Lecture Notes in Computer Science   14049

Founding Editors

Gerhard Goos
Juris Hartmanis

The series Lecture Notes in Computer Science (LNCS), including its subseries Lecture Notes in Artificial Intelligence (LNAI) and Lecture Notes in Bioinformatics (LNBI), has established itself as a medium for the publication of new developments in computer science and information technology research, teaching, and education.

LNCS enjoys close cooperation with the computer science R & D community, the series counts many renowned academics among its volume editors and paper authors, and collaborates with prestigious societies. Its mission is to serve this international community by providing an invaluable service, mainly focused on the publication of conference and workshop proceedings and postproceedings. LNCS commenced publication in 1973.

Heidi Krömker

Editor

HCI in Mobility, Transport, and Automotive Systems

5th International Conference, MobiTAS 2023
Held as Part of the 25th HCI International Conference, HCII 2023
Copenhagen, Denmark, July 23–28, 2023
Proceedings, Part II

 Springer

Editor
Heidi Krömker
Technische Universität Ilmenau
Ilmenau, Germany

ISSN 0302-9743 ISSN 1611-3349 (electronic)
Lecture Notes in Computer Science
ISBN 978-3-031-35907-1 ISBN 978-3-031-35908-8 (eBook)
https://doi.org/10.1007/978-3-031-35908-8

This Springer imprint is published by the registered company Springer Nature Switzerland AG
The registered company address is: Gewerbestrasse 11, 6330 Cham, Switzerland

Foreword

Human-computer interaction (HCI) is acquiring an ever-increasing scientific and industrial importance, as well as having more impact on people's everyday lives, as an ever-growing number of human activities are progressively moving from the physical to the digital world. This process, which has been ongoing for some time now, was further accelerated during the acute period of the COVID-19 pandemic. The HCI International (HCII) conference series, held annually, aims to respond to the compelling need to advance the exchange of knowledge and research and development efforts on the human aspects of design and use of computing systems.

The 25th International Conference on Human-Computer Interaction, HCI International 2023 (HCII 2023), was held in the emerging post-pandemic era as a 'hybrid' event at the AC Bella Sky Hotel and Bella Center, Copenhagen, Denmark, during July 23–28, 2023. It incorporated the 21 thematic areas and affiliated conferences listed below.

A total of 7472 individuals from academia, research institutes, industry, and government agencies from 85 countries submitted contributions, and 1578 papers and 396 posters were included in the volumes of the proceedings that were published just before the start of the conference, these are listed below. The contributions thoroughly cover the entire field of human-computer interaction, addressing major advances in knowledge and effective use of computers in a variety of application areas. These papers provide academics, researchers, engineers, scientists, practitioners and students with state-of-the-art information on the most recent advances in HCI.

The HCI International (HCII) conference also offers the option of presenting 'Late Breaking Work', and this applies both for papers and posters, with corresponding volumes of proceedings that will be published after the conference. Full papers will be included in the 'HCII 2023 - Late Breaking Work - Papers' volumes of the proceedings to be published in the Springer LNCS series, while 'Poster Extended Abstracts' will be included as short research papers in the 'HCII 2023 - Late Breaking Work - Posters' volumes to be published in the Springer CCIS series.

I would like to thank the Program Board Chairs and the members of the Program Boards of all thematic areas and affiliated conferences for their contribution towards the high scientific quality and overall success of the HCI International 2023 conference. Their manifold support in terms of paper reviewing (single-blind review process, with a minimum of two reviews per submission), session organization and their willingness to act as goodwill ambassadors for the conference is most highly appreciated.

This conference would not have been possible without the continuous and unwavering support and advice of Gavriel Salvendy, founder, General Chair Emeritus, and Scientific Advisor. For his outstanding efforts, I would like to express my sincere appreciation to Abbas Moallem, Communications Chair and Editor of HCI International News.

July 2023   Constantine Stephanidis

HCI International 2023 Thematic Areas and Affiliated Conferences

Thematic Areas

- HCI: Human-Computer Interaction
- HIMI: Human Interface and the Management of Information

Affiliated Conferences

- EPCE: 20th International Conference on Engineering Psychology and Cognitive Ergonomics
- AC: 17th International Conference on Augmented Cognition
- UAHCI: 17th International Conference on Universal Access in Human-Computer Interaction
- CCD: 15th International Conference on Cross-Cultural Design
- SCSM: 15th International Conference on Social Computing and Social Media
- VAMR: 15th International Conference on Virtual, Augmented and Mixed Reality
- DHM: 14th International Conference on Digital Human Modeling and Applications in Health, Safety, Ergonomics and Risk Management
- DUXU: 12th International Conference on Design, User Experience and Usability
- C&C: 11th International Conference on Culture and Computing
- DAPI: 11th International Conference on Distributed, Ambient and Pervasive Interactions
- HCIBGO: 10th International Conference on HCI in Business, Government and Organizations
- LCT: 10th International Conference on Learning and Collaboration Technologies
- ITAP: 9th International Conference on Human Aspects of IT for the Aged Population
- AIS: 5th International Conference on Adaptive Instructional Systems
- HCI-CPT: 5th International Conference on HCI for Cybersecurity, Privacy and Trust
- HCI-Games: 5th International Conference on HCI in Games
- MobiTAS: 5th International Conference on HCI in Mobility, Transport and Automotive Systems
- AI-HCI: 4th International Conference on Artificial Intelligence in HCI
- MOBILE: 4th International Conference on Design, Operation and Evaluation of Mobile Communications

List of Conference Proceedings Volumes Appearing Before the Conference

1. LNCS 14011, Human-Computer Interaction: Part I, edited by Masaaki Kurosu and Ayako Hashizume
2. LNCS 14012, Human-Computer Interaction: Part II, edited by Masaaki Kurosu and Ayako Hashizume
3. LNCS 14013, Human-Computer Interaction: Part III, edited by Masaaki Kurosu and Ayako Hashizume
4. LNCS 14014, Human-Computer Interaction: Part IV, edited by Masaaki Kurosu and Ayako Hashizume
5. LNCS 14015, Human Interface and the Management of Information: Part I, edited by Hirohiko Mori and Yumi Asahi
6. LNCS 14016, Human Interface and the Management of Information: Part II, edited by Hirohiko Mori and Yumi Asahi
7. LNAI 14017, Engineering Psychology and Cognitive Ergonomics: Part I, edited by Don Harris and Wen-Chin Li
8. LNAI 14018, Engineering Psychology and Cognitive Ergonomics: Part II, edited by Don Harris and Wen-Chin Li
9. LNAI 14019, Augmented Cognition, edited by Dylan D. Schmorrow and Cali M. Fidopiastis
10. LNCS 14020, Universal Access in Human-Computer Interaction: Part I, edited by Margherita Antona and Constantine Stephanidis
11. LNCS 14021, Universal Access in Human-Computer Interaction: Part II, edited by Margherita Antona and Constantine Stephanidis
12. LNCS 14022, Cross-Cultural Design: Part I, edited by Pei-Luen Patrick Rau
13. LNCS 14023, Cross-Cultural Design: Part II, edited by Pei-Luen Patrick Rau
14. LNCS 14024, Cross-Cultural Design: Part III, edited by Pei-Luen Patrick Rau
15. LNCS 14025, Social Computing and Social Media: Part I, edited by Adela Coman and Simona Vasilache
16. LNCS 14026, Social Computing and Social Media: Part II, edited by Adela Coman and Simona Vasilache
17. LNCS 14027, Virtual, Augmented and Mixed Reality, edited by Jessie Y. C. Chen and Gino Fragomeni
18. LNCS 14028, Digital Human Modeling and Applications in Health, Safety, Ergonomics and Risk Management: Part I, edited by Vincent G. Duffy
19. LNCS 14029, Digital Human Modeling and Applications in Health, Safety, Ergonomics and Risk Management: Part II, edited by Vincent G. Duffy
20. LNCS 14030, Design, User Experience, and Usability: Part I, edited by Aaron Marcus, Elizabeth Rosenzweig and Marcelo Soares
21. LNCS 14031, Design, User Experience, and Usability: Part II, edited by Aaron Marcus, Elizabeth Rosenzweig and Marcelo Soares

22. LNCS 14032, Design, User Experience, and Usability: Part III, edited by Aaron Marcus, Elizabeth Rosenzweig and Marcelo Soares
23. LNCS 14033, Design, User Experience, and Usability: Part IV, edited by Aaron Marcus, Elizabeth Rosenzweig and Marcelo Soares
24. LNCS 14034, Design, User Experience, and Usability: Part V, edited by Aaron Marcus, Elizabeth Rosenzweig and Marcelo Soares
25. LNCS 14035, Culture and Computing, edited by Matthias Rauterberg
26. LNCS 14036, Distributed, Ambient and Pervasive Interactions: Part I, edited by Norbert Streitz and Shin'ichi Konomi
27. LNCS 14037, Distributed, Ambient and Pervasive Interactions: Part II, edited by Norbert Streitz and Shin'ichi Konomi
28. LNCS 14038, HCI in Business, Government and Organizations: Part I, edited by Fiona Fui-Hoon Nah and Keng Siau
29. LNCS 14039, HCI in Business, Government and Organizations: Part II, edited by Fiona Fui-Hoon Nah and Keng Siau
30. LNCS 14040, Learning and Collaboration Technologies: Part I, edited by Panayiotis Zaphiris and Andri Ioannou
31. LNCS 14041, Learning and Collaboration Technologies: Part II, edited by Panayiotis Zaphiris and Andri Ioannou
32. LNCS 14042, Human Aspects of IT for the Aged Population: Part I, edited by Qin Gao and Jia Zhou
33. LNCS 14043, Human Aspects of IT for the Aged Population: Part II, edited by Qin Gao and Jia Zhou
34. LNCS 14044, Adaptive Instructional Systems, edited by Robert A. Sottilare and Jessica Schwarz
35. LNCS 14045, HCI for Cybersecurity, Privacy and Trust, edited by Abbas Moallem
36. LNCS 14046, HCI in Games: Part I, edited by Xiaowen Fang
37. LNCS 14047, HCI in Games: Part II, edited by Xiaowen Fang
38. LNCS 14048, HCI in Mobility, Transport and Automotive Systems: Part I, edited by Heidi Krömker
39. LNCS 14049, HCI in Mobility, Transport and Automotive Systems: Part II, edited by Heidi Krömker
40. LNAI 14050, Artificial Intelligence in HCI: Part I, edited by Helmut Degen and Stavroula Ntoa
41. LNAI 14051, Artificial Intelligence in HCI: Part II, edited by Helmut Degen and Stavroula Ntoa
42. LNCS 14052, Design, Operation and Evaluation of Mobile Communications, edited by Gavriel Salvendy and June Wei
43. CCIS 1832, HCI International 2023 Posters - Part I, edited by Constantine Stephanidis, Margherita Antona, Stavroula Ntoa and Gavriel Salvendy
44. CCIS 1833, HCI International 2023 Posters - Part II, edited by Constantine Stephanidis, Margherita Antona, Stavroula Ntoa and Gavriel Salvendy
45. CCIS 1834, HCI International 2023 Posters - Part III, edited by Constantine Stephanidis, Margherita Antona, Stavroula Ntoa and Gavriel Salvendy
46. CCIS 1835, HCI International 2023 Posters - Part IV, edited by Constantine Stephanidis, Margherita Antona, Stavroula Ntoa and Gavriel Salvendy

47. CCIS 1836, HCI International 2023 Posters - Part V, edited by Constantine Stephanidis, Margherita Antona, Stavroula Ntoa and Gavriel Salvendy

https://2023.hci.international/proceedings

Preface

Human-computer interaction in the highly complex field of mobility and intermodal transport leads to completely new challenges. A variety of different travelers move in different travel chains. The interplay of such different systems, such as car and bike sharing, local and long-distance public transport, and individual transport, must be adapted to the needs of travelers. Intelligent traveler information systems must be created to make it easier for travelers to plan, book, and execute an intermodal travel chain and to interact with the different systems. Innovative means of transport are developed, such as electric vehicles and autonomous vehicles. To achieve the acceptance of these systems, human-machine interaction must be completely redesigned.

The 5th International Conference on HCI in Mobility, Transport, and Automotive Systems (MobiTAS 2023), an affiliated conference of the HCI International (HCII) conference, encouraged papers from academics, researchers, industry, and professionals, on a broad range of theoretical and applied issues related to mobility, transport, and automotive systems and their applications.

For MobiTAS 2023, a key theme with which researchers were concerned was autonomous and assisted driving, as well as intelligent transportation systems. Topics covered in this area included designing driver training applications for adaptive cruise control, designing automation and intelligent car features, designing and evaluating intelligent assistants, exploring the cooperation between humans and agents, and understanding user behavior towards automation. Other papers focused on urban and sustainable mobility, exploring various aspects of urban transportation, such as autonomous shuttles, delivery robots, heavy vehicles in underground mines, urban air mobility, and electric vehicles and e-scooters, as well as sustainability perspectives, such as service ecosystems, vehicle sharing, route optimization, and traffic planning. In addition, a considerable number of contributions reflected the need for understanding driver behavior and performance, offering insights into the effect of technology on driving experience, exploring driver preferences for automated driving styles, and classifying driving styles based on driver behavior. The design of in-vehicle experiences for drivers and passengers was another key theme of research addressed in this year's proceedings, covering a range of topics, such as interface design and evaluation of alert systems and dashboards, data-driven design, game-based design, multimodal design featuring ultrasound skin stimulation and music, as well as design to alleviate motion sickness for passengers. Finally, a number of papers focused on the topic of accessibility and inclusive mobility, addressing accessibility models, independent mobility for people with intellectual disabilities, and understanding driver behavior in individuals with mild cognitive disabilities.

Two volumes of the HCII 2023 proceedings are dedicated to this year's edition of the MobiTAS conference. The first focuses on topics related to autonomous and intelligent transport systems for urban and sustainable mobility, while the second focuses on topics related to human-centered design of mobility and in-vehicle experiences.

Papers of these volumes are included for publication after a minimum of two single-blind reviews from the members of the MobiTAS Program Board or, in some cases, from members of the Program Boards of other affiliated conferences. I would like to thank all of them for their invaluable contribution, support, and efforts.

July 2023   Heidi Krömker

5th International Conference on HCI in Mobility, Transport and Automotive Systems (MobiTAS 2023)

Program Board Chair:

- Heidi Kroemker, *Technische Universität Ilmenau, Germany*
- Avinoam Borowksy, *Ben-Gurion University of the Negev, Israel*
- Angelika C. Bullinger, *Chemnitz University of Technology, Germany*
- Bertrand David, *Ecole Centrale de Lyon, France*
- Marco Diana, *Politecnico di Torino, Italy*
- Cyriel Diels, *Royal College of Art, UK*
- Chinh Ho, *University of Sydney, Australia*
- Christophe Kolski, *Université Polytechnique Hauts-de-France, France*
- Josef F. Krems, *Chemnitz University of Technology, Germany*
- Roberto Montanari, *RE:LAB srl, Italy*
- Matthias Rötting, *Technische Universität Berlin, Germany*
- Frank Ritter, *Penn State University, USA*
- Philipp Rode, *Volkswagen Group, Germany*
- Thomas Schlegel, *Hochschule Furtwangen University, Germany*
- Felix W. Siebert, *Technical University of Denmark, Denmark*
- Ulrike Stopka, *Technische Universität Dresden, Germany*
- Tobias Wienken, *CodeCamp:N GmbH, Germany*
- Xiaowei Yuan, *Beijing ISAR User Interface Design Limited, P.R. China*

The full list with the Program Board Chairs and the members of the Program Boards of all thematic areas and affiliated conferences of HCII2023 is available online at:

http://www.hci.international/board-members-2023.php

5th International Conference on ICT in Mobility, Transport and Automotive Systems (MOTAS 2023)

HCI International 2024 Conference

The 26th International Conference on Human-Computer Interaction, HCI International 2024, will be held jointly with the affiliated conferences at the Washington Hilton Hotel, Washington, DC, USA, June 29 – July 4, 2024. It will cover a broad spectrum of themes related to Human-Computer Interaction, including theoretical issues, methods, tools, processes, and case studies in HCI design, as well as novel interaction techniques, interfaces, and applications. The proceedings will be published by Springer. More information will be made available on the conference website: http://2024.hci.international/.

General Chair
Prof. Constantine Stephanidis
University of Crete and ICS-FORTH
Heraklion, Crete, Greece
Email: general_chair@hcii2024.org

https://2024.hci.international/

Contents – Part II

Accessibility and Inclusive Mobility

Contents – Part I

Urban Mobility

Sustainable Mobility

Driver Behavior and Performance

FIGCONs: Exploiting FIne-Grained CONstructs of Facial Expressions for Efficient and Accurate Estimation of In-Vehicle Drivers' Statistics

Zhuoran Bi, Xiaoxing Ming, Junyu Liu, Xiangjun Peng, and Wangkai Jin[✉]

User-Centric Computing Group, Hangzhou, China
wangkaijin00@gmail.com

Abstract. Recent advances in-vehicle monitoring of drivers' statistics attempt to leverage Machine-Learning (ML)-based techniques for prediction, credited to its high accuracy and less cost for deploying redundant sensors/cameras inside the vehicle. However, existing approaches for in-vehicle ML-based predictors heavily rely on the raw driver data for processing, which mcan push the burdens on multiple components, ranging from on-vehicle devices to Internet-of-Vehicles (e.g., on-vehicle computation and storage units). To obtain a better balance between the performance and cost, we propose FIne-Grained CONstructs (**FIGCONs**) of collected drivers' statistics, to further exploit its applicability. We do so by integrating it with state-of-the-art ML-based approaches to predict in-vehicle statistics. The experimental results deliver three key findings, which showcase the benefits of FIGCONs and the potentials for practical deployments of in-vehicle ML-based predictors. We hope our work can stimulate more future works and potential practices in similar manners.

Keywords: Face Segmentation & Localization · Image Processing · Human-Computer Interaction

1 Introduction

In-vehicle statistics are critical to support intelligent Human-Vehicle Interaction systems [12, 16, 22–25, 28]. The naive way of collecting in-vehicle statistics is by equipping sensors and cameras inside vehicles, each with specific monitoring targets. For example, [8] uses physiological sensors to detect drivers' stress levels in real-world driving scenarios, [11] uses cameras and GPS to learn to predict and fuse multiple sensory streams, and [7] uses a combination of physiological, environmental, and visual sensors to assess drivers' attention. These methods have multiple drawbacks that prevent them from being widely used in real-world as deploying these sensors brings high cost, reduced user experience, and the complexity of handling all the data streams.

With the surge of applying Machine Learning techniques for intelligent vehicles, there is a growing trend to use Machine-Learning (ML) -based solutions as

H. Krömker (Ed.): HCII 2023, LNCS 14049, pp. 3–17, 2023.
https://doi.org/10.1007/978-3-031-35908-8_1

alternatives can achieve minimum sensor/camera deployments, while maintaining high in-vehicle statistic prediction accuracy. There is a line of work which takes driver facial images (or run-time driver video) and outputs pre-configured in-vehicle statistics based on the input driver facial images. (e.g., Face2Multi-Modal [9], Face2Statistics [27,30], etc.). These works rely on one facial camera for monitoring and have high prediction accuracy for configured metrics, which showcase their practical values. However, these works still suffer from three major limitations, resulting from leveraging full drivers' facial images throughout the whole pipeline for these applications. **First, it brings a significant burden on the limited hardware resources on vehicles and on the network traffic within the Internet-of-Vehicles.** Due to the limited storage and computation resources on each vehicle, processing drivers' full facial images might take up all the available storage and computation units and even need communications with central servers for assisted computations for predictions [4,5,9,12,16,17,22]. **Second, the granularity of input features is coarse (i.e., only at face-level),** which can still contain noises that can lead to prediction degradation. For example, illumination changes on the face may hide certain features that are useful for prediction, thus resulting in accuracy loss. **Third, the throughput is lowered and the latency is increased** since extra transmissions and computations of noise data are also carried out in the pipeline.

To this end, we argue that a fine-grained exploitation of input data can be beneficial for ML-based solutions. We make the key observation that relying on full facial images for ML-based in-vehicle monitoring still has multiple limitations that can bottleneck the system in the real world. To overcome this limitation, we propose a <u>FI</u>ne-<u>G</u>rained <u>CON</u>structs (FIGCONs) of drivers' facial images as the input data for ML-based predictors. Specifically, we leverage a lightweight Deep Neural Network (DNN) to extract key features on human faces (e.g., eyes, mouths, etc.) as we notice organs can contain important features for monitoring and can serve as alternatives to whole facial images. Then, we integrate FIGCONs into a state-of-the-art in-vehicle ML-based multi-modal predictor, Face2Statistics, to evaluate the performance of FIGCONs by measuring the prediction accuracy, throughput, and latency of Face2Statistics with and without FIGCONs on three standard in-vehicle statistics. We extract three key findings from the experimental results and the experimental results show that FIGCONs can boost prediction accuracy, increase pipeline throughput, and lower latency to a great extent. We believe FIGCONs can serve as a starting point for more fine-grained exploitation of input data for in-vehicle processing, where the hardware resources are limited, and can be deployed widely to support lightweight intelligent in-vehicle applications.

To sum up, We make four key contributions:

- We address three limitations of existing in-vehicle Machine-Learning-based predictors, which all resulted from using raw input data throughout the whole pipeline.
- We propose a <u>FI</u>ne-<u>G</u>rained <u>CON</u>structs (FIGCONs) of drivers' facial images which exploit the input driver facial image at the organ level.

- We integrate FIGCONs into a state-of-the-art in-vehicle multi-modal predictor - Face2Statistics, where FIGCONs serve as a lightweight pre-processing module for efficient input data exploitation.
- We conduct extensive experiments to evaluate the benefits of FIGCONs. Our experimental results show that FIGCONs can boost prediction accuracy, increase pipeline throughput and lower latency. We also extract three key findings from the experimental results which can guide future research.

The rest of this paper is organized as follows. Section 2 introduces the background and motivation of our work. Section 3 presents our design and methodology for exploiting FIGCONs. Section 4 showcases the experimental results and analysis. Section 6 discusses FIGCONs' applicability to other usages. Section 7 concludes our paper.

2   Background and Motivation

Recent techniques for in-vehicle statistics collection rely heavily on vision-based systems, which mount cameras in the cabin to capture video streams and store data for analysis. Among all possible data streams that can be collected, drivers' facial images are essential and widely used data streams for understanding driver status. To obtain such streams, one standard method is to mount a camera on the driving dashboard (e.g., [9, 26, 30]) and exploit drivers' facial images for different monitoring purposes. Some common use cases are: 1) driving behaviors and styles classifications [15, 21, 31], 2) driver drowsiness/inattention detection [18] and 3) driver distraction tracking [7, 8]. Considering the variety of purposes that facial images can support, it is intuitive when recent researchers attempt to apply Machine Learning (ML) techniques on intelligent vehicles and develop multi-modal predictor which uses drivers' facial images only.

However, we observe that ML-based in-vehicle predictors suffer from several limitations that may bottleneck their deployment in the real world. First, current intelligent vehicles have limited hardware resources to support long-running ML-based predictors. Although there is recent hardware advancement on edge sides [10], further exploration on minimizing computation/storage of ML-based applications are still necessary to boost performance and reduce costs. Second, the performance concern of these predictors is also critical as high accuracy and low latency are the primary requirements to guarantee immediate and accurate responses. Such a goal might be compromised by using raw data throughout the whole pipeline, which may introduce redundant computation or unexpected noises. Third, when ML-based predictors use drivers' full facial images, privacy can also be a potential issue. There are also existing works trying to tackle this problem in the context of Internet-of-Vehicles (e.g., blockchain-based methods [2], privacy-preserving-based methods [4, 5, 17, 29], processing tradeoffs [13, 20]). However, these methods still incur certain overheads and better alternatives are still under active research and development. To this end, our goal is to address the aforementioned limitations by putting more focus on find-grained constructs of input data for ML-based applications and comparing the performance of FIGCONs with raw data on modern ML-based applications.

3   Methodology

In this section, we will introduce our methodology by first presenting the designs and implementations of building FIne-Grained CONstructs of drivers' facial expressions (Sect. 3.1) and then showcase how FIGCONs can be integrated into existing in-vehicle applications, which we use Face2Statistics in this work (Sect. 3.2).

3.1   FIne-Grained CONstructs of Facial Expressions

The key insight of FIne-grained CONstructs (FIGCONs) of driver's facial expressions is extracting facial features from a finer granularity level (i.e., different organs) than full faces, to improve the performance of using such features in ML-based in-vehicle applications like multi-modal predictors. Our primary focus of building FIGCONs lies in reconstructing organs on drivers' faces, since recent studies prove that pictures of facial organs can represent important information for human bio-signals, which can be learned/predicted by ML with sufficient training (e.g., [7,9,11,30]). Specifically, we aim to construct FIGCONs with either separated facial organs or a combination of facial organs, characterize their performance on in-vehicle ML-based applications, and compare the results with the baseline solution (i.e., using drivers' full facial images).

Building FIGCONs also needs to overcome three challenges that are critical to the performance of real-world in-vehicle applications: First, the process of constructing FIGCONs should have a minimal delay. Otherwise, it can lead to applications halting for inputs and delay in decision-making. Second, FIGCONs should remain invariant to drivers' facial angle changes. Ideally, FIGCONs are expected to be centered images that are easier for further processing. Third, the prediction results of ML-based applications using FIGCONs should have comparable or better accuracy than that of full driver images. In the rest of the section, we will present the building process of FIGCON in the following four steps, which are 1) Face Detection, 2) Landmark Localization, 3) Face Segmentation, and 4) Combination Selection.

Step1: Face Detection. The first step of FIGCONs is face detection, which includes a coarse-grained selection of video frames with human faces presented, a point-of-interest crop, and an alignment of cropped images with preliminary detected facial landmarks.

The reason for cropping the face accurately is that the facial video frames usually contain a considerable portion of useless information like backgrounds. Cropping could reduce this interference of irrelevant information and concentrate the computation power on drivers' faces. This is crucial for landmark prediction and localization in Step 2 as eliminating the interference of background noises can improve landmark prediction accuracy and faster processing speed by concentrating more computation power on generating landmarks.

To enable quick construction and rotation invariance, we leverage BlazeFace [1], a lightweight and high-throughput Neural Network model for face detection. BlazeFace was originally designed for fast and run-time face detection and cropping point-of-interest regions on edge devices like Augmented Reality (AR) applications. It can run at a speed of 200–1000+ FPS on mobile devices and present several landmarks that can be used for rotation estimation. Figure 1 shows the outputs of BlazeFace with our data, where the right frame shows the generated bounding box and some preliminary key landmarks on the driver's face (e.g. eye centers, ear tragions, and nose tip) that can be used to re-transform the rotated driver image to be centered and aligned.

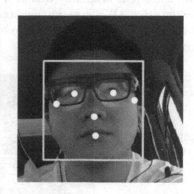

Fig. 1. An original driver facial image (left) and the corresponding output of Step 1 with face bounding rectangle and key landmarks (right).

Step2: Landmark Localization. The second step of FIGCONs is landmark localization, which takes the detected faces in Step 1 as input and generates fine-grained facial landmarks. Based on the preliminary landmarks generated in Step 1 (e.g. eye centers, ear tragions, and nose tips), we explore more detailed landmark localization to further exploit fine-grained facial features, which is beneficial for Step 3 when implementing face segmentation. To this end, we first introduce a mesh prediction neural network for landmark localization, then we introduce an optimized usage of the face detector at the application level.

Mesh Prediction Neural Network. To generate facial landmarks, we leverage a mesh prediction neural network [14], which uses a custom but relatively lightweight residual neural network architecture. This model produces 468 landmarks with 3D coordinates information on detected faces. Each landmark has its unique ID and a uniquely correlated landmark position, which ensures that the position of a landmark remains invariant on different faces for easy mapping between abstracted landmarks features and original images. Taking landmarks with ID 4 (denote the landmarks on the nose tips) as an example, it is intuitive to acquire information about the landmarks on the nose tips for all different faces by calling ID 4. With such invariance of mesh IDs, it is easier for us to map back a landmark to the original face image by using its coordinates and ID, which lay the groundwork for us to get the location on the original face image according to specified landmarks in the next step.

We choose [14] for mesh prediction for two main reasons. Firstly, this network could provide robust landmark detection with good generalization and limited overfitting, since it is trained on a large dataset containing about 30,000 face images gathered by different sensors under various lighting conditions and image quality. Secondly, this network could provide fine-grained detection on various facial regions (e.g. eyebrows, eyes, nose, mouth) with different movements (shown in Fig. 2), since it is originally designed for complicated real-world applications such as expressive AR effects, virtual accessories, and apparel try-on and makeup [14]. This matches our initial objective to explore the fine-grained facial features at the organ level.

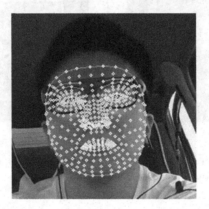

Fig. 2. A frame of drive with 468 landmarks generated by mesh prediction neural network.

Optimize the Usage of Face Detector. At the application level, we introduce an optimization design to reduce redundant face detection and save computation power, since many edge devices require fast and efficient on-the-fly detection. In traditional face detection pipelines on the single-frame level with video frames as input, perform face detection is performed on every frame before using the mesh prediction. However, we notice that if a previously detected frame provides an excellent facial crop and localization, it could be unnecessary to re-detect and re-localize the face in the following few frames. Instead, we could directly apply the detection model of the previous frames on the following frames to get decent detection and localization results. Therefore, we introduce a distinct scalar network output (face flag) that produces the probability that measures if a face is suitably detected and localized in the current crop with a controllable threshold. The detection model will be reused to save computation power when the measured probability is over an acceptable threshold.

Step3: Face Segmentation. In this step, we use the landmark coordinates information to select different facial organs and their combinations. Figure 3 shows a template face mesh with 468 landmarks and a frame of a driver's face where the template mesh is applied on. Landmarks cluster densely around the detailedly divided regions of facial organs and thus can provide fine-grained segmentation. To select and crop the segmented regions, various shapes could be used, including straightforward rectangular bounding boxes and complicated irregular cropping along the exact contour of the organs. We choose the rectangle bounding box for cropping since it could generate regularly cropped images that are more suitable to implement the combination selection and feed into neural networks like Face2Statistics [30] (which will be illustrated in detail in Sect. 3.2). The 4 vertices of bounding boxes are located by the 4 outermost landmarks of a facial organ.

For implementation, we use OpenCV library [3], which is a state-of-the-art image processing library, to handle this face segmentation. There are mainly two steps for cropping. First, we identify the four landmarks and use their unique IDs as the index to acquire their coordinates information. We use the X- and Y- coordinated to accurately locate the landmark and specify the bounding box. Then, we pass the coordinates of four landmarks to the OpenCV library and crop out the facial region within the specified bounding box.

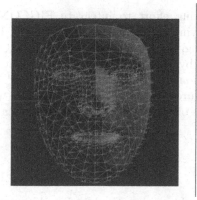 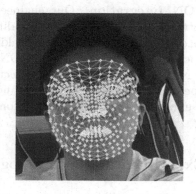

Fig. 3. A template face mesh with 468 landmarks and their IDs (left) [6] and a frame of a driver's face with face mesh (right).

Step4: Combination Selection. In this step, we combine individual facial organs obtained from the previous step into a few combinations. Our goal is to find out whether the combination of different organs could lead to better-predicting accuracy than only using individual ones, and how different combinations could affect the accuracy. We first use only individual facial organ images (e.g. eyes, nose, mouth) to test the prediction model performance under the minimal storage occupation and use the least computational power. Then we perform various combination of different organ images (e.g., mouth and nose, eyes and nose) to test how prediction performance vary.

3.2   Integrating FIGCON Method with Face2Statistics

To evaluate the performance of FIGCONs in real-world applications, we integrate FIGCON design into state-of-the-art in-vehicle multi-modal predictors. We choose Face2Statistics as a proof-of-concept, the methodology to design and implement is also expected to be compatible to other ML-based applications. In this section, we present a brief overview of Face2Statistics and showcase our engineering effort to integrate FIGCON into Face2Statistics.

Face2Statistics Overview: Face2Statistics is designed for user-friendly and cost-effective alternatives of sensors/cameras to collect driver statistics. The end goal of Face2Statistics is to predict multiple in-vehicle statistics (i.e., both bio-signals and situational statistics) on driver facial expressions only. The working pipeline of Face2Statistics starts with raw video input, which is streamed from the camera mounted on the vehicle dashboards. Then, after several image pre-processing steps, it applies Deep-Neural-Network-based Predictors to predict pre-configured statistics and visualize the predictions results. Face2Statistics is suitable for FIGCON evaluation as it only relies on input facial expressions for prediction. Therefore, the impact of FIGCON can be directly reflected by the prediction outputs of Face2Statistics.

FIGCON Integration: Our engineering effort to integrate FIGCON into Face2Statistics mainly focuses on modifying the front-end pipeline (i.e., Image Preprocessing part). Specifically speaking, we implement automatic FIGCON construction for each possible FIGCON by incorporating the lightweight Deep Neural Networks and the cropping/selection algorithms introduced in Sect. 3 into the pipeline. Therefore, users can configure different types of FIGCON, which can be either solo organs or combinations of organs for in-vehicle statistics monitoring.

4   Experimental Methodology

In this section, we will introduce our experimental methodologies. We start with the details of our implementations (Sect. 4.1). Then we introduce our selected dataset (Sect. 4.2) and evaluation metrics (Sect. 4.3) respectively.

4.1   Implementation Details

We implement FIGCONs using: (1) Mediapipe library [19] for face detection and landmark localization; and (2) OpenCV library [3] for face segmentation and combination selection. We integrate FIGCON into Face2Statsitics by using the open-source version. We train and evaluate different variants of FIGCONs with Face2Statistics using a machine with Intel(R) Xeon(R) CPU @ 2.30 GHz CPU and an Nvidia Tesla T4 GPU.

4.2   Dataset

We use the BROOK dataset [12,16,22], an open-source and multi-modal dataset with facial videos from 34 drivers over the duration of a 20-min driving session, to evaluate the integration of Face2Statistics with different sets of FIGCONs. BROOK covers various dimensions of driving statistics including physiological status (e.g., skin conductivity, heart rate, and eye tracking) and vehicle status (e.g., vehicle speed, vehicle acceleration, brake status, .etc). We choose drivers' facial images as input, and use Face2Statistics to predict three driving statistics: vehicle speed, skin conductivity, and heart rate. Vehicle speed is an important and indicative type of data for showing the vehicle status, whereas skin conductivity and heart rate are representative types of data for monitoring drivers' physiological status.

4.3   Evaluation Metrics

We use three evaluation metrics to comprehensively understand the performance benefits of FIGCON on ML-based applications. We first compare the validation accuracy of different FIGCONs in three selected datatypes with the validation accuracy of full facial expressions in Face2Statistics. Then we compare the end-to-end throughput and latency results when Face2Statistics uses FIGCON and full facial expressions during inference.

5   Experimental Results

In this section, we present the experimental results and our corresponding observations from these results. We first showcase the performance results of validation accuracy for the three selected types of data in Scct. 5.1. Then, we present our throughput and latency analysis of FIGCON in Sect. 5.2.

5.1   Performance Analysis

Figure 4, Fig. 5 and Fig. 6 show the validation accuracy of Face2Statistics using four sets of FIGCONs (i.e., Mouth and Nose, Mouth, Eyes and Nose, Eyes) and full facial expression respectively. From these three figures, we draw three key observations:

Observation 1: Using FIGCONs can boost validation accuracy. From Fig. 4 and Fig. 5, we observe that by fine-graining features on Mouth + Nose, Face2Statistics achieves an average of 3.80% and 3.70% validation accuracy improvement than using full facial images to explore features. Also, when predicting Vehicle Speed (VC) and Skin Conductivity (SC), the Mouth + Nose FIGCON results in validation accuracy increase for three out of fivers drivers for VC and all drivers for SC. This is because by using fine-grained features as the input for training and inference, Face2Statistics can learn more detailed information that can be beneficial to the prediction. Therefore, FIGCONs have the potential to optimize the performance of current ML-based in-vehicle predictors.

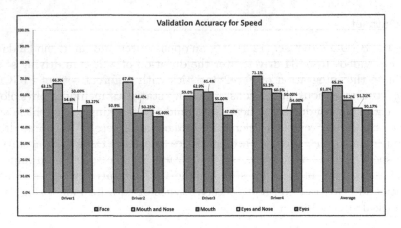

Fig. 4. Validation Accuracy of four sets of FIGCONs and face for Vehicle Speed (VC).

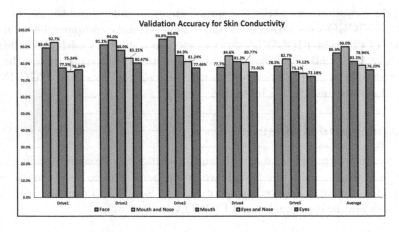

Fig. 5. Validation Accuracy of four sets of FIGCONs and face for Skin Conductivity (SC).

Observation 2: Combining different FIGCONs sets can provide performance benefits than using FIGCONs using solo features (i.e., organs). In most of the reported cases, we observe that FIGCON which uses a combination of organs outperforms FIGCON using only a single organ. For example, using *Mouth+Nose* FIGCON set outperforms the *Mouth* FIGCON set and using *Eyes+Nose* FIGCON set outperforms the *Eyes* FIGCON set in Fig. 4, Fig. 5 and Fig. 6. This is because the combination of different FIGCONs can be viewed as fusing multiple fine-grained features together, which can contribute to the performance gain. A special observation we need to point out is that this observation mainly applies to FIGCON sets that contain a sole organ and its combinations with other organs, which means that not all of the combinations of FIGCON sets can outperform any sole organ FIGCON sets. For instance, the *Mouth* FIGCON has better performance than the *Eyes+Nose* FIGCON set.

Observation 3: The performance of different FIGCONs vary greatly for different drivers and prediction targets. Following the previous observation, we notice that the performance variation is relatively large for different FIGCONs on different drivers and different prediction targets. For example, when predicting Heart Rate, the *Mouth* FIGCON set is the best-performing FIGCON for Driver 3. But when predicting for SC, the *Mouth + Nose* FIGCON sets become the best FIGCOn among the four FIGCON sets. One possible explanation for this phenomenon is that the current granularity of FIGCON can be enough to fully exploit the features of driver facial expressions and further study is still needed to understand the benefits of FIGCON in the future.

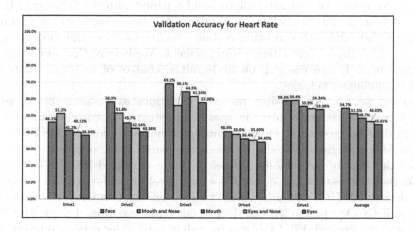

Fig. 6. Validation Accuracy of five sets of FIGCONs for Heart Rate.

5.2   Throughput and Latency Analysis

Table 1 shows the normalized throughput and latency of selected FIGCONs sets and full images. For throughput, we conclude that selected FIGCONs sets can significantly increase the throughput by at most 25× when using *Eyes* FIGCON set, and at least 17× when using *Mouth+Nose* and *Eyes+Nose* FIGCONs sets. This is because the sizes of all FIGCONs sets are much smaller than full images: a full facial image occupies about 17 KB storage on our machine, a *Mouth+Nose* FIGCONs image occupies about 1 KB, and a *Eyes* FIGCON image occupies about 750 B. For latency, we conclude that all these selected FIGCON sets can reduce the latency by around 1/3 compared to full images. This is because using FIGCON sets enables Face2Statistics to eliminate transmissions and computations of noisy data.

Table 1. The normalized throughput and latency for selected FIGCONs sets and full images.

	Full image	Mouth+Nose	Eyes+Nose	Mouse	Eyes
Throughput	1	17X	17X	20X	25X
Latency	1	0.37X	0.35X	0.33X	0.29X

6 Discussion and Future Work

To advocate for more fine-grained control of input data for intelligent vehicle systems, we make the first attempt to build a FIne-Grained CONstructs (FIG-CON) of driver facial expressions for in-vehicle ML-based applications. Our findings show that FIGCON can bring certain benefits such as validation accuracy increase, throughput, and latency improvement. We believe that our work can stimulate more future research on find-grained control of real-world data collected in intelligent vehicles.

Hereby, we list two possible research directions as possible future works. First, it is potential to study the spatial and temporal correlations of FIG-CON in different scenarios. Our current experiments do not include FIGCONs such as *Mouth + Eyes* as a simple concatenation of these organs together can destroy the spatial correlations. Therefore, more advanced FIGCON construction methodologies that can incorporate spatial relationships are still needed for more diversified FIGCON sets. Also, a study on the temporal correlations of different FIGCONs (e.g., [27]) can deepen our understanding of the FIGCON in the time scale. Second, FIGCON can be widely applied for other studies such as privacy-related works (e.g., [4,5,17]). As FIGCON only contain parts of the bio-information of drivers, it makes them ideal for privacy-related studies as applying privacy protections on FIGCON is expected to have multiple advantages such as low computation overhead and less privacy leakage risks.

7 Conclusions

In this paper, we propose FIn-Grained CONstructs (FIGCONs) of drivers' facial expressions and exploit its applicability on Face2Statisics, a state-of-the-art in-vehicle Multi-modal predictor. Specifically, we build FIGCON by reconstructing facial features at organ-level and integrate FIGCON into Face2Statistics pipeline by incorporating lightweight DNN models and corresponding cropping/selection algorithms. We evaluate FIGCON with the support of Face2Statistics and the experimental results show that FIGCON can boost validation accuracy and improve system throughput while maintaining low latency. We believe this work can serve as a starting point to re-think data representation for in-vehicle ML-based applications and our methodologies can stimulate more future research.

Acknowledgement. We thank the anonymous reviewers from HCI 2023 and all members from User-Centric Computing Group for their valuable feedback and comments.

References

1. Bazarevsky, V., Kartynnik, Y., Vakunov, A., Raveendran, K., Grundmann, M.: BlazeFace: sub-millisecond neural face detection on mobile GPUs. CoRR, abs/1907.05047 (2019)
2. Blockchain-based security attack resilience schemes for autonomous vehicles in industry 4.0: a systematic review. Comput. Electr. Eng. **86**, 106717 (2020)
3. Bradski, G.R., Kaehler, A.: Learning OpenCV - Computer Vision with the OpenCV Library: Software that Sees. O'Reilly, Sebastopol (2008)
4. Duan, Y., Liu, J., Ming, X., Jin, W., Song, Z., Peng, X.: Characterizing and optimizing differentially-private techniques for high-utility, privacy-preserving internet-of-vehicles. In: International Conference on Human-Computer Interaction (2023)
5. Duan, Y., Liu, J., Jin, W., Peng, X.: Characterizing differentially-private techniques in the era of internet-of-vehicles (2022)
6. Google. Augmented faces developer guide for AR foundation. https://developers.google.com/ar/develop/unity-arf/augmented-faces/developer-guide?hl=en
7. Haouij, N.E., Poggi, J.-M., Sevestre-Ghalila, S., Ghozi, R., Jaïdane, M.: AffectiveROAD system and database to assess driver's attention. In: Proceedings of the 33rd Annual ACM Symposium on Applied Computing, pp 800–803 (2018)
8. Healey, J.A., Picard, R.W.: Detecting stress during real-world driving tasks using physiological sensors. IEEE Trans. Intell. Transp. Syst. **6**(2), 156–166 (2005)
9. Huang, Z., et al.: Face2Multi-modal: in-vehicle multi-modal predictors via facial expressions. In: Adjunct Proceedings of the 12th International Conference on Automotive User Interfaces and Interactive Vehicular Applications, AutomotiveUI 2020, Virtual Event, Washington, DC, USA, 21–22 September 2020, pp. 30–33. ACM (2020)
10. Primeo Inc., Hardware acceleration: Edgeboost nodes (2021)
11. Jain, A., Koppula, H.S., Soh, S., Raghavan, B., Singh, A., Saxena, A.: Brain4Cars: car that knows before you do via sensory-fusion deep learning architecture. arXiv preprint arXiv:1601.00740 (2016)
12. Jin, W., Duan, Y., Liu, J., Huang, S., Xiong, Z., Peng, X.: BROOK dataset: a playground for exploiting data-driven techniques in human-vehicle interactive designs. Technical report-Feb-01 at User-Centric Computing Group, University of Nottingham Ningbo China (2022)
13. Jin, W., Ming, X., Song, Z., Xiong, Z., Peng, X.: Towards emulating internet-of-vehicles on a single machine. In AutomotiveUI 2021: 13th International Conference on Automotive User Interfaces and Interactive Vehicular Applications, Leeds, United Kingdom, 9–14 September 2021 - Adjunct Proceedings, pp. 112–114. ACM (2021). https://doi.org/10.1145/3473682.3480275
14. Kartynnik, Y., Ablavatski, A., Grishchenko, I., Grundmann, M.: Real-time facial surface geometry from monocular video on mobile GPUs. CoRR, abs/1907.06724 (2019)
15. Khodairy, M.A., Abosamra, G.: Driving behavior classification based on oversampled signals of smartphone embedded sensors using an optimized stacked-LSTM neural networks. IEEE Access **9**, 4957–4972 (2021)
16. Liu, J., et al.: BROOK dataset: a playground for exploiting data-driven techniques in human-vehicle interactive designs. In: International Conference on Human-Computer Interaction (2023)
17. Liu, J., et al.: HUT: enabling high-UTility, batched queries under differential privacy protection for internet-of-vehicles. Technical Report-Feb-02 at User-Centric Computing Group, University of Nottingham Ningbo China (2022)

18. Mao, H., Tang, J., Zhao, X., Tang, M., Jiang, Z.: A driver drowsiness detection scheme based on 3D convolutional neural networks. Int. J. Pattern Recogn. Artif. Intell., 36(2), 2252007:1–2252007:21 (2022)
19. MideaPipe (2020). https://google.github.io/mediapipe/
20. Ming, X., et al.: Enabling efficient emulation of internet-of-vehicles on a single machine: practices and lessons. In: International Conference on Human-Computer Interaction (2023)
21. Moukafih, Y., Hafidi, H., Ghogho, M.: Aggressive driving detection using deep learning-based time series classification. In: Koprinkova-Hristova, P.D., Yildirim, T., Piuri, V., Iliadis, L.S., Camacho, D. (eds.) IEEE International Symposium on INnovations in Intelligent SysTems and Applications, INISTA 2019, Sofia, Bulgaria, 3–5 July 2019, pp. 1–5. IEEE (2019)
22. Peng, X., Huang, Z., Sun, X.: Building BROOK: a multi-modal and facial video database for human-vehicle interaction research, pp. 1–9 (2020). https://arxiv.org/abs/2005.08637
23. Song, Z., Wang, S., Kong, W., Peng, X., Sun, X.: First attempt to build realistic driving scenes using video-to-video synthesis in OpenDS framework. In: Adjunct Proceedings of the 11th International Conference on Automotive User Interfaces and Interactive Vehicular Applications, AutomotiveUI 2019, Utrecht, The Netherlands, 21–25 September 2019, pp. 387–391. ACM (2019). https://doi.org/10.1145/3349263.3351497
24. Song, Z., Duan, Y., Jin, W., Huang, S., Wang, S., Peng, X.: Omniverse-OpenDS: enabling agile developments for complex driving scenarios via reconfigurable abstractions. In: HCI in Mobility, Transport, and Automotive Systems: 4th International Conference, MobiTAS 2022, Held as Part of the 24th HCI International Conference, HCII 2022, Virtual Event, June 26-July 1, 2022, Proceedings, pp. 72–87. Springer, Cham (2022). https://doi.org/10.1007/978-3-031-04987-3_5
25. Sun, X., et al.: Exploring personalised autonomous vehicles to influence user trust. Cogn. Comput. **12**(6), 1170–1186 (2020). https://doi.org/10.1007/s12559-020-09757-x
26. Wahlstrom, E., Masoud, O., Papanikolopoulos, N.: Vision-based methods for driver monitoring. In: Proceedings of the 2003 IEEE International Conference on Intelligent Transportation Systems, vol. 2, pp. 903–908 (2003)
27. Wang, J., Xiong, Z., Duan, Y., Liu, J., Song, Z., Peng, X.: The importance distribution of drivers' facial expressions varies over time!. In: 13th International Conference on Automotive User Interfaces and Interactive Vehicular Applications, pp. 148–151 (2021)
28. Wang, S., et al.: Oneiros-OpenDS: an interactive and extensible toolkit for agile and automated developments of complicated driving scenes. In: HCI in Mobility, Transport, and Automotive Systems: 4th International Conference, MobiTAS 2022, Held as Part of the 24th HCI International Conference, HCII 2022, Virtual Event, June 26–July 1, 2022, Proceedings, vol. 13335, pp. 88–107. Springer, Cham (2022). https://doi.org/10.1007/978-3-031-04987-3_6
29. Xiong, J., Bi, R., Zhao, M., Guo, J., Yang, Q.: Edge-assisted privacy-preserving raw data sharing framework for connected autonomous vehicles. IEEE Wirel. Commun. **27**(3), 24–30 (2020)
30. Xiong, Z., et al.: Face2statistics: user-friendly, low-cost and effective alternative to in-vehicle sensors/monitors for drivers. In: Krömker, H. (ed.) HCI in Mobility, Transport, and Automotive Systems – 4th International Conference, MobiTAS 2022, Held as Part of the 24th HCI International Conference, HCII 2022, Virtual

Event, June 26–July 1, 2022, Proceedings, Lecture Notes in Computer Science, vol. 13335, pp. 289–308. Springer, Cham (2022). https://doi.org/10.1007/978-3-031-04987-3_20

31. Zhang, Yu., Jin, W., Xiong, Z., Li, Z., Liu, Y., Peng, X.: demystifying interactions between driving behaviors and styles through self-clustering algorithms. In: Krömker, H. (ed.) HCII 2021. LNCS, vol. 12791, pp. 335–350. Springer, Cham (2021). https://doi.org/10.1007/978-3-030-78358-7_23

Using Eye Tracking to Guide Driver's Attention on Augmented Reality Windshield Display

Elham Fathiazar[1]([✉]) [iD], Bertram Wortelen[2] [iD], and Yvonne Kühl[1]

[1] German Aerospace Center (DLR), Institute of Systems Engineering for Future Mobility, Oldenburg, Germany
elham.fathiazar@uol.de
[2] Humatects GmbH, Oldenburg, Germany

Abstract. Situation awareness is a critical component for driving performance and safety, especially when the driver is out of the loop and should take over an automated vehicle. To enhance safety in the driving task, head-up displays with augmented reality are being considered as a key technology. In this paper, we propose a novel augmented reality-based attention guidance system for full windshield head-up displays for in-view and out-of-view objects. The proposed system takes the current gaze data into account to guide the user dynamically to the important objects rather than just highlighting the target. We conducted a user study to assess the attention guidance system on a simulated full windshield display. We explored how the gaze behavior of locating relevant objects changed when using the attention guidance system compared to the system with no visual cues as a baseline. Our results suggest that the proposed system significantly reduces the time to perceive the objects of interest.

Keywords: Attention Guidance · Windshield Display · Head-up display · Driver Situation Awareness · Augmented Reality

1 Introduction

Situation awareness is a critical component for driving safety because it directly influences the decision-making process and subsequent human performance [1]. Endsley defines situation awareness (SA) as "the perception of the elements in the environment within a volume of time and space, the comprehension of their meaning and the projection of their status in the near future" [2]. According to Endsley definition of awareness, the first level in gaining situation awareness involves the perception of elements in the current situation. To obtain a sufficient situation awareness, the driver needs to continuously observe his environment and perceive all relevant information for the driving task, e.g., signs and signals, unexpected hazards, and nearby traffic. In takeover situations of an automated vehicle, gaining a sufficient level of situation awareness in a short amount of time can be a challenging task, especially if the driver is out of the driving loop at the time of the takeover request by the autonomous car. Studies on takeover scenarios show that the driver needs at least 7 s to safely take over the driving task [3, 4]. The time needed depends on how quickly the driver can perceive all the information from the environment and develop sufficient situation awareness. Since the acquisition of situation awareness in takeover situations can be time- and safety-critical, it is important to design interfaces that support the driver in acquiring sufficient situation awareness, e.g., by directing the driver's attention to critical components in the driving scene to accelerate the perception of relevant information in less time.

Augmented reality head-up displays (HUD) are becoming a common feature in modern cars. These displays normally show relevant information like speed or navigation information on a small display in front of the driver. Augmented reality windshield displays extend head-up displays to use entire windshield for display of information [5–8]. These displays make a larger zone accessible to display information and can overlay objects in the environment with additional information and cues. The augmented reality-based technologies can also be used to assist the driver detecting important parts of the environment, like potentially dangerous events, by directing the drivers' attention. Some research has evaluated methods for directing driver attention with augmented reality cues. Most of them used augmented reality cues for highlighting potential hazardous objects to increase detection rate or reaction time. In these studies, the augmented reality cues were directly mapped over the object, for example, with a frame [9–13] or silhouette [14] around the object, a line below the object [15], or lines to the object's corners [16]. The experiments with these techniques show that the participants detected critical situations with visual warning cues earlier, this led to a safer and smoother driving experience and a better situation awareness. Further studies in take over from self-driving cars provided additional information about the current traffic situation or the reason for the impending handover to support the driver [17–19]. The problem here is that these visualizations only provide abstract information about the situation and often direct the attention to an extra display, not to the driving environment. Moreover, all these techniques only work when the hazardous objects are in-view and on-screen. But important or safety critical information is often out of the view of the driver. Research about visual attention guidance to out-of-view objects in driving context are limited. The work from Tönnis et al. [20], presented two visualization schemes: a 3D arrow pointing in the direction of danger and a bird's eye perspective. The results showed that with the 3D arrow as a warning indicator, reaction times were significantly shorter. These results suggests that a simple technique of directing attention in the target direction, as is done with the arrows, is a promising approach that we adapt in our work.

In this paper, we propose an augmented reality-based attention guidance system on a full windshield head-up display in takeover situations to direct the driver's gaze directly to the important information in the environment. The system is designed to attract the driver's attention near to his/her foveal field of view and guide it toward critical information in the driving scenery. The proposed attention guidance, therefore, actively directs attention to objects inside and outside the field of view of the driver. In a user study we tested our attention guidance strategy on a virtual full windshield display (simulated using simulator software SILAB), in a high-fidelity driving simulator from German Aerospace Center (DLR). A three lane highly automated driving scenario with various traffic conditions were used to guide driver attention towards traffic elements such as traffic signs or other traffic participants. To simulate the takeover process of inattentive drivers, the participants were engaged with a display located inside the vehicle. Acoustics tones were used to inform the participants about the intervals to attend to the driving scene. We examined how the detection time of relevant objects and the drivers' gaze behavior changed when using the proposed attention guidance system compared to a baseline where no attention guidance system was used. Our results shows that the detection time for relevant objects reduced significantly using attention guidance system compared to the baseline condition.

2   Attention Guidance System Design

In this section we describe the proposed attention guidance system that leads the user to perceive relevant objects during a takeover task from automated to manual driving. We implemented the attention guidance system on a simulated full windshield head-up display. Head-up displays are becoming a common feature in modern cars. These displays normally show relevant information like speed or navigation information on a small display in front of the driver. However, as theses displays usually cover only a small zone of the windshield, the displayed information are not always located close to foveal field of view (FoV).

The use of a head-up display that utilizes the entire windshield area, along with an eye tracking system, and advanced environmental sensors such as cameras and lidars, enables the vehicle with two distinct features. First, knowing the position of objects in the vehicle's environment enables the vehicle to display information about these objects close to them on the windshield. Additionally knowing the position of the driver's eyes and the line of sight between object and eyes allows for a more precise placement of the information on the object. Thus, augmenting the driver's environment. For example, it could place warning symbols on vehicles that are breaking or getting close to the lane boundaries. Second, knowing the position of the driver's eyes and the direction of the gaze enables the vehicle to display information on the windshield close to the line of sight within the driver's foveal FoV.

Our proposed attention guidance system uses both features but does not directly display information or augment objects. Instead, it displays an animation that guides the driver's attention away from the current focus of attention to a target object, which the vehicles consider critical for the driver to know about. The attention guidance system actively directs the user's gaze to the target object rather than just highlighting the object. Thus, these objects can be both in and out of the view.

We consider such a system especially useful for highly automated driving. In situations, in which the vehicle is driving autonomously but reaches its operational limits and has to hand over control to the driver by issuing a takeover request (TOR). The driver now has to perceive all relevant elements and information about the current driving situation in a limited amount of time. If the vehicle identifies that the driver missed important aspects of the situation, e.g., a traffic participant or a traffic sign, it can guide the driver's attention to it using the proposed system.

To this end, the attention guidance system consists of a circle that appears very close to the foveal FoV of the driver and moves toward the target object. The circle is not displayed directly in the foveal FoV to avoid obstructing the elements that the driver is currently looking at. However, the movement of the circle close to the foveal FoV is easily detectable by the driver. The circle attracts the attention of the driver and guides it from its current focus toward the target object. The circle is displayed in white as a neutral color not to be overloaded by other colors of traffic elements. The start position of the animated circle is calculated from the current gaze position of driver, obtained from a real time eye tracking system mounted on the driver's head.

To have a smooth transition, animation includes a fade in and fade out phase in the beginning and at the end of the animation. The circle fades in close to the current gaze position of the driver within 200 ms. The circle then moves to the position of the target object in about 700 ms, depending on the distance of the detected gaze position to the target area of interest. The speed of moving is relatively fast to avoid that the driver fixates on the animation but directs the driver to look to the target object. When the target is reached, the circle fades out (duration of about 150 ms).

Fig. 1. Guidance system designed to actively direct the driver attention to a target object, consist of a moving white circle, starting close to current gaze position in green. The circle moves toward a speed sign as target object in this example. The arrow depicts the path for the moving circle.

To simulate a full windshield HUD, we implemented the animation directly in the simulator software. Figure 1 depicts schematic guidance system on the projection screen. The green filled circle depicts the gaze position at the beginning of the animation. The small white circle close to gaze position depicts the start of animation. The circle moves towards the target object (here a speed limit sign on the right side of the road). The white arrow indicates the path of the circle motion. The circle fades out at the position of the sign, depicted with a white circle around the speed sign.

3 Experiment

To evaluate the attention guidance system described in the previous section we conducted a user study in a high-fidelity driving simulator. We simulated a takeover situation during autonomous driving and analyzed whether the attention guidance system reduces the detection time of the target object during the takeover preparation.

Fig. 2. Sample driving scenarios for testing attention guidance system. In each scenario, an element was selected to which attention should be directed, e.g., traffic signs and other road participants slowing down or cutting the lane. These elements are highlighted by white circles.

3.1 Study Design

The study was designed as a within-subjects controlled driving simulator study. We investigated if the dependent variable detection time (time it takes the driver to look at a specific element within the driving environment) is influenced by the independent variable (no attention guidance vs. attention guidance on a simulated full windshield head-up display). Various traffic scenarios were simulated on a three-lane highway. The vehicle drove in autonomous mode and no interaction with the vehicle was required by the participants. The different scenarios were driven through by the autonomous vehicle one after the other without vehicle stopping. The scenarios included a variety of situations and contained varying environmental objects, including low, medium and high traffic, speeding vehicles, overtaking maneuvers, accidents and traffic jams on the oncoming lanes, emergency vehicles, different traffic signs, warning triangles, and construction vehicles next to the road. In each scenario, a traffic element was selected as a target object to which attention should be drawn, e.g., hazard road signs, speed limit signs, cutting vehicles ahead, and braking of vehicles ahead. These elements were used to guide attention in 14 different traffic scenarios. Figure 2 shows six of the driving scenarios, with the elements of interests highlighted by a white circle around them.

3.2 Experimental Setup and Procedure

The study was performed in a high-fidelity driving simulator in the German Aerospace Center (DLR) in Oldenburg. The driving simulator consists of a full vehicle mockup with a 200° field of view cylindrical projection wall at a focal distance of about 3 m in front of the driver (see Fig. 3b). In addition, the side mirrors and rearview mirror show the traffic scene on planar displays. The driving simulator was programmed to drive autonomously during the whole study. The SILAB software was used for the driving simulation.

We recruited 10 participants (3 males and 8 females) from the university of Oldenburg. The participants aged between 19 and 36 (mean = 24.8, SD = 5.4). All participants had a valid driver license and drove on average 6138 km per year.

The experiment procedure started with a briefing session and the participants were handed information about the experiment. They signed a consent form to attend to the experiment followed by filling out a demographic questionnaire. Afterwards, they sat in the driver's seat of the driving simulator and drove autonomously two rounds of scenarios, once with no attention guidance system (baseline) and once with attention guidance system. To avoid any order effect, half of the participants started with the baseline condition (no attention guidance) while the other half started with the use of the attention guidance system.

To simulate a TOR situation where the driver has to get back into the loop, the drivers were instructed to not look at the environment for some time and instead solve an in-vehicle task. This represents a non-driving related task (NDRT) a driver might conduct during autonomous driving. A touch display was located inside the car in the center console for this purpose. Acoustic signals were used to inform the participants about when to attend to the road and when to perform the NDRT. When the vehicle entered a new scenario an audio signal – representing the takeover request – was played, instructing the driver to focus attention to the driving environment and assess the situation, but the drivers did not have to take over control. Instead, a second audio signal instructed the driver to focus attention back to the NDRT and away from the environment. The participants replied to the questions from NDRT until the audio signal for the next scenario was played, instructing them to focus attention back to the environment. Figure 3 shows how these two phases (NDRT phase and attention phase) were alternated during the experiment.

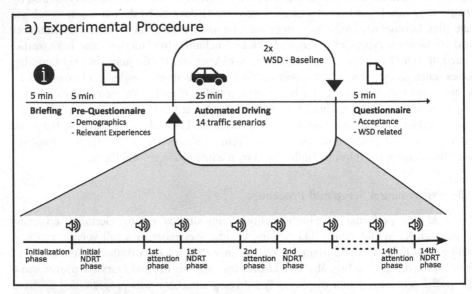

Fig. 3. (a) The procedure followed during experiments. Each participant drove the scenarios two times, once with attention guidance system and once no attention guidance system as a baseline. Half of the participants started with baseline session first and the other half with attention guidance system. In each session participants were instructed to alternately attend to the road or reply to a NDRT questionnaire on a display inside the vehicle. Acoustic tones were used to inform participants about the intervals to attend to the road or perform NDRT questionnaire about their last observed driving scenes. (b) The physical set up used to perform experiments.

The duration of the attention phases was much shorter (mean duration = 9.3 s, SD = 4.5 s) than the NDRT phases (mean duration = 104.4 s, SD = 19.2 s). However, the duration of the attention phase was intentionally very short in order to put time pressure on the participants during the attention phase. To give a high incentive to the drivers to make a thorough situation assessment in the short period of attention phase, the NDRT questioned the drivers about the last observed driving situation. The NDRT questions asked about a variety of different aspects of the driving situation. This was done to stimulate the participants to try to pay attention to every aspect of the driving situation. Table 1 shows the list of questions. The NDRT phases always started with different questions and slightly varying order. After each answer the drivers had to rate the confidence in their answer on a four-point scale of high confidence to no confidence. On average drivers needed 5.5 s (SD = 2.6 s) to answer a question.

Table 1. The list of questions drivers had to answer during the NDRT phases (translated from German).

	Question	Answer options	
1	In which lane is a car approaching from behind?	• Left lane • Middle lane	• Right lane • No vehicle
2	Did you notice anything unusual from the following list?	• Flashing blue lights • Warning triangle • Hazard lights	• Traffic rule violation • Construction site • Nothing
3	What is your estimate for the speed of the vehicle in front?	• Slower than me • Same speed	• Faster than me • No vehicle
4	How many vehicles are behind you in the right lane?	• 0 • 1	• More than 1
5	Is your indicator on?	• Yes	• No
6	Is there a vehicle to your left?	• Yes	• No
7	How many vehicles are ahead of you in the right lane?	• 0 • 1	• More than 1
8	Which traffic sign was just displayed?	• 80 km/h limit • 100 /h limit • 120 km/h limit	• 130 km/h limit • Other • None
9	How many vehicles are in front of you in the middle lane?	• 0 • 1	• More than 1
10	How far ahead is the next exit?	• 300 m • 700 m	• 1500 m • No exit
11	How many vehicles are ahead of you in the left lane?	• 0 • 1	• More than 1
12	Did a vehicle just switch lanes?	• No	• Yes, ahead of me • Yes, behind me
13	Which city was shown on the blue navigation sign?	• No sign • Hamburg • Berlin	• Dresden • Oldenburg • Ulm
14	Is there a traffic jam ahead of you?	• Yes	• No

(*continued*)

Table 1. (*continued*)

	Question	Answer options	
15	How many vehicles are behind you in the middle lane?	• 0 • 1	• More than 1
16	Which traffic sign was just displayed?	• No-passing zone • Traffic jam waning • Redirection	• Warning of strong wind • Other • None
17	How many vehicles are behind you in the left lane?	• 0 • 1	• More than 1
18	Is another vehicle about to change lanes?	• No	• Yes, ahead of me • Yes, behind me
19	Did you notice anything unusual from the following list?	• Accident • Speeder • Traffic jam	• Overtaking maneuver • No
20	Which traffic sign was just displayed?	• Curvy road ahead • Construction site • No speed limit	• Traffic jam ahead • Other • None
21	How do you estimate the distance to the vehicle behind you?	• Large distance • Normal distance	• Tailgating • No vehicle
22	Is another vehicle driving unusually close to the lane boundary?	• No	• Yes, ahead of me • Yes, behind me
23	How is the traffic flow on the oncoming lane?	• No traffic • Low traffic	• Heavy traffic • Traffic jam
24	What is your speed?	• 80 km/h • 90 m/h • 100 km/h	• 110 km/h • 120 km/h • 130 km/h
25	What was on the hard shoulder?	• Warning triangle • Normal car	• Construction site vehicle • Nothing

As the last step of the process, the participants filled out acceptance questionnaire of Van der Laan [21].

4 Results

We examined the performance of drivers during both driving attention phases and NDRT phases. For driving attention phase, we recorded gaze data and examined whether attention guidance leads to a better performance in detecting the target object (Sect. 4.2). For NDRT phase we analyzed the responses to the questions and examined whether the attention guidance leads to a better situation awareness with respect to the target object (Sect. 4.3). We also measured user acceptability and satisfaction with the van der Laan questionnaire. The results were a mean score of 0.85 (±0.7) for usefulness and a mean score of 0.52 (±0.9) for satisfaction.

4.1 Gaze Data Processing

Gaze data was recorded through the entire experiment. As a first preprocessing step, we extracted the intervals in which drivers were paying attention to the road. The 14 scenarios yielded a total number of 140 paired gaze data sequences, collected from 10 participants for two conditions (with attention guidance and baseline). We then examined the extracted data for valid gaze data. Non-valid data points were flagged by the eye tracker during recording. Therefore, for consistent data analysis, we first filtered out gaze data sequences that were flagged as non-valid. Because we recorded paired data for each participant (one for the baseline and one for the attention guidance system), we also performed the filtering process in pairs and therefore removed a paired sequences if one of the gaze data sequences in the pair was not valid. This preprocessing step resulted in a total of 46 paired valid gaze data sequences that were used for statistical analysis.

To detect gaze fixations from saccades, we used the velocity thresholding method. We set the angular velocity threshold of 10 degrees per second and the minimum gaze duration of 120 ms as fixation detection criteria. We analyzed the number of gaze fixations for the area in front of driver where the eye tracker could provide eye movement information, but we did not find any statistical difference between the baseline and attention guidance system in the number of gaze fixations.

We also examined whether participants failed to perceive the target object. Our analysis shows that in the condition with attention guidance, participants perceived the target object in all 46 data sequences. However, in the case of the baseline without attention guidance, participants did not perceive the target object in 7 data sequences (from 4 different participants in 4 different scenarios) out of 46 data sequences.

4.2 Detection Time

The detection time is defined as the time span between the appearance of the target object in the driving scene and the time when the gaze is fixated on the target object for the first time. The driver has time to fixate the target object until the object disappears from the scene, which means that the disappearance of the object is the latest time at which fixation can occur. Therefore, we analyzed the data to calculate the detection time until the object disappeared. Accordingly, the maximum possible detection time is the time difference between the appearance and disappearance of the target object. We used the maximum detection time for the data samples, in which drivers failed to detect the target object. Figure 4 shows the boxplots for detection times for the condition with attention guidance and no attention guidance (baseline). Our results show that the detection time reduced with attention guidance system. We performed independent samples t-test to examine whether the differences in detection times between the two groups were significant. The results show a significant difference ($t = 3.984$, $p = 0.0001$) for detection time between the condition with attention guidance system and baseline.

4.3 Analysis of NDRT Questions

We analyzed the drivers' performance on the NDRT questions between the driving attention phases. The main purpose of the questions was to give the drivers a strong

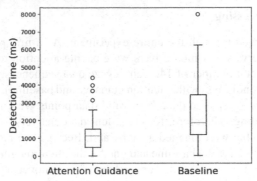

Fig. 4. – Boxplot depicting the time to fixate to target object when using attention guidance system and when no attention guidance system was used (baseline). A significant difference is found between the two groups using independent samples t-test (p = 0.0001).

incentive to try to perceive as much of the driving situation as possible between the two audio signals. However, it also reflects how much the drivers were able to perceive during the short attention phase.

We analyzed the rate of correct answers. For this analysis, we excluded all questions in scenarios where no clear correct answer could be determined. This was mainly the case when the traffic situation changed during the scenario. Of the remaining 3123 answers, 2280 (73%) were correct and 843 (27%) were incorrect. This high error rate indicates that the driving attention phase was often shorter than needed to fully grasp the traffic situation. Drivers seemed to be aware of this. The results show that the confidence level for correct answers were higher than the confidence level for incorrect answers (on average 0.5 points higher on the 4-point confidence scale). An independent samples t-test showed that this difference was significant, $t(3121) = 14.9549$, $p < 0.01$.

Another finding that emerged from the questions is whether the attention guidance leads to a better situation awareness with respect to the target object. In this case, drivers should be more likely to answer questions about the target object correctly if attention guidance was used. To test this, we determined the questions about the target object for each scenario and analyzed the proportion of correct answers. Without attention guidance, only 41.0% of the answers were correct, while with attention guidance, 66.2% were correct. A two proportions z-test showed that the difference is significant ($z = 2.9615$, $p = 0.002$). Note that both rates are lower than the previously reported overall rate of correct answers. This is because the target objects were only visible for a short period of time and were generally hard to spot. Therefore, the error rate is higher than for questions about other elements in the environment.

5 Discussion

When the vehicle informs the driver to take over control, this task of the driver consists of an active search for all relevant information. The driver performs this task by looking around in the environment in a sequence of fixations in areas with a high expected value. Based on Wickens et al. [22] expected value model of attention, these are areas that have a statistically high probability of providing new information and where it would be critical for the driver to miss the information, e.g., the behavior of the vehicle ahead. Our proposed attention guidance system automates parts of the active search task as it already indicates to the driver where to look. If the driver perceives the system to be reliable, i.e., if it reliably directs the driver's attention to new information of high value, then this should increase the expected value of the area highlighted by the system. Therefore, it should strongly increase the probability that the driver will choose the highlighted area as the next area to look at.

Our results showed that the time to detect the target object was significantly reduced when attention guidance was used. In addition, analysis of the NDRT showed that directing attention led to better situation awareness with respect to the target object. This was consistent with our expectation that the participant would look more quickly at the relevant object when the gaze was directed with a visual guidance system, especially when the information is outside the field of view.

Analysis of participants' gaze behavior showed that they scanned the windshield and mirrors during the attention phase of driving, indicating an active search for relevant information to take over the vehicle. However, this extensive search could be an additional effort of active search due to the NDRT questions that participants answered between attention phases. We speculate that the questions may have motivated participants to become overly attentive to the driving environment and mirrors in order to memorize more detailed information about the driving environment to better answer the questions. Despite this motivation, the attention guidance system was able to direct participants' attention to reduce the time needed to perceive the target object by giving it priority in observation.

Even though we were able to show a clear improvement for the detection time of the target objects, we believe that the benefit of the system highly depends on the selection of target objects. Human attention is a limited resource. Guiding attention to a target object means "stealing" attention from other objects. The attention guidance system results in safety gains in takeover situations only if the target object is actually very critical and the driver does not notice it. Otherwise, the system might even reduce safety by diverting attention from other critical information.

6   Conclusion and Future Work

In this work, we presented an augmented reality-based attention guidance system that actively directs the attention to in-view and out-of-view objects. We implemented the attention guidance system on a simulated full windshield HUD. Our results show that the system reduces the time for the driver to perceive relevant objects to gain required driving situation awareness.

In the future, we plan to combine the HUD with a system that predicts the current situation awareness based on eye tracking data. This should ensure that the user's attention is only directed to important objects in the environment that they have not yet noticed. In the next study the situation awareness questionnaire should be designed with a narrower focus to the current situation to question for a more goal-oriented situation awareness. Furthermore, we desire to study our approach in real takeover situations and during manual driving.

Acknowledgement. This work was funded by the German Federal Ministry of Economics and Energy, via national project, SituWare, with the topic of assessing driver situation awareness for adaptive cooperative handover strategies in highly automated driving. Our special thanks to Lars Webers for his valuable contribution to implementation and execution of the experiments.

References

1. Gugerty, L.: Situation awareness in driving. Handb. Driv. Simul. Eng. Med. Psychol. **1**, 265–272 (2011)
2. Endsley, M. R.: Design and evaluation for situation awareness enhancement. In Proceedings of the Human Factors Society Annual Meeting, vol. 32, no. 2, pp. 97–101. Sage Publications, Los Angeles (1988)
3. Gold, C., Damböck, D., Lorenz, L., Bengler, K.: "Take over!" How long does it take to get the driver back into the loop?. In: Proceedings of the Human Factors and Ergonomics Society Annual Meeting, vol. 57, no. 1, pp. 1938–1942. Sage Publications, Los Angeles (2013)
4. Lu, Z., Coster, X., De Winter, J.: How much time do drivers need to obtain situation awareness? a laboratory-based study of automated driving. Appl. Ergon. **60**, 293–304 (2017)
5. Takaki, Y., Urano, Y., Kashiwada, S., Ando, H., Nakamura, K.: Super multi-view windshield display for long-distance image information presentation. Opt. Express **19**, 704–716 (2011)
6. Doshi, A., Cheng, S.Y., Trivedi, M.M.: A novel active heads-up display for driver assistance. IEEE Trans. Syst. Man Cybern. Part B (Cybern.) **39**(1), 85–93 (2008)
7. Sun, T.X.: 47.4: a novel full windshield heads-up display technology. In: SID Symposium Digest of Technical Papers, vol. 46, no. 1, pp. 712–715 (2015)
8. Road & Track Homepage. https://www.roadandtrack.com/news/a42594729/your-windshield-is-the-next-big-screen/. Accessed 02 Feb 2022
9. Frémont, V., Phan, M.T., Thouvenin, I.: Adaptive visual assistance system for enhancing the driver awareness of pedestrians. Int. J. Hum.-Comput. Interact. **36**(9), 856–869 (2020)
10. Phan, M.T., Thouvenin, I., Frémont, V.: Enhancing the driver awareness of pedestrian using augmented reality cues. In: IEEE 19th International Conference on Intelligent Transportation Systems (ITSC), pp. 1298–1304. IEEE (2016)
11. Rusch, M.L., et al.: Directing driver attention with augmented reality cues. Transport. Res. F: Traffic Psychol. Behav. **16**, 127–137 (2013)
12. Damböck, D., Weißgerber, T., Kienle, M., Bengler, K.: Evaluation of a contact analog head-up display for highly automated driving. In: 4th International Conference on Applied Human Factors and Ergonomics, San Francisco, USA (2012)
13. Schall Jr., M.C., et al.: Augmented reality cues and elderly driver hazard perception. Hum. Factors **55**(3), 643–658 (2013)
14. Bozkir, E., Geisler, D., Kasneci, E.: Assessment of driver attention during a safety critical situation in VR to generate VR-based training. In: ACM Symposium on Applied Perception 2019, pp. 1–5 (2019)
15. Hwang, Y., Park, B.J., Kim, K.H.: Effects of augmented-reality head-up display system use on risk perception and psychological changes of drivers. ETRI J. **38**(4), 757–766 (2016)
16. Pomarjanschi, L., Dorr, M., Barth, E.: Gaze guidance reduces the number of collisions with pedestrians in a driving simulator. ACM Trans. Interact. Intell. Syst. (TiiS) **1**(2), 1–14 (2012)
17. Weidner, F., Broll, W.: Stereoscopic 3D dashboards: an investigation of performance, workload, and gaze behavior during take-overs in semi-autonomous driving. Pers. Ubiquit. Comput. **26**(3), 697–719 (2022)

18. Walch, M., Lange, K., Baumann, M., Weber, M.: Autonomous driving: investigating the feasibility of car-driver handover assistance. In: Proceedings of the 7th International Conference on Automotive User Interfaces and Interactive Vehicular Applications, pp. 11–18 (2015)
19. Naujoks, F., Forster, Y., Wiedemann, K., Neukum, A.: A human-machine interface for cooperative highly automated driving. In: Stanton, N.A., Landry, S., Di Bucchianico, G., Vallicelli, A. (eds.) Advances in Human Aspects of Transportation, pp. 585–595. Springer, Cham (2017). https://doi.org/10.1007/978-3-319-41682-3_49
20. Tönnis, M., Klinker, G.: Effective control of a car driver's attention for visual and acoustic guidance towards the direction of imminent dangers. In: IEEE/ACM International Symposium on Mixed and Augmented Reality 2006, pp. 13–22. IEEE (2006)
21. Van Der Laan, J.D., Heino, A., De Waard, D.: A simple procedure for the assessment of acceptance of advanced transport telematics. Transp. Res. Part C: Emerg. Technol. 5(1), 1–10 (1997)
22. Wickens, C.D., Helleberg, J., Goh, J., Xu, X., Horrey, W.J.: Pilot task management: Testing an attentional expected value model of visual scanning. Savoy, IL, UIUC Institute of Aviation Technical Report (2001)

The Effect of Implicit Cues in Lane Change Situations on Driving Discomfort

Konstantin Felbel[✉], André Dettmann, and Angelika C. Bullinger

Chair for Ergonomics and Innovation, Chemnitz University of Technology, Chemnitz, Germany
konstantin.felbel@mb.tu-chemnitz.de

Abstract. A high driving comfort is one of the key prerequisites for the use of automated driving functions. Therefore, it is important to understand how possible discomfort while driving could be reduced. The presence of implicit cues used by manual drivers to anticipate lane changes could effect perceived discomfort in lane change situations. The present study investigated the effect of implicit cues in a driving simulation study. The test scenario was a two-lane motorway where participants were automatically driven in the left lane. During the automated drive, 20 lane-change situations occurred. A vehicle in the right lane in front of the participant's vehicle either changed into the participant's lane due to a slow-moving truck or did not. Three implicit cues were systematically varied. Participants were asked to continuously indicate their discomfort using a handset control.

The results show that implicit cues effect the anticipation of a lane change as well as the perceiving of discomfort. In particular, speed difference and Time To Collision have an high effect on perceived discomfort. This may be due to the relation-ship between speed and distance. Small speed differences result in a smaller optical distance between the ego-vehicle and the lead-vehicle. The reduced distance between the ego-vehicle and the lead-vehicle increases the feeling of insecurity in case of emergency braking. A small effect could be observed by pre-shifting before the actual lane change. By pre-shifting within the lane, the driver may be able to prepare for an upcoming lane change.

1   Theoretical Background

By 2050, a mixed traffic scenario of manually driven and highly automated vehicles (HAFs) is expected [1–3]. By then, HAFs must be able to handle interactions with manually driven vehicles [4, 5]. According to [6] and [4], road traffic can be considered as a social system with the goal of efficient, smooth, safe and comfortable interaction between agents [7, 8]. Current automated driving systems, such as Lane Departure Warning or Active Lane Keeping Assistant, try to take these goals into account, but due to their less predictive function, this is often not optimal, especially in interactions with other road users. For example, these systems keep the vehicle as close to the centre of the lane as possible, resulting in a trajectory that is clearly different from a manual (natural) trajectory [9]. In oncoming traffic, for example, the manual trajectory is shifted laterally towards the edge of the lane, which is seen as an implicit cue to the driver's

H. Krömker (Ed.): HCII 2023, LNCS 14049, pp. 32–41, 2023.
https://doi.org/10.1007/978-3-031-35908-8_3

intention (e.g. to create more safety distance) [10, 11]. Human drivers can recognise these implicit cues (i.e. lateral and longitudinal trajectories/vehicle movements) and infer early driving intentions, as well as use them themselves to communicate their own driving intentions to other road users [4, 12–16]. However, current automation systems are not yet able to do this. Two problems arise. Firstly, if the driven passenger sees conflicting automated driving behaviours in the presence of implicit cues (e.g. lateral and longitudinal trajectories/vehicle movements), this makes it difficult for the passenger to correctly understand the automated behaviour. Secondly, due to poor interpretation of natural driving behaviour or failure to recognise implicit cues, current automated systems may also fail to adequately predict the driving intentions of manually driven vehicles [5]. As a result, automated systems may detect a manual lane change too late, resulting in inadequate response from the HAF. This significantly reduces the driving comfort for the occupants. According to [17], discomfort is a state of driving-related psychological tension or stress in moments of a limited harmony between the driver and the environment. Therefore discomfort must to be reduced in order to experience comfort, which [18] defines as relaxation. Failure to mitigate discomfort may result in the technology not being used and the positive effects of automated driving being lost. It is therefore important to understand which implicit cues we, as human drivers, particularly use to anticipate the driving intentions of other road users and whether the presence of implicit cues with a specific characteristic can help to reduce the feeling of discomfort in a lane change situation. This knowledge could be used in appropriately weighted way in automated driving functions to improve predictive performance and reduce discomfort in the automated driving.

2 Methodology

To investigate this issue, a driving simulator study was conducted with 15 participants (1 female, 14 male). The mean age was 26.1 years (SD = 3.8 years). 13 participants. Had held a driving licence for 5.4 years (SD = 3.3) and had driven approximately 50.407 km (SD = 115.900 km) since receiving their licence. Two participants did not have a driving licence. In a series of questions about subjective driving style, the majority of drivers described their driving style as anticipatory, relaxed, cautious but also courageous.

2.1 Design and Apparatus

At the beginning of the study, each participant was informed about the study procedure. Participants were told that they could choose to withdraw at any time without negative consequences. No compensation was given for participation. The total duration of the study was 30 min. After completing a demographic questionnaire, participants were seated in the driving simulator. A 5-min test drive allowed participants to familiarise themselves with the simulator. The task was explained and any remaining questions were answered. After the test drive, during which any signs of simulator sickness were observed, the experimental drive was started. A driving simulator study based on a mixed $2 \times 2 \times 2 \times 2$ design was used to investigate driver discomfort during 20 more or less predictable lane change situations. The test scenario was a two-lane motorway where the participants were automatically driven in the left lane. The participant's vehicle (ego-vehicle) stayed in the left lane and maintained a speed of 130 km/h throughout the drive. In each situation, a lead-vehicle in the right lane either changed lanes due to a slow-moving truck or did not. According to [15] three implicit cues used in manual driving to anticipate a possible lane change were varied. These were the relative speed difference between the ego-vehicle and the lead-vehicle (10 km/h or 30 km/h), the Time To Collision (TTC) between the lead-vehicle and the slow-moving truck (5 s or 10 s), and a visible lateral shift to the left or not of the lead-vehicle (0.40 m to the left). The shift to the left was performed in the lane change and no lane change conditions. By combining these, 16 situations were programmed in which the lead-vehicle changed lanes eight times and did not change lanes eight times. A further four situations were programmed with the lead-vehicle's indicators activated while not changing lanes or with the indicators deactivated while changing lanes. Table 1 gives an overview of all constructed situations. In all situations, the TTC from the ego-vehicle to the lead-vehicle was 10 s at the moment when the lead-vehicle changed or did not change its lane, i.e. braked. Figure 1 shows a schematic of the observed lane-changing behaviour of the lead-vehicle. For easier interpretation of the situation, the different road users are defined in Table 2. As the lead-vehicle approaches the slow-moving truck, it first either shifts or does not shift to the left, which takes 2 s. Immediately afterwards, the lead-vehicle maintained its lateral offset position for 2.33 s. After a total of 4.33 s, the lead-vehicle either changed lanes to overtake the truck, which took 1.16 s, or braked with its lights on. Lane changing without pre-shifting took an additional 300 ms.

Table 1. Situations characterisation. Marked situation were constructed with turned off indicators of the lead-vehicle while changing the lane (*) and turned on indicators while not changing the lane (")

Situation	Rel. Speed difference ego-vehicle to lead-vehicle	Distance lead-vehicle to slow-moving truck	Lateral pre-shift by lead-vehicle	Lane change lead-vehicle
1	10 km/h	5 s	yes	yes
2	10 km/h	5 s	yes	no
3	10 km/h	5 s	no	yes
4	10 km/h	5 s	no	no
5	10 km/h	10 s	yes	yes
6	10 km/h	10 s	yes	no
7	10 km/h	10 s	no	yes
8	10 km/h	10 s	no	no
9"	10 km/h	5 s	yes	yes
10*	10 km/h	5 s	yes	no

(*continued*)

Table 1. (*continued*)

Situation	Rel. Speed difference ego-vehicle to lead-vehicle	Distance lead-vehicle to slow-moving truck	Lateral pre-shift by lead-vehicle	Lane change lead-vehicle
11	30 km/h	5 s	yes	yes
12	30 km/h	5 s	yes	no
13	30 km/h	5 s	no	yes
14	30 km/h	5 s	no	no
15	30 km/h	10 s	yes	yes
16	30 km/h	10 s	yes	no
17	30 km/h	10 s	no	yes
18	30 km/h	10 s	no	no
19"	30 km/h	10 s	no	yes
20*	30 km/h	10 s	no	no

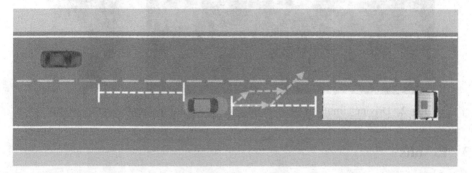

Fig. 1. Schematic of the lane-changing situation. The distance between the slow-moving truck and the lead-vehicle (brown) was dichotomous (TTC = 5 s & 10 s). The distance between ego-vehicle (blue) and the lead-vehicle was constant (TTC = 5 s). The relative speed difference between the ego-vehicle and the lead-vehicle was dichotomous (10 km/h & 30 km/h)

During the drive, participants were asked to state out loud if they expected the lead-vehicle to change lanes. Participants could give an answer and change it as often as they wished. The response was only counted and analysed further if the response was given before the lead-vehicle started its final lane change or until the lead-vehicle braked. In addition, participants were asked to rate their discomfort continuously using a handset controller (see Fig. 2). As mentioned in the introduction, implicit cues could reduce discomfort. Therefore, the lane changes were either very expected, i.e. the implicit cues indicated an upcoming lane change strongly or very unexpected, i.e., the implicit cues did not indicate an upcoming lane change.

Table 2. Road user definition in accordance with [19]

Description	Corresponding icon
Ego-Vehicle The vehicle that approaches the lead-vehicle	
Lead-Vehicle The vehicle that does or does not carry out the lane change	
Slow-moving truck Surrounding vehicle that is likely affect the ego-vehicle or the lead-vehicle	

Fig. 2. Driving simulator (left), handset controller for discomfort rating (right)

3   Results

Due to the small sample size, no statistics were calculated. Instead, the absolute number of participants who experienced discomfort in a given interval of a situation was calculated. This was calculated as follows: if a participant pressed the trigger on the handset controller in a given situation, this was counted as one discomfort rating. If the same participant pressed the trigger again in the same situation, it was not counted as a new discomfort rating. The start of the interval was defined as two seconds before the lead-vehicle pre-shifted laterally and ended when the lead-vehicle's lane change ended. In the no lane change condition, the exact same interval was taken, except that the lead-vehicle did not change lanes. One participant was excluded from the evaluation due to simulation errors.

In the following Table 3 all lane change situations with all implicit cue variations and all no lane change situations with all implicit cue variations are aggregated separately. It shows that more than three times as much discomfort was expressed in the lane change condition. This negates the effect of implicit cues. Without a lane change, implicit cues do not affect the discomfort rating. As discomfort was expressed only eight times in the no lane change situation, further analysis will only focus on the lane change situation.

Table 3. Comparison between aggregated lane change and no lane change situations

Number of participants which expressed discomfort	Rel. Speed difference ego-vehicle to lead-vehicle	Distance lead-vehicle to slow-moving truck	Lateral pre-shift by lead-vehicle	Lane change by lead-vehicle
29	10 kmh & 30 kmh	5 s & 10 s	yes & no	yes
8	10 kmh & 30 kmh	5 s & 10 s	yes & no	no

In the following Table 4 shows a comparison of lateral pre-shift, relative speed and distance on expressed discomfort. Here, relative speed difference has an effect on perceived discomfort. This could be due to the relationship between speed and distance. Small speed differences result in a smaller optical distance between the ego-vehicle and the lead-vehicle. The reduced distance between the ego-vehicle and the lead-vehicle increases the feeling of insecurity in case of emergency braking. In addition, the lateral displacement shows as small effect on discomfort. Therefore, Table 5 compares the effect of the lateral pre-shift in combination with the relative speed difference between the ego-vehicle and the lead-vehicle and the distance from the lead-vehicle to the slow-moving truck.

Table 4. Comparison of lateral pre-shift, relative speed and distance on expressed discomfort

Comparison	Number of participants which expressed discomfort	Rel. Speed difference ego-vehicle to lead-vehicle	Distance lead-vehicle to slow-moving truck	Lateral pre-shift by lead-vehicle
#1	10	10 kmh	5 s & 10 s	yes
	11	10 kmh	5 s & 10 s	no
#2	2	30 kmh	5 s & 10 s	yes
	6	30 kmh	5 s & 10 s	no

The results show that the lateral pre-shift has a small effect on the discomfort rating. Only in the comparison #3 did participants report higher discomfort.

Table 5. Comparison between situations with and without a lateral pre-shift

Comparison	Number of participants which expressed discomfort	Rel. Speed difference ego-vehicle to lead-vehicle	Distance lead-vehicle to slow-moving truck	Lateral pre-shift by lead-vehicle
#1	5	10 kmh	5 s	yes
	8	10 kmh	5 s	no
#2	1	30 kmh	5 s	yes
	3	30 kmh	5 s	no
#3	5	10 kmh	10 s	yes
	3	10 kmh	10 s	no
#4	1	30 kmh	10 s	yes
	3	30 kmh	10 s	no

Participants were asked to verbally indicate whether they think the lead-vehicle would make a lane change. As a lane change was not predetermined by the presence or absence of implicit cues no correct answer could be given by the participant. The question whether a lane change was anticipated was aimed at the general anticipation of a lane change. Table 6 shows the distribution of anticipated lane changes in all situations. Lane change and no lane change situations have been aggregated.

The results are inconclusive. There is no visible correlation. Only the 30 km/h condition appears to have a slight effect on anticipating an upcoming lane change.

Table 6. Participants anticipation of lane changes. Marked situation were programmed with turned off indicators of the lead-vehicle while changing the lane (*) and turned on indicators while not changing the lane (")

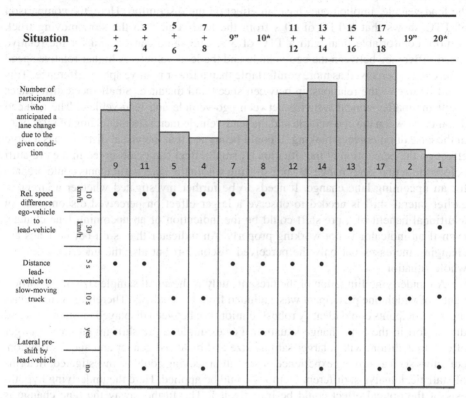

4 Conclusion and Outlook

The current study investigated the effect of implicit cues in lane change situations to reduce perceived discomfort during automated driving. Along with trust and acceptance, comfort is the most important factor in the use of an automated vehicle [20]. If passengers of an automated vehicle experience too much discomfort, the likelihood of not using the automated driving function increases [21]. Therefore, it is important to understand how to reduce discomfort during an automated driving. One possible solution is to understand and use of implicit cues as indicators of upcoming lane changes by lead-vehicles. [15] showed that manual drivers use implicit cues to understand the evolving driving situations, and therefore the presence of implicit cues could have an effect on the perceived discomfort. In the present study, participants rated discomfort in different situations. In these situations, implicit cues were systematically varied. Participants were asked to provide feedback on perceived discomfort using a handset controller, and to express their anticipation of the lead-vehicle changing into the participants's lane.

The results show a strong effect of a lane change performed by a lead-vehicle. When a lane change is performed, discomfort is rated higher than when no lane change is performed. When analysing only situations where a lane change was performed by the lead-vehicle, implicit cues have an effect on the discomfort. Here, the comparison of TTC shows that a TTC of 10 s from the lead-vehicle to the slow-moving truck is more comfortable than with a TTC of 5 s. The same can be said for the relative speed difference between the ego-vehicle and the lead-vehicle. A higher relative speed difference is perceived as more comfortable than a lower relative speed difference. This could be due to the relationship between speed and distance. Small speed differences result in smaller optical distance between ego-vehicle and lead-vehicle. The reduced distance between the ego-vehicle and the lead-vehicle increases the feeling of insecurity in the case of emergency braking. It could be argued that the visual distance has a large effect on the perception of the situation. A small effect can be observed in the pre-shift before the actual lane change. By pre-shifting within the lane, participants could prepare for an upcoming lane change. It needs to be further investigated whether a larger or earlier lateral shift is needed to observe a larger effect on perceived discomfort. An additional benefit of a pre-shift could be the indication of an upcoming lane change, even if an indicator is not working properly. An indicator that is on but no lanes are changing increases not only the perceived discomfort but also the uncertainty of the whole situation.

An underlying limitation of the present study is the small sample size of 15 participants, of which one participant was excluded from the analysis. Therefore, a study with more participants could identify robust relationship between displayed implicit cues and discomfort in the lane change situation. For example, a pre-shift might have a larger effect. In addition, with a larger sample size and balanced gender and age distribution, cofactors such as driving experience or sensation seeking could be investigated. In terms of statistical analysis, different methods could be applied. Here the underlying hypothesis of the optical effect could be investigated. The further away the lane change is optically, the less discomfort it causes. One implication of this may be that lane changes should not be based on TTC but rather on absolute distance. On the contrary, discomfort may not be the best measure of visual distance. Here, the probability of lane changing or how unsafe the lane change was could be assessed.

Acknowledgements. The research was funded by the Deutsche Forschungsgemeinschaft (DFG, German Research Foundation) – Project-ID 416228727 – SFB 1410.

References

1. Ghiasi, A., Hussain, O., Qian, Z., Li, X.: A mixed traffic capacity analysis and lane management model for connected automated vehicles: a Markov chain method. Transp. Res. Part B: Methodol. **106**, 266–292 (2017). https://doi.org/10.1016/j.trb.2017.09.022
2. Patel, R.H., Härri, J., Bonnet, C.: Braking strategy for an autonomous vehicle in a mixed traffic scenario. In: 3rd International Conference on Vehicle Technology and Intelligent Transport Systems; Porto, Portugal: SCITEPRESS - Science and Technology Publications; 2017, pp. 268–275. https://doi.org/10.5220/0006307702680275

3. Altenburg, S., Kienzler, H.-P., auf der Maur, A.: Einführung von Automatisierungsfunktionen in der Pkw-Flotte: Auswirkungen auf Bestand und Sicherheit (2018)
4. Rasouli A, Kotseruba I, Tsotsos JK. Agreeing to cross: How drivers and pedestrians communicate. 2017:264–9. doi:https://doi.org/10.1109/IVS.2017.7995730
5. Schwarting, W., Pierson, A., Alonso-Mora, J., Karaman, S., Rus, D.: Social behavior for autonomous vehicles. Proc. Natl. Acad. Sci. U S A. **116**, 24972–24978 (2019). https://doi.org/10.1073/pnas.1820676116
6. Müller, L., Risto, M., Emmenegger, C.: The social behavior of autonomous vehicles. In: Lukowicz, K., et al. (eds.) Proceedings of the 2016, pp. 686–689. ACM (2016). https://doi.org/10.1145/2968219.2968561
7. ERTRAC. Connected Automated Ariving Roadmap (2019). https://www.ertrac.org/uploads/documentsearch/id57/ERTRAC-CAD-Roadmap-2019.pdf
8. Dettmann, A.: Eignung autostereoskopischer Displays im Fahrzeugkontext. Springer Fachmedien, Wiesbaden (2021)
9. Rossner, P., Bullinger, A.C.: Do you shift or not? influence of trajectory behaviour on perceived safety during automated driving on rural roads. In: Krömker, H. (ed.) HCII 2019. LNCS, vol. 11596, pp. 245–254. Springer, Cham (2019). https://doi.org/10.1007/978-3-030-22666-4_18
10. Schick, B., Seidler, C., Aydogdu, S., Kuo, Y.-J.: Fahrerlebnis versus mentaler Stress bei der assistierten Querführung. ATZ Automobiltech Z. **121**, 70–75 (2019). https://doi.org/10.1007/s35148-018-0219-9
11. Voß, G.M., Keck, C.M., Schwalm, M.: Investigation of drivers' thresholds of a subjectively accepted driving performance with a focus on automated driving. Transport. Res. F: Traffic Psychol. Behav. **56**, 280–292 (2018). https://doi.org/10.1016/j.trf.2018.04.024
12. Simon, K., Bullinger, A.C.: Was stresst, ärgert und beunruhigt Fahrer? Emotionale Reaktionen auf alltägliche Fahrsituationen bei jüngeren und älteren Fahrern. Braunschweig (2017)
13. Simon, K., Bullinger, A.C.: To change or not to change – that is the question: detecting lane change signals for anticipatory highly automated driving, Berlin, 08–10 October 2018 (2018)
14. Sommer, K.C.: Vorausschauendes Fahren: Erfassung Beschreibung und Bewertung von Antizipationsleistung im Straßenverkehr [Dissertation]: Universität Regensburg (2012)
15. Felbel, K., Dettmann, A., Lindner, M., Bullinger, A.C.: Communication of intentions in automated driving – the importance of implicit cues and contextual information on freeway situations. In: Krömker, H. (ed.) HCII 2021. LNCS, vol. 12791, pp. 252–261. Springer, Cham (2021). https://doi.org/10.1007/978-3-030-78358-7_17
16. Dey, D., Terken, J.: Pedestrian interaction with vehicles: roles of explicit and implicit communication. In: Proceedings of the 9th ACM International Conference on Automotive User Interfaces and Interactive Vehicular Applications (AutomotiveUI 2017), pp. 109–13 (2017). https://doi.org/10.1145/3122986.3123009
17. Hartwich, F.: Hartwich (2018)
18. Engelbrecht, A.: Fahrkomfort und Fahrspaß bei Einsatz von Fahrerassistenzsystemen [Dissertation]. Humboldt-Universität zu Berlin, Berlin (2013)
19. Lee, S.E., Olsen, E.C., Wierwille, W.W.: A Comprehensive Examination of Naturalistic Lane-Changes (2004)
20. Dettmann, A., et al.: Comfort or not? automated driving style and user characteristics causing human discomfort in automated driving. Int. J. Human–Comput. Interact., 331–339 (2021). https://doi.org/10.1080/10447318.2020.1860518
21. Hartwich, F., et al.: Fahrstilmodellierung im hochautomatisierten Fahren auf Basis der Fahrer-Fahrzeuginteraktion: Abschlussbericht "DriveMe" (2016). https://doi.org/10.2314/GBV:870302329

Design Study on the Effect of Intelligent Vehicles Interaction Mode on Drivers' Cognitive Load

Miao Liu[ID] and Bo Qi[(✉)]

East China University of Science and Technology, Shanghai 200237, People's Republic of China
qibo1998@126.com

Abstract. The cognitive load of drivers is examined based on Multiple-Resource Theory by building the Human-Machine Interaction model of intelligent vehicles in order to improve the driving experience of this group and prevent traffic accidents as much as possible. The Multiple-Resource Theory is used to analyze the interaction modes of intelligent vehicles in this study, and the role of Physical Stuff interaction, Touch-screen Interaction, Voice Interaction, System-initiative, and Multimodal Interaction on the driver's cognitive load is evaluated using simulated driving experiments and a Likert 5-point Likert scale. Among them, Multimodal Interaction evaluates the effect of physical Stuff interaction with Touch-screen Interaction and Physical Stuff Interaction with Voice Interaction. The principles of HMI design for Intelligent vehicles are investigated and deduced. Finally, an intelligent vehicle HMI system is developed based on the research and analysis findings, and the system is evaluated once again to demonstrate that the research findings may give relevant design ideas for intelligent vehicle HMI development.

Keywords: Human-Machine Interaction · Intelligent Vehicles · Multiple-Resource Theory · Interaction Modes

1 Intelligent Vehicle and Interaction Design Research

1.1 A Brief Description of Intelligent Vehicle Interaction

The Human-Machine Interaction (HMI) design study is a systemic problem, and Intelligent vehicles are a complex system that is integrated and tightly entwined, necessitating examination of the interaction between the vehicle system and the driver.

Intelligent vehicles vary greatly from ordinary vehicles in terms of function and technology, and the whole system grows more complicated. It covers, at the functional level, environmental awareness, planning and decision-making, automated aided driving, and so on. On a technological level, it combines many high-tech technologies, including big data, artificial intelligence, the Internet of Things, and contemporary sensing [1].

The interaction design of intelligent vehicles must be revisited in light of improvements in function and technology.

Traditional vehicles are often based on the functioning of solid mechanical mechanisms with a single mode of interaction in the vehicle's interaction system. The steering

H. Krömker (Ed.): HCII 2023, LNCS 14049, pp. 42–57, 2023.
https://doi.org/10.1007/978-3-031-35908-8_4

wheel, knobs, and paddles have been kept as the primary mode of interaction. When building the interaction system of conventional vehicles, the emphasis is on ergonomic features such as interior layout and operating buttons. Screens and speech modules were progressively added as the hardware evolved. The variety of functional carriers allows for a variety of interaction types. The steering wheel and other mechanical structures as the principal job carrier for the driver in today's intelligent vehicles age; the center control screen as the secondary task carrier for the driver; and the integration of additional interaction modes to build a full interaction system.

1.2　A Brief Description of Intelligent Vehicle Interaction

Interaction design may be classified into distinct interaction modes based on the interaction media used, and the interaction mode is the method of human-system interaction. The key interaction medium of intelligent vehicles' interaction system is the steering wheel, center screen, center console, doors and windows, and other components. The interaction mode of intelligent vehicles may be split into the following components depending on the interaction medium and natural human-machine interaction.

– Steering wheel, doors, and windows: Physical Stuff Interaction (PI).
– Center console: Touch-screen Interaction (TI), Voice Interaction (VI), System-initiative Interaction (SI) [2].

Because single-mode interaction has limits, designers in the age of intelligent vehicles have begun to integrate several interaction modes to develop Multimodal Interaction. For example, the current intelligent vehicles model in the Chinese market, the Xiaopeng G9, blends object Physical Stuff Interaction with touch interaction and voice interaction (see Fig. 1).

Fig. 1. Interior of Xiaopeng G9 (from the official website of Xiaopeng)

2   Interaction Design and Cognitive Load Research Framework

2.1   Multiple-Resource Theory and Cognitive Load

Multiple-Resource Theory. Wickens' Multiple-Resource Theory is one of the most prominent ideas for researching psychological cognition. Wickens' Multiple-Resource Theory model (see Fig. 2) in *Human Factors Engineering* is a well-known model that describes human cognition and m Multiple-Resource Theory in four dimensions [3]. The Multiple-Resource Theory model architecture is made up of three parts: demand, resource overlap, and allocation mechanisms. The idea of mental load, which describes the task's demand on finite human mental resources, either single or multiple, is most closely connected to the first component [4].

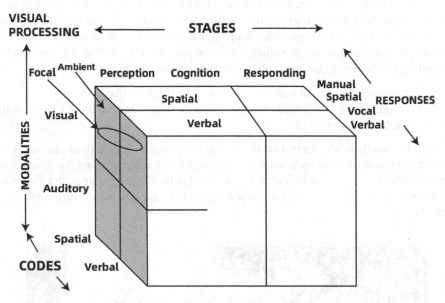

Fig. 2. Multiple-Resource Theory Model [3]

The model is separated into three dimensions: the dimension of "CODES" (Verbal and Spatial), the dimension of "MODALITIES" (Visual and Auditory), and the dimension of "STAGES" (perception, cognition, and Responding).

According to Multiple-Resource Theory, human attention and cognitive load are finite, and when a person's attention capacity is surpassed, it is possible for human mistakes to occur. If a driver is overburdened with cognitive load as a result of many driving responsibilities while driving, the likelihood of a traffic collision increases.

The primary task of driving requires the driver to devote a significant amount of attention resources to the visual channel at all times, and the driver must always observe road conditions and react in a timely manner, so the amount of information shared by both the secondary and primary tasks, as well as the visual channel's information processing capacity, must be considered when analyzing the interaction model.

Cognitive Load. One of the most important concerns researched in the field of driver psychocognition is cognitive load. The amount of mental effort expended in working memory is referred to as cognitive load. Working memory is the mechanism that stores and processes information momentarily as we learn or solve issues.

Processing and remembering new information may be more challenging when cognitive load is high since the brain is attempting to keep track of what it is currently doing. A multitude of factors can impact cognitive load, including the difficulty of the content being acquired, the learner's past knowledge, and the learning environment. Learning efficiency and efficacy can be improved by reducing cognitive load.

Cognitive load can be divided into Internal Cognitive Load, External Cognitive Load, and Related Cognitive Load, where Internal Cognitive Load refers to the load that all learning or working processes bring, which depends on the difficulty of the task and the level of the performer; External Cognitive Load is caused by the way information is organized and presented, and improper design can add additional cognitive load; Related Cognitive Load refers to the load associated with facilitating the process of schema construction and schema automation, which is about establishing associations between information, packaging information, and automated information processing, and related cognitive load can over reduce the cognitive load in learning or working processes [5]. The cognitive load discussed in this study generally refers to External Cognitive Load.

Scholars have conducted extensive research on cognitive load, and Paas et al. discuss the contribution of cognitive load measurement techniques to cognitive load theory (CLT) [6]. Since the 1990s, interactions have begun to emerge as an explicit area of research. Paas et al. describe the rationale for cognitive load theory, outline the origins of instructional meaning, and these studies represent a representative sample of current research in the field and discuss overall results in a theoretical context [7]. Plass et al. propose a view that describes emotions as External Cognitive Load that competes for limited resources by requiring the processing of off-task or task-irrelevant information to compete for the limited resources of working memory [8]. Castro-Meneses et al. aim to address the issue of high cognitive load can impede performance and educational outcomes by examining the validity of Electroencephalography (EEG) as an objective measure of cognitive load in the context of educational video [9].

He et al. investigated the effects of high cognitive load on vehicle driver Electroencephalographic (EEG) signals in the intersection of cognitive load and HMI [10], and Li et al. conducted a simulator study in which participants drove on a straight urban roadway while completing cognitive tasks of varying difficulty [11]. These findings imply that a high cognitive load might affect a driver's driving condition. When a driver's driving condition is damaged, his or her driving experience and even safety are jeopardized.

This study investigates the effect of different interaction modes on the driver's cognitive load, using the interaction mode as the main object of study. The main objective is to reduce cognitive load.

2.2 Analysis of Vehicle HMI Based on Multiple-Resource Theory Model

Multi-Resource Theory suggests that human information processing sources are usually divided into 4 parts: visual, auditory, cognitive, and motor responses any task can be composed of 28 behavioral elements under these 4 processing sources [12] (see Table 1).

Table 1. Behavior element of multi-resource model [12].

Visual	Auditory	Cognitive	Psychomotor
Detect	Detect	Simple Reaction	Speak
Text Symbology	Verity Feedback	Recognize	Reach/Switch
Search	Locate	Choice Reaction	Manipulate
Inspect/Check	Interpret Speech	Calculate	Adjust/Move
Compare Identity	Verity &Locate	Decide	Control
Trace	Compare	Recall/prepare	Keying
Align Track	Analyse Pattern	Judge	Write/Draw

Based on the Multiple-Resource Theory, the interaction pattern of intelligent vehicles is analyzed from four information-processing sources: visual, auditory, cognitive, and action as shown in Fig. 3.

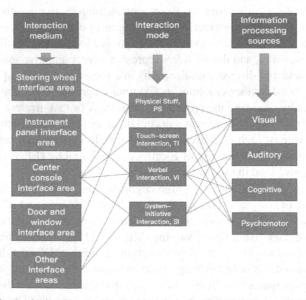

Fig. 3. Analysis of interaction patterns based on Multiple-Resource Theory

3 Method

In this study, driver experience data were collected by means of driving simulations, and subjective perception survey statistics were administered to the experimenters quickly after they performed the driving simulations.

The questionnaire survey uses the Likert 5-point Likert scale to assess the driver's cognitive load for different interaction modes in three dimensions: visual subjective rating, usability rating, and cognitive load [13].

The visual intuitive rating is an intuitive evaluation of the cockpit, and usability is an assessment of the effectiveness, efficiency, and satisfaction of the interaction system. The cognitive load score is the main evaluation index, and the visual and usability subjective scores are used as references. A comparison experiment was set up to compare the differences in cognitive load brought about by traditional vehicles' HMI and intelligent vehicles' HMI systems. The scales were all scored positively. The scale data were analyzed for reliability and validity, and the Cronbach. α coefficient value was 0.958, indicating that the scale data were credible, and the results of the KMO test showed a value of 0.771, while the results of Bartlett's sphericity test showed a significance p-value of 0.000. It showed significance at the level, the variables were associated, and the factor analysis is correct.

According to Multiple--Resource Theory, the degree of mental load can be measured in four ways:

(i) the number of information being processed simultaneously;
(ii) the total information demanded by the task at a given moment (difficulty of the task);
(iii) whether the simultaneous tasks require the same processing source or processing channel;
(iv) the degree of interference of information located in the same processing stage with each other [13].

Mental load generally includes cognitive, emotional, and volitional load [14]. In this study, the cognitive load during driving is analyzed from these four aspects.

Physical Stuff Interaction, Touch-screen Interaction, Voice Interaction, and System-initiative Interaction are used as research objects in single interaction mode, while Physical Stuff Interaction combined with Touch-screen Interaction and Physical Stuff Interaction combined with Voice Interaction are used as research objects in Multimodal Interaction.

4 Result

4.1 Analysis and Evaluation of Physical Stuff Interaction Mode

Figure 4 depicts the subjective assessment of physical Stuff Interaction engagement. In terms of usability, this interaction is somewhat higher than the visual rating and cognitive load since it is reliant on mechanical components such as buttons, knobs, and toggles. In the general layout, the interaction has a distinct recognition area and encompasses multiple surfaces such as the center console, steering wheel, doors, and windows. The driver can clearly identify the role of each physical stuff through training. At the same time, due to the characteristics of its mechanical structure, the Physical Stuff can achieve the purpose of precise operation and can achieve haptic-based feedback, the driver can feel the response even when operating them.

Based on the Multiple-Resource Theory model, the driver needs to process through Visual and Psychomotor processing sources when only the Physical Stuff is used, and not even through the visual channel in the dimension of "MODALITIES" when the driver can master the function and position of each physical stuff. The dimension of "CODES" requires the simultaneous encoding of language and spatial graphics. This operation does not require the simultaneous processing of several pieces of information, the total human information requirements for the corresponding tasks are not high, the same processing sources and channels do not need to be used for plural tasks, and the information in the same processing stage does not interfere with each other to a great extent.

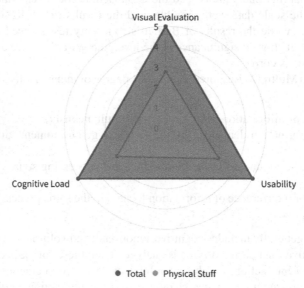

Fig. 4. Evaluation of Physical Stuff Interaction mode

Because Physical Stuff is restricted in the information it can represent owing to physical operating features, the Physical Stuff interaction method cannot suit the different demands of drivers. To fulfill the requirements of more users, the vehicle has an excessive number of Physical Stuff, which increases the driver's cognitive load.

The most common interaction mode in conventional automobile HMI is Physical Stuff Interaction, and the scores of Physical Stuff Interaction are utilized as comparison indicators in the ensuing interaction mode analysis.

4.2 Evaluation and Analysis of Touch-Screen Interaction Mode

Overall Evaluation of Touch Screen Interaction. Figure 5 depicts the results of the Touch-screen Interaction ratings, with Touch-screen Interaction exceeding Physical Stuff Interaction contact in all three dimensions. The usability dimension has a higher grade. This interaction option, which is one of the most used in intelligent vehicles, frequently employs the center console screen as a carrier.

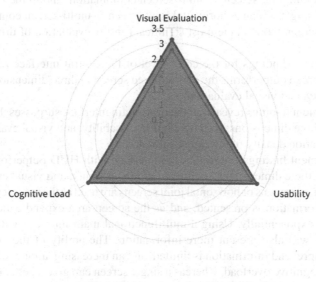

Fig. 5. Evaluation of Touch-screen Interaction mode

The Touch-screen Interaction modality necessitates Visual and Psychomotor processing sources, according to the Multiple-Resource theoretical paradigm. In comparison to the Physical Stuff interaction mode, the Touch-screen Interaction area is fixed in the center screen area, thus the recognition area is more clear. The driver must parse both the language and spatial dimensions in the "CODES" dimension, as well as the Visual channel in the "MODALITIES" dimension while engaging with the touch screen. At the same time, the touch screen's function is quite similar to that of today's smartphones. According to Jacob's law [15], Consistent performance, symbols, and a consistent conceptual model can all help to lower driver learning costs. Visual evaluation of Touch-screen Interaction and cognitive load reduction is superior to physical button contact.

There are several types of center control displays available in today's intelligent vehicle industry, including multi-screen combinations, single screens, Head-Up Displays (HUD), and joint screens. At the same time, because a center control screen may convey far more information than physical stuff, the design of different types of center control interfaces must be explored independently. The center control screen will be examined on three levels in this study: screen arrangement, center control interface color and center control interface information density.

Evaluation and Analysis of Screen. Further evaluation results about the center control screen are shown in Fig. 6.

The evaluation results show that, in the cognitive load dimension single screen > multi-screen combination > horizontal joint screen, in the usability dimension single screen > horizontal joint screen > multi-screen combination, and in the visual evaluation dimension single screen > horizontal joint screen > multi-screen combination, the overall single screen mode is the most significant in the evaluation of driver cognitive load reduction.

The assessment findings for the color level of the central interface reveal that the light-colored screen outperforms the dark-colored screen in three dimensions: cognitive burden, usability, and visual evaluation.

The assessment findings reveal that the basic main interface surpasses the multifunction screen in three dimensions: cognitive burden, usability, and visual evaluation at the level of information density of the center interface.

The assessment findings reveal that the interface with HUD outperforms the non-HUD screen in three dimensions: cognitive burden, usability, and visual evaluation.

When multiple screens or horizontal joint screens are used as the main screen arrangement, more information is presented, and as the screen area expands, the recognition range expands exponentially. Using a multifunctional main interface with high information density will also present more information. The ability of the driver's visual channel to comprehend information is limited, and an increasing amount of information might lead to cognitive overload, whereas a single screen can give a better driving experience. Interaction with a touch screen necessitates coding of Verbal and Spatial, and as the amount of graphic information rises, so does the amount of Verbal and Spatial that the driver must code, increasing the cognitive burden on the driver. Because the dark-colored interface is more difficult to perceive than the light-colored interface, more attention resources must be given to identification and judgment in the dimension of "STAGES".

The investigation of the HUD's cognitive load variables necessitates a comprehensive examination of driving states. The driver must devote the majority of his attention to the primary task while driving, must constantly monitor road conditions, and the visual processing channel is always active in the MRT model, whereas the HUD does not require the driver to process multiple pieces of information simultaneously in the visual processing channel.

4.3 Evaluation and Analysis of Voice Interaction Mode

Concerning the assessment of the Voice Interaction mode depicted in Fig. 7, because the Voice Interaction module is frequently embedded into the vehicle construction, there is no need to analyze the visual evaluation. Due to technology restrictions, voice interaction in the realm of usability seldom achieves rapid response. According to the questionnaire survey, among the reasons for poor driving conditions, 65.89% of drivers chose slow system response (see Fig. 8), when the driver issues a voice command and must wait for

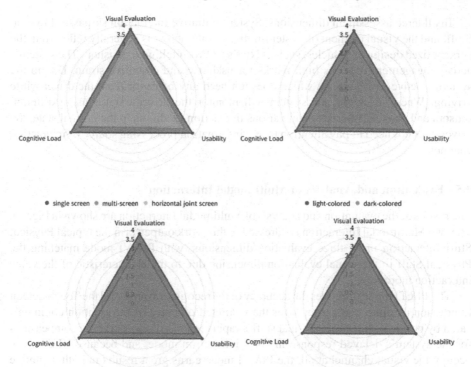

Fig. 6. Evaluation and analysis of screen

the system to respond before the next operation, during this time the driver must devote a lot of attention to cognitive and auditory processing sources. Because the driver must devote a significant amount of attention to cognitive and auditory processing sources during this period, voice interaction is not as convenient as Physical Stuff Interaction.

The driver performs the act of voice interaction only on the auditory channel, which has no effect on the visual channel where the primary task is performed and does not use the same processing source or processing channel when processing the relevant task, so voice interaction is much more subjectively evaluated in terms of cognitive load than Physical Stuff Interaction.

4.4   Evaluation and Analysis of System-Initiative Interaction Mode

The System-initiative Interaction mode emphasizes active interaction with the user, which is based on different types of sensors to actively identify the driver's behavior, expressions, and language, judge the relevant behavior, and generate different responses to different states of the driver. It frequently displays free dialogue, tiredness recognition, attention reminder, emotion identification, autonomous driving, and other capabilities in terms of function.

The evaluation was performed for the System-initiative Interaction mode, and the evaluation results are shown in Fig. 9.

In all three assessment dimensions, System-initiative Interaction surpasses Physical Stuff, and the visual element of System-initiative Interaction is generally reflected in the personalized design of vehicles, such as NIO's NOMI intelligent assistant. The system's active engagement receives high marks for usability and cognitive strain. Due to the system's active involvement, which does not need any behavior from the driver while driving. When necessary, the system will monitor the driver's status using different sensors and actively interact using various interaction mediums in the normal state, the driver does not need to pay attention to any information processing source or interaction channel.

4.5 Evaluation and Analysis of Multi-modal Interaction

The results of the evaluation and analysis of Multimodal Interaction are shown in Fig. 10. The two Multimodal Interactions addressed in this work outperform the typical Physical Stuff Interaction in all three evaluation dimensions, with PI&VI mode matching the Physical Stuff in the visual evaluation dimension due to the characteristic of the voice interaction module.

The PI&TI mode improves the accuracy of the recognition area, and the Touch-screen Interaction medium compensates for the restricted quantity of information communicated by physical stuff. The physical stuff's rapid responsiveness perfectly compensates for the system's delayed response in voice interaction mode, and because VI does not occupy the visual channel at all, the PI&VI mode earns great results in both cognitive load and usability criteria.

Multi-modal interaction can minimize excessive information processing at the same time, reduce task complexity, and avoid information processing in the same processing source, processing channel, and processing stage of the information interference as much as feasible.

5 Intelligent Vehicle HMI Design Based on the Analysis Result

5.1 Summary and Design Solutions

According to the analysis results, the following principles of intelligent vehicle HMI design are summarized.

(i) Physical Stuff interaction are limited to their own structure and can only convey a limited amount of information, but they have the characteristics of timely response and precise operation, so they must retain a certain number of Physical Stuff when designing interactions, and they can also be combined with a variety of interaction modes such as TI and VI to form Multimodal Interactions.

(ii) Touch-screen Interaction is the most common way of contact in the age of Intelligent vehicles, and its recognition area is more precise, but the design of the center control interface must be carefully studied. Based on the subjective assessment results, a single-screen configuration with a light-colored interface and reduced information density of the main interface can be employed. The adoption of HUD can also significantly lessen the driver's cognitive strain.

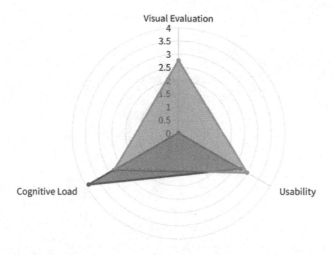

Fig. 7. Evaluation of Voice Interaction mode

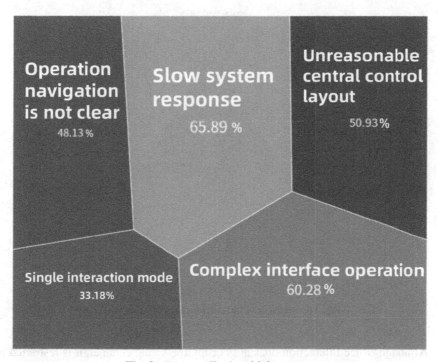

Fig. 8. Factors affecting driving status

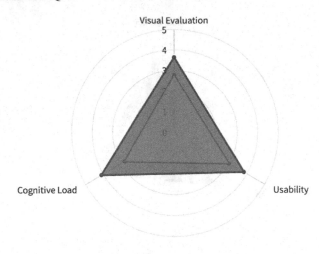

Fig. 9. Evaluation of System-initiative Interaction mode

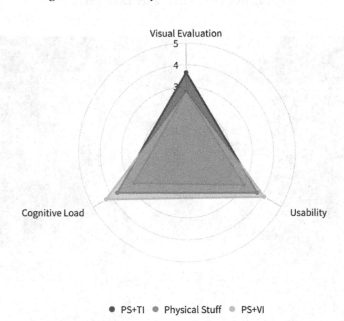

Fig. 10. Evaluation of S Multi-modal Interaction

(iii) Although Voice Interaction does not occupy the visual channel, it is restricted by hardware and mechanism and can be created in conjunction with other interaction modes.

(iv) System-initiative Interaction is a way of engagement based on new technology and theory that is one of the intelligent qualities of intelligent vehicles. System-initiative Interaction is a critical component of intelligent vehicle HMI.

Based on the findings of the preceding research, this study creates a set of schematic diagrams (see Fig. 11) and a minimalist effect diagram (see Fig. 12) of an intelligent vehicle interaction system.

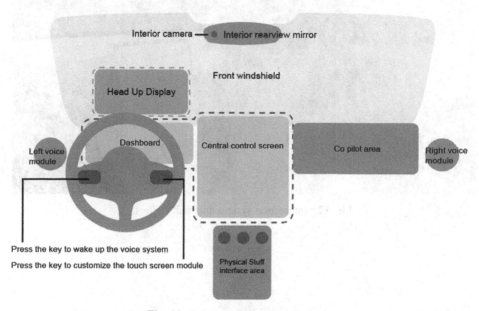

Fig. 11. Interaction system schematic

5.2 Solution Validation and Discussion

The interactive system was put through its paces, and the experimental crew refilled the load scale following the simulation. The evaluation of the interactive system is shown in Fig. 13.

The findings of the evaluation reveal that the interaction system has a substantial effect on lowering cognitive load.

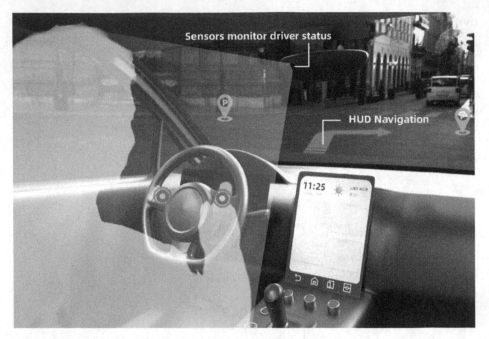

Fig. 12. Interaction system minimalist effect

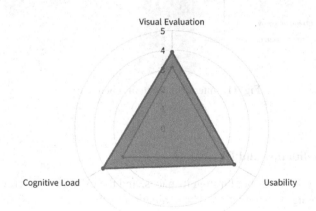

● Interaction System Solution ● Traditional physical stuff interaction

Fig. 13. Interaction system program evaluation

6 Discussion

This study is conducted in the context of the rapidly expanding intelligent vehicles market, discussing the driver's cognitive load by analyzing different interaction modes, designing simulation experiments using the Multiple-Resource Theory model, and analyzing the Scoring Results to derive the role of Physical Stuff Interaction, Touch-screen Interaction, Voice Interaction, System-initiative Interaction, and Multimodal Interaction on the driver's cognitive load, which provides some reference values for intelligent vehicles HMI design. Finally, an interaction system was designed and evaluated based on the research findings to demonstrate the validity of the findings. This study has some limitations, and objective experiments can be designed in future studies to validate the findings and investigate other multimodal combinations.

References

1. Tang, Z.: Research on the interaction design of automobile on-board control HMI in intelligence. Southeast University (2017)
2. Wang, R., Dong, S., Xiao, J.: Research on human-machine natural interaction of intelligent vehicle interface design. J. Mach. Des. **36**(2), 132–136 (2019)
3. Wickens, C.D.: Multiple resources and mental workload. Hum. Factor **50**(3), 449–455 (2008)
4. Moray, N., Pew, R., Rasmussen. J., Sanders, A., Wickens, C.D.: Final report of experimental psychology group. In: Moray, I.S. (ed.) Mental Workload: Its Theory and Measurement, pp. 101–116. Plenum Press, New York (1979)
5. Sweller, J.: Cognitive load during problem solving: effects on learning. Cogn. Sci. **12**(2), 275–285 (1988)
6. Paas, F., Tuovinen, J.E., Tabbers, H., Van Gerven, P.W: Cognitive load measurement as a means to advance cognitive load theory. Educ. Psychol. **38**(1), 63–71 (2003)
7. Paas, F., Renkl, A., Sweller, J.: Cognitive load theory: instructional implications of the interaction between information structures and cognitive architecture. Instr. Sci. **32**, 1–8 (2004)
8. Plass, J.L., Kalyuga, S.: Four ways of considering emotion in cognitive load theory. Edu. Psychol. Rev. **31**, 339–359 (2019)
9. Castro-Meneses, L.J., Kruger, J.-L., Doherty, S.: Validating theta power as an objective measure of cognitive load in educational video. Educ. Technol. Res. Dev. **68**, 181–202 (2020)
10. He, D., Donmez, B., Liu, C., Plataniotis, K.N.: High cognitive load assessment in drivers through wireless electroencephalography and the validation of a modified N-back task. IEEE Trans. Hum. Mach. Syst. **49**(4), 362–371 (2019)
11. Li, P., Markkula, G., Li, Y., Merat, N.: Is improved lane keeping during cognitive load caused by increased physical arousal or gaze concentration toward the road center. Accid. Anal. Prev. **117**, 67–74 (2018)
12. Cullen, L.: Validation of a methodology for predicting performance and workload. Eurocontrol Experimental Centre (1999)
13. Wang, J., Fang, W., Li, G.: Mental workload evaluation method based on multi-resource theory mode. J. Beijing Jiaotong Univ. **34**(6), 107–110 (2010)
14. Fang, X.: School mental health education work under the concept of positive psychology. Knowl. Econ. **166**(04), 126 (2010)
15. Yablonski, J.: Jakobs-law Homepage (2021). https://lawsofux.com/jakobs-law

Consistency Analysis of Driving Style Classification Based on Subjective Evaluation and Objective Driving Behavior

Xiaodong Xu[1], Qidi Zhang[1], Yeping Mao[2], Zehui Cheng[2], and Liang Ma[1]([✉]) [iD]

[1] Department of Industrial Engineering, Tsinghua University, Beijing, China
liangma@tsinghua.edu.cn

[2] Intelligent Driving R&D Institute, Chongqing Changan Automobile Co., Ltd., Chongqing, China

Abstract. The aim of this study was to evaluate the consistency of driving style assessments from self-reported information and actual driving behavior. 32 participants participated in the study and completed the MDSI-C questionnaire and drove in a simulator for 50 min, which involved seven different types of driving scenarios on a highway. The data obtained from the questionnaire scores on six dimensions and thirteen driving behavior data were used to perform clustering respectively. The results showed that the consistency of the results obtained from the two methods reached 50%, suggesting that the MDSI-C has some degree of predictive value for driving behavior in simulators. The inconsistency could be attributed to differences in the information used for clustering, limitations of self-reported methods, and the impact of simulator distortion.

Keywords: Driving style · MDSI-C · Driving behavior · Consistency

1 Introduction

Driving style is defined as a driver's or a group of drivers' habitual way of driving (Sagberg et al. 2015). It was initially proposed to measure and predict the driver's driving performance and safe driving ability, so as to reduce the occurrence of traffic accidents and driving crimes (Elander et al. 1993). With the development of autonomous driving technology, the recognition and classification of driving styles has gradually become a crucial aspect in enhancing the automation, safety, and personalization of autonomous vehicles (AVs), as well as improving the driver's driving experience (Yan et al. 2019). Adaptively adjusting the parameters of the advanced driver assistance system (ADAS) based on the driver's manual driving style, thus forming an automatic driving style that matches the owner, can not only reduce the interference of invalid warnings to the driver and avoid wrong takeover behaviors (Yang et al. 2018); it can also reduce the discomfort caused by loss of controllability (Elbanhawi et al. 2015), enhance the driver's trust (Ekman et al. 2019), acceptance and enjoyment of ADAS (Hartwich et al. 2018), and thus effectively improve the ADAS's reliability and ensure the driver's driving experience and safety.

H. Krömker (Ed.): HCII 2023, LNCS 14049, pp. 58–71, 2023.
https://doi.org/10.1007/978-3-031-35908-8_5

Currently, there are two main approaches to driving style classification: knowledge-driven and data-driven. The former involves using subjective questionnaires to define different driving styles. Commonly used scales include Driving Behavior Inventory (DBI) (Gulian et al. 1989), Driver Behavior Questionnaire (DBQ) (Reason et al. 1990), Driving Style Questionnaire (DSQ) (French et al. 1993), Driving Vengeance Questionnaire (DVQ) (Wiesenthal et al. 2000), etc. However, all of these scales typically consider only one or two aspects of driving style, such as distracted, stressful, or aggressive driving styles. Taubman-Ben-Ari et al. developed the Multidimensional Driving Style Inventory (MDSI) in 2004, which describes driving style from four facets at once: reckless and careless, anxious, angry and hostile, patient and careful (Taubman-Ben-Ari et al. 2004). Several studies have validated the validity and utility of the scale (Taubman-Ben-Ari and Skvirsky 2016), and the scale has also been translated and validated for use in many different cultural contexts (Freuli et al. 2020; Long and Ruosong 2019; Poó et al. 2013). However, as is common with all self-reporting methods, drivers may conceal their risk tendencies and may also forget some violations when answering questionnaires, thus bringing into question the reliability of self-reported information (Af Wåhlberg and Dorn 2015).

The latter approach primarily uses driving simulator data or naturalistic driving data to distinguish driving styles among different groups through clustering methods. A variety of clustering methods have been reported in the literature, including K-means (Peng et al. 2021), GMM (Zhu et al. 2019), Fuzzy C-means (Wang and Wang 2020), Random Forest Clustering (Chen and Chen 2019), and DNN (Wang et al. 2020). These methods are typically based on data such as speed, acceleration, jerk, lateral position, time headway, and even physiological signals. While previous studies have shown that the different driving style classes obtained through clustering methods are significantly different from each other, it should be noted that driving behavior is not synonymous with driving habits. Driving behavior can vary systematically depending on traffic conditions and the driving environment (Sagberg et al. 2015), and drivers may exhibit different driving behaviors in different situations (Dörr et al. 2014; Ma et al. 2021). As a result, the driving styles of drivers identified through clustering methods may be unstable or inconsistent. Additionally, there may be a lack of interpretability for the clustering results, making it difficult to understand the differences between categories from a perspective other than the data.

Thus, it is essential to establish a link between the two classification methods. Currently, two main types of literature focus on both self-reporting methods and objective driving behavior data. The first one is to verify the validity of the self-reported results through driving behavior data (Helman and Reed 2015; Van Huysduynen et al. 2018). Such studies are relatively scarce and mostly rely on correlation analysis, without defining the population and categorizing it. Additionally, the driving scenarios used in these studies are often homogeneous, failing to reflect systematic differences between groups of drivers. The second type of literature uses the self-reported driving style results as labels to train machine learning models for driving behavior data (Hong et al. 2014). This type of literature defaults to the correctness of the scale classification results, but if the two classification methods are inherently inconsistent or even contradictory to each other, the trained model will lose its interpretability and practical value.

In this study, we aimed to assess the consistency of the two classification methods by categorizing the driving population in two independent ways. On the one hand, we used the Chinese version of the MDSI scale (MDSI-C) (Long and Ruosong 2019) to determine the drivers' self-reported driving styles. On the other hand, we categorized drivers based on their driving behavior data, which were collected through an experiment where seven different driving scenarios were designed to comprehensively and systematically depict driving styles. Finally, we compared the similarities and differences between the two classification results and investigated the reasons for agreement and disagreement.

2   Method

2.1   Scenario Design

The driving experiment was conducted on a highway with favorable weather conditions and low traffic density. The experiment scene included ramps on and off the highway, three regular lanes, and an emergency lane. The total length of the experimental scene was 42 km and each lane was 3.75 m wide. The speed limit on the ramps was 40 km/h and on the expressway, it was between 60 and 120 km/h. The emergency lane was only to be used in exceptional circumstances.

In order to collect more systematic and comprehensive data on the driving behavior of the participants and to gain a more general understanding of their driving styles, we designed seven different driving scenarios. Each type of driving scenario was repeated twice, and their order of occurrence was disrupted, as shown in Fig. 1. These scenarios included various driving tasks such as leading, following, cornering, free lane changes, and forced lane changes. Participants had to deal with the corresponding driving tasks and make corresponding driving responses according to their driving habits.

Among them, Events 1 and 12 were scenarios of neighboring vehicles jumping ahead in a queue, where the vehicle in front of the participant would suddenly accelerate and the vehicle in an adjacent lane would signal a lane change. Participants could either accelerate to pass the vehicle and refuse the queue jump or slow down to yield to it. Events 6 and 10 were curved driving scenarios and there were vehicles passing quickly in adjacent lanes in Events 3 and 11. In both scenarios, participants simply needed to continue driving normally in their original lane. These three scenarios mainly reflected the longitudinal control style of the participants. Events 2 and 7, and Events 4 and 8, were car-following scenarios where the vehicle in front of the participant would continuously decelerate until 60 km/h. Participants independently decided the timing of their lane change and only needed to change lanes once. The difference between the two scenarios was the presence or absence of vehicles in the adjacent lanes. Events 5 and 14, and Events 9 and 13, were mandatory lane change scenarios where participants had to change lanes twice in a row due to lane closures caused by construction, traffic accidents, or leaving the expressway. The difference between the two scenarios was again the presence or absence of vehicles in the adjacent lane. These four scenarios mainly reflected the lateral control style of the participants.

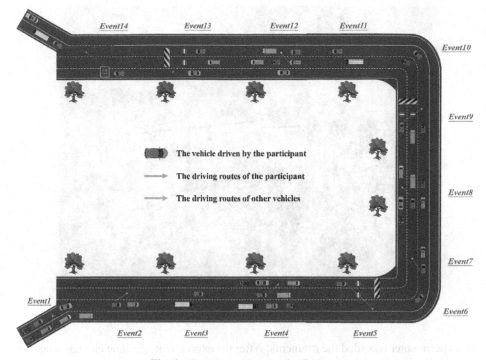

Fig. 1. Experimental scenario diagram

2.2 Apparatus

The experimental scenarios were developed using SCANeR studio and displayed on three combined monitors, which respectively showed the forward view and the side-view mirror view. Participants operated the simulator via Beitong steering wheel and pedals with a visual angle of 120° (see Fig. 2). The simulator was set to automatic transmission, so participants did not need to perform gear shifting operations. The participants' current driving speed was displayed at the top of the center screen, and the top left of the center screen intermittently displayed a message such as "Exit – 2000 m" or "Left Two Lanes Closed – 1600 m" to simulate road signs or warning signs.

2.3 Data Collection

With 100 Hz as the sampling rate, the driving dynamics data of speed (m/s), longitudinal and lateral acceleration (m/s^2), steering wheel angle (degree), acceleration pedal percentage, space headway (m), time headway (s), travel time and car locations were automatically recorded by the simulation software. The lane change was a relatively subjective operation, and the definition and understanding of the beginning and end of a lane change might differ among participants. For example, some participants considered the end of the lane change to be when they straightened the steering wheel while others to be when they adjusted the car body to be parallel to the lane. Therefore, we requested the participants to give a verbal signal at the beginning and end of the lane change, while

Fig. 2. Experimental instruments diagram.

the experimenter recorded the moments. After the experiment, the lane change time was calculated manually.

2.4 Procedure

Participants first filled in the electronic version of the MDSI-C questionnaire, along with relevant demographic information and driving history. After arriving at the experimental site, they signed the informed consent form and then were introduced to the relevant requirements of the experiment and the specific meaning of the prompt information. In the practice session, the participants familiarized themselves with the simulator's operation and specific experimental procedures through practice scenarios. When the participants felt no problem, they would enter the formal experimental session, during which the experimenter would prompt the participants if they exceeded the speed limit for a long time or drove in a wrong lane. The duration of the whole experimental process was about 60 min.

2.5 Participants

A total of 32 drivers (including 28 males and 4 females, M = 29.1, SD = 4.0, years of age) were recruited for the experiment. These participants were all employees of Changan Automobile Co., Ltd and had a valid driving license. Their average effective driving time was 5.2 years (SD = 3.9), and 75% of them had an effective driving mileage of over 3000 km. They all had normal or corrected to normal vision ability and did not suffer from motion sickness. After completing the experiment, each participant would receive ¥150 (about $25) as guerdon.

2.6 Data Processing

The MDSI-C consisted of 6 dimensions and 32 items. After each participant filled out the questionnaire, the average score of each dimension was calculated manually. Since the scale considered the driving style to be multidimensional and lacked specific criteria for group classification, we used the K-means method to divide the participants. The number of clusters was set to 4 according to Taubman-Ben-Ari and Skvirsky (2016).

For the collected driving behavior data, it was suggested that the mean and standard deviation of some related variables could respectively indicate the two main aspects of driving behavior classification (Yang et al. 2018). Thus, the mean and standard deviation of speed, lateral and longitudinal acceleration, steering wheel angle and pedal percentage were calculated to characterize the driving skills and driving stability of the participants. It was important to note that since the acceleration and steering wheel angle had some negative values (indicating direction), their mean was actually replaced by the mean of the absolute value of that variable. In addition, the mean of space headway and time headway were used to evaluate the relationship between the vehicle and the vehicle ahead when following, and the mean lane change time was used to reflect the lane change style of the participants. Due to the large number of variables, we first used principal component analysis (PCA) to simplify the data. Six principal component factors were retained with the criterion of explaining 95% of the variance. Then based on these six principal components, the participants were clustered into 4 clusters using the K-means method.

The Kruskal-Wallis test was used to compare differences among the 4 clusters, and Dunn's test with Bonferroni adjustment was used for post hoc analysis.

3 Result

3.1 MDSI-C Scale

Figure 3 illustrates the results of clustering based on MDSI-C scores, with the number of participants included in the four clusters being 8, 7, 7, and 10, respectively. Except for the distress reduction dimension, the scores of the four clusters in the other five dimensions showed significant differences (all $p < .001$, see Table 1), which indicates that the four clusters were distinguishable and meaningful. Table 2 demonstrates the part of statistically significant differences in the post hoc test results for the five dimensions.

Cluster 4 had the highest mean score on the risky dimension among the four clusters, and was significantly larger than cluster 2 and 3. In contrast, in the careful dimension, its score was significantly lower than the other three clusters. This fact indicates that the participants in cluster 4 had a strong tendency to violate traffic rules and seek thrills through dangerous driving. Hence, cluster 4 was labeled as the "Risky style".

Cluster 3 scored highest on the anxious and dissociative dimensions, and was significantly higher than cluster 1 and 2. Compared to cluster 4, it was more careful and less risky. This suggests that the participants in cluster 3 were very prone to nervousness and distraction when driving. Although they would not intentionally drive dangerously, they were also likely to unconsciously engage in some wrong or risky driving behaviors. Thus, cluster 3 was labeled as the "Anxious style".

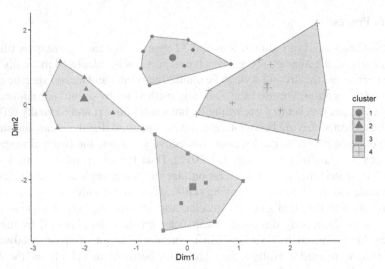

Fig. 3. Cluster plot based on MDSI-C

Cluster 1 scored highest on the angry dimension, and was significantly higher than cluster 2 and 3. Additionally, it scored significantly higher on the careful dimension compared to cluster 4. This implies that the participants in cluster 1 tended to exhibit hostile and aggressive attitudes and behaviors towards other drivers, such as verbal abuse and frequent flashing of lights. However, they were less likely to engage in dangerous driving for the sake of thrill-seeking. As a result, cluster 1 was labeled as the "Angry style".

Cluster 2 had the highest score on the careful dimension, and was generally the lowest scoring cluster in the other four dimensions. This reflects a well-adjusted driving style, which means the participants in cluster 2 were characterized by attentiveness, patience, politeness and calmness. Therefore, cluster 2 was labeled as the "Careful style".

Table 1. Mean scores and K-W test results of the MDSI-C

Dimension	Cluster 1	Cluster 2	Cluster 3	Cluster 4	statistic	p
Anxious	1.16	1.36	3.29	1.98	20.28	<.001**
Angry	4.04	1.86	2.33	3.40	21.02	<.001**
Risky	2.31	1.05	1.48	3.43	21.68	<.001**
Careful	5.33	5.76	5.48	3.80	22.17	<.001**
Dissociative	1.73	1.45	2.95	2.50	20.96	<.001**
Distress Reduction	4.84	5.07	3.71	4.68	6.68	.083

*: $p < .05$; **: $p < .001$

Table 2. The significant part of post hoc test results of MDSI-C

Dimension	Cluster Number	Cluster Number	statistic	p
Anxious	1	3	4.17	<.001
	2	3	3.41	.004
Angry	1	2	−4.10	<.001
	1	3	−3.13	.011
	2	4	3.07	.013
Risky	1	2	−2.88	.024
	2	4	4.32	<.001
	3	4	3.06	.013
Careful	1	4	−3.00	.016
	2	4	−4.31	<.001
	3	4	−3.35	.005
Dissociative	1	3	3.26	.007
	2	3	3.77	<.001
	2	4	3.17	.009

3.2 Driving Behavior Data

Figure 4 displays the results of clustering of driving behavior data, with the number of participants included in the four clusters being 6, 16, 7, and 3, respectively. The four clusters were significantly different on all thirteen dimensions (all $p < .001$, see Table 3), suggesting that the four clusters were also distinguishable and meaningful. Table 4 shows the part of statistically significant differences in the post hoc test results for the thirteen dimensions.

It was suggested drivers with angry or risky style would have higher average speed and greater speed variability, and maintain a shorter distance from the vehicle in front (Van Huysduynen et al. 2018). Cluster 4 demonstrated the highest mean speed and standard deviation of speed, as well as the lowest space headway and time headway. Furthermore, this cluster also had the highest mean and standard deviation of lateral acceleration and the highest mean and standard deviation of steering wheel angle, as well as the shortest lane change time. These findings suggest that the participants in cluster 4 preferred to drive at high speeds, follow vehicles closely, and make sudden lane changes, which aligns with the characteristics of adventurous drivers. Thus, cluster 4 was designated as the "Risky style". Cluster 1 was second only to cluster 4 in mean speed, standard deviation of speed, space headway, time headway, and mean and standard deviation of lateral acceleration, but had the longest lane change time. This indicates that the participants in cluster 1, while also favoring high-speed driving and close following, had a lower level of risk-taking behavior compared to those in cluster 4. Thus, cluster 1 was designated as the "Angry style".

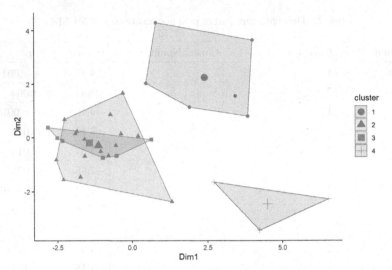

Fig. 4. Cluster plot based on driving behavior data

Increases in lane position variability were associated with dissociative driving (Just et al. 2008), which also means people with dissociative driving styles have less stability in lateral control. Calm and cautious drivers rarely brake suddenly (Murphey et al. 2009), and thus their longitudinal acceleration would be lower. And they also tend to maintain a longer distance from the vehicle in front when following. Compared to cluster 2, cluster 3 had a significantly higher mean and standard deviation for both steering wheel angle and lateral acceleration. In contrast, cluster 2 had a smaller mean for longitudinal acceleration, and a shorter space headway and time headway. In conclusion, cluster 3 was designated as the "Anxious style", while cluster 2 was designated as the "Careful style".

Table 3. Mean and K-W test results of the driving behavior data

Dimension		Cluster 1	Cluster 2	Cluster 3	Cluster 4	statistic	p
Speed (m)	mean	102.51	95.02	84.12	110.37	21.96	<.001**
	SD	28.47	26.52	22.97	28.88	14.15	.003*
Longitudinal acceleration (m/s²)	mean	1.08	0.48	0.55	0.61	15.78	.001*
	SD	1.65	0.86	0.93	1.07	16.43	<.001**
Lateral acceleration (m/s²)	mean	0.19	0.15	0.15	0.26	11.31	.010*
	SD	0.51	0.38	0.40	0.66	10.37	.016*

(*continued*)

Table 3. (*continued*)

Dimension		Cluster 1	Cluster 2	Cluster 3	Cluster 4	statistic	p
Pedal percentage	mean	0.82	0.82	0.80	0.84	15.77	.001*
	SD	0.16	0.11	0.11	0.12	16.04	.001*
Steering wheel angle (degree)	mean	0.11	0.09	0.11	0.12	14.88	.002*
	SD	0.14	0.11	0.15	0.16	12.59	.006*
Space headway (m)		109.34	119.27	117.18	72.81	9.78	.021*
Time headway (s)		6.28	7.30	7.53	3.82	12.33	.006*
Lane change time (s)		5.78	4.82	4.49	2.87	9.34	.025*

*: p < .05; **: p < .001

Table 4. The significant part of post hoc test results of the driving behavior data

Dimension	Cluster Number	Cluster Number	statistic	p
Speed (mean)	1	3	−3.84	<.001
	2	3	−2.86	.026
	3	4	3.91	<.001
Speed (SD)	1	3	−3.36	.005
	3	4	2.81	.030
Longitudinal acceleration (mean)	1	2	−3.97	<.001
Longitudinal acceleration (SD)	1	2	−4.00	<.001
	1	3	−2.73	.039
Lateral acceleration (mean)	2	4	2.84	.027
Lateral acceleration (SD)	2	4	2.67	.046
Pedal percentage (mean)	2	3	−2.79	.032
	3	4	3.77	<.001
Pedal percentage (SD)	1	2	−3.93	<.001
	1	3	−2.96	.018
Steering wheel angle (mean)	2	3	2.84	.027
	2	4	2.72	.040
Steering wheel angle (SD)	2	3	2.68	.045
Space headway	2	4	−2.94	.020
Time headway	2	4	−3.01	.016
	3	4	−2.80	.030
Lane change time	1	4	−2.91	.021

4 Discussion

The results of the two clustering methods were compared based on the number of the participants. It was found that only 16 participants were in agreement, making up 50% of the total sample. This proportion is higher than 25%, which would be the expected overlap ratio if the sample was randomly classified in each of the two conditions. This suggests that the outcomes of the MDSI-C do have some degree of predictive value for driving behavior in a simulator. However, the remaining 50% of the participants did not show consistent results, which may be mainly due to the following reasons:

The information contained in the scores obtained from the MDSI-C and the driving behavior data is fundamentally different. The questionnaire measures the driving tendencies and perceptions formed by the driver's long-term driving experience and diverse driving scenarios, which are closely related to the driver's personality traits (Taubman-Ben-Ari et al. 2004) and are relatively stable. Conversely, driving behavior is a short-term and dynamic phenomenon. It is influenced not only by the driver's personal conditions (such as personality, gender, and age), but also by the driving environment, traffic conditions, and the driver's current driving goals and motives (Van Huysduynen et al. 2018). Thus, the clustering results obtained from the driving behavior data collected during different scenarios are likely to be inconsistent (Dörr et al. 2014). Although our study aimed to capture a comprehensive and representative picture of the driving behavior of our participants by including as many diverse driving scenarios as possible, the limited one-hour driving time, the single low-density high-speed scenario, and the simple experimental motives of our participants may still not be sufficient to fully reflect their stable driving habits. Rather, the clustering results likely reflect only a partial aspect of the participants' driving styles. The differences in the information used as the basis for the clustering are the underlying cause of the differences in the clustering results.

The potential discrepancy between self-reported information and actual driving behavior is another significant concern. Firstly, there may be a gap between what someone thinks and what they do. For instance, the MDSI-C contains some questions such as "like to take risks while driving". Different participants may have different understandings of what constitutes risk-taking (Van Huysduynen et al. 2018). Some individuals may exhibit what is commonly perceived as risky behavior while driving, but may not consider themselves as taking risks when filling out the questionnaire. Meanwhile, others may have a strong inclination towards risk-taking at the cognitive level, but may not exhibit such behavior while driving. The inconsistency between cognition, tendency, and actual behavior can lead to incongruence in the clustering results obtained from the two methods. Another important factor is the influence of social desirability. For example, the MDSI-C includes questions such as "when someone tries to skirt in front of me on the road, I drive in an assertive way to prevent it" in the dimension of anger. Participants may consciously or unconsciously evaluate themselves in a socially acceptable manner, but may exhibit completely opposite behavior during actual driving, also leading to incongruence in the clustering results obtained from the two methods.

The driving experience on a simulator diverges greatly from that of actual driving, which further leads to differences in driving behavior. During the brief post-experiment interview, several participants noted that driving on the simulator was unnatural and unfamiliar. Firstly, the absence of vestibular and proprioceptive input affects the participants' ability to judge and maintain speed as they can only rely on auditory and visual cues. Some participants are not sensitive to the changes in simulator sound, and do not have the habit of constantly monitoring the instrument panel. This makes it challenging for them to maintain a stable speed and accelerate or decelerate at a relatively stable rate. Secondly, the steering wheel on the simulator is smaller and therefore more sensitive, which often results in excessive control inputs from participants when driving in a lateral direction. The discrepancy between the driving experience on the simulator and the actual vehicle has a considerable impact on participants' stability while driving during the experiment and contributes to some extent to the disparities between the clustering results obtained from the two methods.

5   Conclusion

This experiment aimed to evaluate the driving styles of participants from both knowledge-driven and data-driven perspectives using clustering methods, and compare the consistency of the results. The experiment found that the consistency between the two methods reached 50%, which was higher than the overlap ratio of randomly classified samples. This indicates that the results of MDSI-C have some predictive value for driving behavior in simulators, but cannot be entirely accurate. Three main reasons were identified for the inconsistency: first, the difference in the information contained in the data used for clustering, second, the limitations of self-reported methods, and third, the impact of simulator distortion. Additionally, the small sample size may also have affected the quality of the clustering. Therefore, it is important to further study the relationship between driving style and behavior using long-term, multi-scenario, and large-sample real-vehicle data in the future.

The experiment also highlighted that driving style is a multi-dimensional, dynamic, and complex concept, and dividing it into a limited number of categories may not accurately capture its complexity and also be unreasonable. Currently, driving style is mainly used to set relevant parameters of personalized assistive driving systems. With the advancement of autonomous driving technology and the increasing complexity of driving scenarios, generic and homogeneous driving modes may not meet the personalized driving requirements of each driver. It is therefore necessary to explore data-driven methods to design single-person single-class and single-event single-class models and experimental paradigms. This experiment can serve as a reference for future explorations in this direction.

Acknowledgement. We appreciated the support from Chongqing Changan Automobile Co., Ltd and the National Natural Science Foundation of China (Grant Nos. 71942005 and 72192824).

References

Af Wåhlberg, A.E., Dorn, L.: How reliable are self-report measures of mileage, violations and crashes? Saf. Sci. **76**, 67–73 (2015). https://doi.org/10.1016/j.ssci.2015.02.020

Chen, K.-T., Chen, H.-Y.W.: Driving style clustering using naturalistic driving data. Transp. Res. Rec. **2673**(6), 176–188 (2019)

Dörr, D., Grabengiesser, D., Gauterin, F.: Online driving style recognition using fuzzy logic. Paper Presented at the 17th International IEEE Conference on Intelligent Transportation Systems (ITSC) (2014)

Ekman, F., Johansson, M., Bligård, L.-O., Karlsson, M., Strömberg, H.: Exploring automated vehicle driving styles as a source of trust information. Transport. Res. F: Traffic Psychol. Behav. **65**, 268–279 (2019)

Elander, J., West, R., French, D.: Behavioral correlates of individual differences in road-traffic crash risk: an examination of methods and findings. Psychol. Bull. **113**, 279–294 (1993). https://doi.org/10.1037/0033-2909.113.2.279

Elbanhawi, M., Simic, M., Jazar, R.: In the passenger seat: investigating ride comfort measures in autonomous cars. IEEE Intell. Transp. Syst. Mag. **7**(3), 4–17 (2015). https://doi.org/10.1109/mits.2015.2405571

French, D.J., West, R.J., Elander, J., Wilding, J.M.: Decision-making style, driving style, and self-reported involvement in road traffic accidents. Ergonomics **36**(6), 627–644 (1993)

Freuli, F., et al.: Cross-cultural perspective of driving style in young adults: psychometric evaluation through the analysis of the multidimensional driving style inventory. Transport. Res. F: Traffic Psychol. Behav. **73**, 425–432 (2020)

Gulian, E., Matthews, G., Glendon, A.I., Davies, D., Debney, L.: Dimensions of driver stress. Ergonomics **32**(6), 585–602 (1989)

Hartwich, F., Beggiato, M., Krems, J.F.: Driving comfort, enjoyment and acceptance of automated driving–effects of drivers' age and driving style familiarity. Ergonomics **61**(8), 1017–1032 (2018)

Helman, S., Reed, N.: Validation of the driver behaviour questionnaire using behavioural data from an instrumented vehicle and high-fidelity driving simulator. Accid. Anal. Prev. **75**, 245–251 (2015). https://doi.org/10.1016/j.aap.2014.12.008

Hong, J.-H., Margines, B., Dey, A.K.: A smartphone-based sensing platform to model aggressive driving behaviors. Paper Presented at the Proceedings of the SIGCHI Conference on Human Factors in Computing Systems (2014)

Just, M.A., Keller, T.A., Cynkar, J.: A decrease in brain activation associated with driving when listening to someone speak. Brain Res. **1205**, 70–80 (2008). https://doi.org/10.1016/j.brainres.2007.12.075

Long, S., Ruosong, C.: Reliability and validity of the multidimensional driving style inventory in Chinese drivers. Traffic Inj. Prev. **20**(2), 152–157 (2019)

Ma, Y., Li, W., Tang, K., Zhang, Z., Chen, S.: Driving style recognition and comparisons among driving tasks based on driver behavior in the online car-hailing industry. Accid. Anal. Prev. **154**, 106096 (2021)

Murphey, Y.L., Milton, R., Kiliaris, L.: Driver's style classification using jerk analysis. Paper Presented at the 2009 IEEE Workshop on Computational Intelligence in Vehicles and Vehicular Systems, 30 March 2009–2 April 2009

Peng, Y., Cheng, L., Jiang, Y., Zhu, S.: Examining Bayesian network modeling in identification of dangerous driving behavior. PLoS ONE **16**(8), e0252484 (2021). https://doi.org/10.1371/journal.pone.0252484

Poó, F.M., Taubman-Ben-Ari, O., Ledesma, R.D., Díaz-Lázaro, C.M.: Reliability and validity of a Spanish-language version of the multidimensional driving style inventory. Transport. Res. F: Traffic Psychol. Behav. **17**, 75–87 (2013)

Reason, J., Manstead, A., Stradling, S., Baxter, J., Campbell, K.: Errors and violations on the roads: a real distinction? Ergonomics **33**(10–11), 1315–1332 (1990)

Sagberg, F., Selpi, Bianchi Piccinini, G.F., Engström, J.: A review of research on driving styles and road safety. Human Factors **57**(7), 1248–1275 (2015)

Taubman-Ben-Ari, O., Mikulincer, M., Gillath, O.: The multidimensional driving style inventory—scale construct and validation. Accid. Anal. Prev. **36**(3), 323–332 (2004)

Taubman-Ben-Ari, O., Skvirsky, V.: The multidimensional driving style inventory a decade later: review of the literature and re-evaluation of the scale. Accid. Anal. Prev. **93**, 179–188 (2016)

Van Huysduynen, H.H., Terken, J., Eggen, B.: The relation between self-reported driving style and driving behaviour. A simulator study. Transport. Res. F: Traffic Psychol. Behav. **56**, 245–255 (2018)

Wang, L., Lin, Q.-F., Wu, Z.-Y., Yu, B.: A data-driven estimation of driving style using deep clustering. In: CICTP 2020, pp. 4183–4194 (2020)

Wang, X., Wang, H.: Driving behavior clustering for hazardous material transportation based on genetic fuzzy C-means algorithm. IEEE Access **8**, 11289–11296 (2020)

Wiesenthal, D.L., Hennessy, D., Gibson, P.M.: The Driving Vengeance Questionnaire (DVQ): the development of a scale to measure deviant drivers' attitudes. Violence Vict. **15**(2), 115–136 (2000)

Yan, F., Liu, M., Ding, C., Wang, Y., Yan, L.: Driving style recognition based on electroencephalography data from a simulated driving experiment. Front. Psychol. **10**, 1254 (2019)

Yang, L., Ma, R., Zhang, H.M., Guan, W., Jiang, S.: Driving behavior recognition using EEG data from a simulated car-following experiment. Accid. Anal. Prev. **116**, 30–40 (2018)

Zhu, B., Jiang, Y., Zhao, J., He, R., Bian, N., Deng, W.: Typical-driving-style-oriented personalized adaptive cruise control design based on human driving data. Transp. Res. Part C Emerg. Technol. **100**, 274–288 (2019)

Do Drivers Vary in Preferences for Automated Driving Styles Across Different Scenarios? Evidence from a Simulation Experiment

Qidi Zhang[1], Yuzheng Wang[1], Gang He[2], David Pongrac[1], Zehui Cheng[2], and Liang Ma[1(✉)]

[1] Department of Industrial Engineering, Tsinghua University, Beijing, China
liangma@tsinghua.eu.cn
[2] Chongqing Changan Automobile Co., Ltd., Intelligent Driving R&D Institute, Chongqing, China

Abstract. The driving experience provided by autonomous vehicles is a crucial factor in determining the level of acceptance and trust that drivers have in the technology. In order to enhance this experience, it is necessary to understand drivers' preferences and attitudes towards the automated driving style. In this study, 32 drivers with 4 manual driving styles were recruited for simulation experiment under 2 driving scenarios. During each scenario, the drivers were presented with 3 different automated driving styles, and their preferences for automated driving style, trust and acceptance for the autonomous vehicle were recorded. The results showed that there were no significant statistical differences between the preferred automated driving styles of different drivers under each scenario. However, the driving style preferences of Risky and Careful drivers differed from those of other drivers. Additionally, driver preferences varied between passive and active driving scenarios. The study found that driver acceptance of autonomous vehicles was more stable and less influenced by the scenario, compared to trust. The findings from this study can be used to improve the design of autonomous driving systems and enhance the overall driving experience. This, in turn, is hoped to drive the widespread adoption of autonomous vehicles and bring us one step closer to realizing the full potential of this transformative technology.

Keywords: Autonomous Vehicles · Driving Style · Acceptance · Trust · Vehicle Control Algorithms

1 Introduction

Autonomous vehicles (AVs) are one of the most intensively investigated technologies within the automotive domain, which are expected to improve travel safety, efficiency, and mobility by taking the driver out of the loop and relying on vehicles to navigate themselves through traffic (Beiker, 2012). However, at the early stage of AV development, the need for human intervention or takeover cannot be ruled out in order to ensure the safety of the overall system. Thus, drivers remain an integral component of the system and

H. Krömker (Ed.): HCII 2023, LNCS 14049, pp. 72–87, 2023.
https://doi.org/10.1007/978-3-031-35908-8_6

their acceptance and use of the automated technology need to be much better understood to optimize system development, testing, operation, and ultimately adoption (Molnar et al., 2018).

Previous research on drivers' acceptance of AVs has considered how macro concerns of safety (Berliner et al., 2019; Zhang et al., 2019), loss of control and driving pleasure (Asgari & Jin, 2019; Baccarella et al., 2020; Rahimi et al., 2020; X. Wang et al., 2020), and privacy (Gurumurthy & Kockelman, 2020) can affect trust and acceptance of this technology. For micro features, rider interface and the algorithms that control vehicle behavior might be equally important, but were often neglected by researchers. Lee et al. pointed out that since fully automated vehicles would have driving styles that may or may not align with the driver's expectations, the automated driving style might bolster or undermine drivers' satisfaction in the automation, as well as their acceptance and use of AVs (Lee et al., 2021). However, there exists inconsistency in the findings of current research on people's driving style preferences for AVs (Berliner et al., 2019; Lee et al., 2021), the results indicated that although there is an overall preference for automated driving styles, people's preferences may vary when faced with different situations. Moreover, previous research on automated driving style are mostly based on fixed parameter settings, there is hardly any research that use driver's own driving style as the benchmark instead of general standards to describe aggressive, moderate, or defensive driving styles.

Therefore, in this research, we aim to investigate the preference of automated driving styles (more aggressive, personal, more defensive comparing to driver him/herself) as well as the trust in and acceptance of AVs by drivers with different manual driving styles (risky, anxious, angry, careful) in 2 driving scenarios: car-following and lane change. This study investigated the relationship among driver's preference, trust and acceptance on AVs, driver's manual driving styles, and automated driving styles under various scenarios, which can contribute to made up the gap between AV design and traffic system construction, and point out that neither blindly improve and demonstrate the driving ability of AVs in order to maximize traffic efficiency nor simply imitating the driver's own driving style will meet the psychological needs of drivers.

2 Literature Review

2.1 Driving Style Evaluation

Measuring manual driving style has been a topic of interest in AV research, and previous studies has used two main approaches: self-report measures and behavioral indicators. For self-report measures, such as the Multidimensional Driving Style Inventory (MDSI) and the Driver Behavior Questionnaire (DBQ), have assessed drivers' personality traits and behavior behind the wheel. For example, Poó and Ledesma used the MDSI to examine personality and driving characteristics in Argentine drivers (Poó & Ledesma, 2013). Wang et al. used the MDSI to examine how Chinese drivers' personality traits affect driving style (Wang et al., 2018). Self-report indicators rely on drivers to report their driving behavior and attitudes. These measures are relatively easy to administer and can provide insight into drivers' perceptions of their behavior.

Behavioral indicators help to measure drivers' actions and performance behind the wheel. Such indicators include reaction delay, time headway, lane change time, or lane change gap tolerance. For example, in a study by Lee et al., the authors used behavioral indicators to assess drivers' trust in automated vehicle driving styles (Lee et al., 2021). They found that drivers had higher confidence in automated driving styles when the reaction delay and time headway were similar to their driving styles. Such measures are less prone to bias and provide a more accurate picture of a driver's actual behavior. However, behavioral indicators can be costlier and more time-consuming to collect and may not provide insight into drivers' attitudes or perceptions of their behavior.

2.2 Automated Driving Style and Human Acceptance

The acceptance and trust of AVs by drivers are essential in successfully implementing these vehicles on public roads. One aspect that has been studied concerning driver acceptance is the automated driving style.

Studies have shown that the automated driving style can significantly impact driver acceptance and trust. For example, a study by Bellem et al. found that participants reported higher acceptance levels when the the AV used a more cautious driving style than when the car used a more aggressive one (Bellem et al., 2018). However, another study by Price et al. found that the automated driving style may or may not meet the driver's expectations, which may either increase or decrease the riders' trust (Price et al., 2016).

Personal manual driving style has been found to influence the evaluation of automated driving style. For example, a study by Sagberg et al. found that more aggressive drivers were more critical of the automated driving style when it was too cautious. On the other hand, more cautious participants were more critical of the automated driving style when it was too aggressive (Sagberg et al., 2015). Similarly, in a study by Ma and Zhang, the authors found that the similarity between drivers' manual and automated driving styles was positively related to their trust in the automated driving styles (Ma & Zhang, 2021).

The evaluation and preference of automated driving style also vary among different driving scenarios. For example, a study by Bellem et al. found that participants perceived driving style based on maneuver metrics, meaning they can prefer it differently in urban areas and on highways (Bellem et al., 2016). Another study by Haghzare et al. found that participants' preferences depend on the traffic conditions, e.g., heavy and light traffic (Haghzare et al., 2021).

Overall, previous studies suggest that the automated driving style used by AVs can significantly impact driver acceptance and trust and that engineers should tailor the driving style to the specific driving scenario to optimize acceptance and trust. Personal driving style and traffic conditions also play a role in evaluating the automated driving style. There are several research gaps that need to be addressed. One the one hand, there is a limited scope of studied driving scenarios, further research is needed to understand how drivers respond to various driving situations or driving tasks. On the other hand, previous research on automated driving style often used fixed parameter settings and general standards to describe aggressive, moderate, or defensive driving styles, hardly any research uses a driver's driving style or behavior as the benchmark for automated driving styles. Therefore, it is essential to understand how drivers respond to automated driving styles customized to their driving behavior parameters. This would give a more accurate understanding of how drivers' acceptance and trust of AVs are affected.

3   Method

3.1   Participants

Participants were recruited from Choqing, China using online questionnaire through WeChat (https://www.wjx.cn/vm/wF3bfFo.aspx#). Those people who filled out the recruitment questionnaire were asked to complete the simplified version of Multidimensional Driving Style Inventory (MDSI) (Huysduynen et al., 2015) in order to help us identify their manual driving style. After retrieving the questionnaire, we selected participants based on their driving style from among all those who filled out the questionnaire. Finally, 32 participants, between the ages of 23 and 36 (M = 29, SD = 4, number of males = 28), participated in this study. According to the manual driving style, these participants can be divided into 4 categories: anxious, angry, risky and careful, with 8 people in each category. For cumulative driving mileage, 8 participants (25%) was less than 3,000 km, 4 participants (12.5%) was between 3,000 km and 19,999 km, 10 participants (31.25%) was between 20,000 km and 49,999 km, 5 participants (15.63%) was between 50,000 km and 99,999 km, and 5 participants (15.63%) was equal or above 100,000 km. Detailed information about the demographic characteristics and driving record information of the respondents is presented in Table 1.

Table 1. Summary of demographic and driving record information

	Frequency	Percentage
Gender		
Male	28	87.50%
Female	4	12.50%
Age		
≤ 25	5	15.63%
26–30	19	59.38%
31–35	5	15.63%
≥ 36	3	9.38%
Manual Driving Style		
Anxious	8	25.00%
Angry	8	25.00%
Risky	8	25.00%
Careful	8	25.00%

(continued)

Table 1. (*continued*)

	Frequency	Percentage
Driving Experience		
≤ 5 years	18	56.25%
6–9 years	8	25.00%
10–13 years	5	15.63%
14–18 years	1	3.13%
Cumulative Driving Mileage		
< 3,000 km	8	25.00%
3,000–19,999 km	4	12.50%
20,000–49,999 km	10	31.25%
50,000–99,999 km	5	15.63%
≥ 100,000 km	5	15.63%

3.2 Apparatus and Vehicle Automation

Driving Simulator
The portable desk-top driving simulator (Beitong steering wheel and pedals) was used. Virtual driving scenarios were driven by the software SCANeR™ studio (AVSimulation Inc., France), and projected on three display screen (1 forward view and 2 side-view mirror view) 150 cm ahead from the participant's seat position (See Fig. 1). During manual driving, SCANeR™ studio automatically records each participants' driving data such as speed, position, braking time at a frequency of 100 Hz.

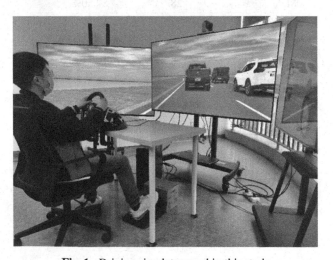

Fig. 1. Driving simulator used in this study

For manual driving data extraction, 32 participants manually drove the simulator on a stretch of highway, two driving scenarios (car following and lane changing) that were exactly the same in the following automated driving tasks were interspersed in this manual driving path, each scenario appeared twice. From their driving data, the driving-style-related indicators selected for this study (Time Headway, Reaction Delay, Lane Change Time, and Gap Tolerance) could be derived.

Time Headway (TH) was used to reflect the time distance between the participant's vehicle and the front vehicle during car following, and was calculated as the distance between the front and participant's vehicles divided by the speed of the participant's vehicle. Reaction Delay (RD) was used to reflect the time lag of the participant's reaction when the front vehicle braked sharply during car following, and was calculated by subtracting the moment the front vehicle started braking from the moment the participant's vehicle started braking. Lane Change Time (LCT) was used to reflect the time it took for the participant to change from the original lane to the adjacent lane, and was calculated by the moment when the participant ended the lane-changing behavior minus the moment when the lane-changing behavior started. Gap Tolerance (GT) was used to reflect the distance between the front and rear vehicles on the target adjacent lane (that is, the gap in the traffic flow that can be inserted by participant's vehicle) that the participant can accept when lane changing. GT was calculated as the position distance of the front and rear vehicles in the target adjacent lane at the moment when the participant starts to change lanes.

Since each of the car following scenario and lane changing scenario appeared twice in the manual driving task, the average of each participant's TH, RD, LCT and GT data could be calculated. These data of the participants themselves were used to guide the automated driving style profiles they would experience in the follow-up experiment.

Subjective Mesurements
Checklist for Trust between People and Automation (Jian et al., 2000): This questionnaire was used to evaluate 12 potential factors of trust between people and automated system, including 'deception', 'underhanded manner', 'suspicion', 'beware', 'harm', 'security', 'integrity', 'dependable', 'reliability', 'entrust', 'familiarity' on a 7-point scale ('1' = not at all to '7' = extremely).

System Acceptance Questionnaire (Van Der Laan et al., 1997): This nine-item questionnaire was designed to measure human's acceptance of new technology with two dimensions, usefulness, and satisfaction. Participants were required to evaluate systems by rating on a 7-point scale from - 3 to + 3 (e.g., '- 3' = useless to ' + 3' = useful).

3.3 Experimental Design

Two automated driving scenarios were simulated in this study: car following (Scenario I) and lane changing (Scenario II). We used each driver's own driving-style-related data (TH, RD, LCT, and GT) as indicators to reflect various automated driving styles through the upward/downward adjustment of indicators for each driver, so as to reduce the variability of results caused by individual differences. Under each scenario, 3 auto-mated driving styles were simulated: (1) More Aggressive; (2) Personal; and (3) More Defensive.

Scenario I: Car following

The scenario was simulated as highway driving with three lanes in each direction of the roadways, and the AV was in the middle lane. The posted speed limits were 100 km/h. The scenario began with the AV (i.e. the vehicle taken by the subject) following a vehicle in front at a constant speed of 90km/h at a certain THW. After 15 s of car following, the front vehicle braked sharply and decelerated to 0 km/h with a deceleration speed of 20 km/s^2. After the front vehicle started to brake, the AV decelerated to 0 after a certain RD.

Under this scenario, each participant would experience 3 kinds of automated driving styles compared to themselves, each style included 2 kinds of parameter settings (see Fig. 2). Three kinds of automated driving styles were different in THW settings: for *More Aggressive*, the THW used by the AV was the participant's own average THW minus 0.5s; for *Personal*, the THW used by the AV was the participant's own average THW; for *More Defensive*, the THW used by the AV was the participant's own average THW plus 0.5 s. For all kinds of automated driving styles, the RD used by the AV was 0 or the participant's own average RD. The reason for setting the RD in this way was that we all knew that the reaction time of future AVs can theoretically converge to zero indefinitely, so its RD must not be longer than that of humans.

Each participant would experience 6 trials of AV driving, to avoid the order effect of style sequence and RD sequence under each style, a Latin square design was used.

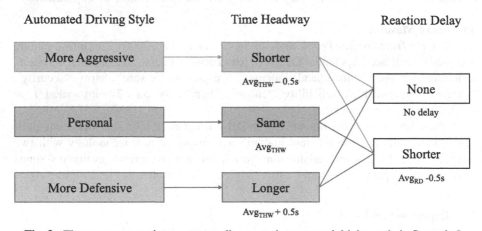

Fig. 2. The parameter settings corresponding to each automated driving style in Scenario I

Scenario II: Lane Changing

The scenario was simulated as highway driving with three lanes in each direction of the roadways, and the AV was in the middle lane. The posted speed limits were 100 km/h.

The scenario began with the AV (i.e. the vehicle taken by the subject) following a vehicle in front at a constant speed of 90km/h at the participant's own average THW. After 15 s of car following, the front vehicle started to slowly decelerate to 60 km/h with a deceleration speed of 4 km/s². At this time, the right lane is not eligible for lane change, while the left lane is eligible for lane change. Therefore, in order to avoid the slowing vehicle ahead, the AV would change to the left lane using a certain LCT when the traffic on the left reaches its desired GT.

Under this scenario, each participant would experience 3 kinds of automated driving styles compared to themselves, each style included 2 kinds of parameter settings (see Fig. 3). Three kinds of automated driving styles were different in LCT and GT settings: for *More Aggressive*, the LCT used by the AV was the participant's own average LCT minus 0.8 s, and the GT used by the AV was participant's own average GT or average GT minus 15 m; for *Personal*, the LCT used by the AV was the participant's own average LCT, and the GT used by the AV was participant's own average GT or average GT plus 15 m; for *More Defensive*, the LCT used by the AV was the participant's own average LCT plus 0.8 s, and the GT used by the AV was participant's own average GT or average GT plus 15 m.

Each participant would experience 6 trials of AV driving, to avoid the order effect of style sequence and GT sequence under each style, a Latin square design was used.

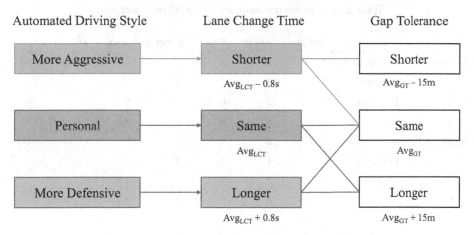

Fig. 3. The parameter settings corresponding to each automated driving style in Scenario II

3.4 Procedure

Upon arrival, participants' driver's licenses were checked, and they filled out the informed consent. Participants then completed 2 questionnaires, including Checklist for Trust between People and Automation reflecting pre-experimental trust, and System Acceptance Questionnaire reflecting pre-experimental acceptance to AV.

A two-phase simulation experiment was conducted. In the first phase, driver's own driving style data toward each specific scenario was measured trough a manual driving task. In the second phase, different driving styles comparing to the driver himself were carried out by the AV simulator toward the specific scenario. Under each scenario, after experiencing 1 driving style (include 2 kinds of parameter settings), the participant was asked to choose which of two parameter settings they preferred. Thus, by the time the participant has experienced all three driving styles, he/she has already chosen a preferred parameter setting for each driving style. The participants then experienced all three chosen parameter settings again and ranked them according to their preferences. Subjective evaluations (post-experimental trust, post-experimental acceptance) were recorded after experiencing each style of driving under each scenario.

4 Results

4.1 Automated Driving Style Preferences Under Different Scenarios

The results showed the insignificant associations between the driver's manual driving styles and their preference of automated driving style under both Scenario I ($\chi^2(6) = 4.556$, $p = 0.609$) and Scenario II ($\chi^2(6) = 4.3778$, $p = 0.626$). The summary of the chi-square tests is presented in Table 2.

Table 2. Chi-square test summary under different scenarios

		Anxious	Angry	Risky	Careful	Total	χ^2	p
Scenario I	Aggressive	2	2	3	1	8	4.556	0.609
	Personal	5	4	5	4	18		
	Defensive	1	2	0	3	6		
	Total	8	8	8	8	32		
Scenario II	Aggressive	2	2	4	1	9	4.378	0.626
	Personal	2	1	2	3	8		
	Defensive	4	5	2	4	15		
	Total	8	8	8	8	32		

4.2 Automated Driving Style Preferences and Different Manual Driving Styles

For drivers of all manual driving styles, the results showed that their preferred choice of automated driving style was not significantly associated with autonomous driving scenario (anxious: $\chi^2(2) = 3.086$, $p = 0.214$; angry: $\chi^2(2) = 3.086$, $p = 0.214$; risky: $\chi^2(2) = 3.429$, $p = 0.180$; careful: $\chi^2(2) = 0.286$, $p = 0.867$). The summary of the chi-square tests was presented in Table 3.

Table 3. Chi-square test summary under different driving styles

Driving Style		Scenario I	Scenario II	Total	χ^2	p
Anxious	Aggressive	2	2	4	3.086	0.214
	Personal	5	2	7		
	Defensive	1	4	5		
	Total	8	8	16		
Angry	Aggressive	2	2	4	3.086	0.214
	Personal	4	1	5		
	Defensive	2	5	7		
	Total	8	8	16		
Risky	Aggressive	3	4	7	3.429	0.180
	Personal	5	2	7		
	Defensive	0	2	2		
	Total	8	8	16		
Careful	Aggressive	1	1	2	0.286	0.867
	Personal	4	3	7		
	Defensive	3	4	7		
	Total	8	8	16		

4.3 Automated Driving Style Preferences for All Drivers

Chi-square test was used to investigate the automated driving style preferences for all drivers. The results showed significant associations between autonomous driving scenarios and driver's choice of automated driving style ($\chi^2(2) = 7.762$, $p = 0.021$) (see Table 4). From the preference distribution of each scenario, results revealed that drivers preferred personal or more aggressive automated driving style in Scenario I, while preferred personal or more defensive automated driving style in Scenario II.

Table 4. Chi-square test summary of all drivers under Scenario I and II

	Scenario I	Scenario II	Total	χ^2	p
Aggressive	8	9	17	7.762	0.021
Personal	18	8	26		
Defensive	6	15	21		
Total	32	32	64		

4.4 Post- Experimental Trust of AVs

The ANCOVA among different scenarios was conducted with the manual driving style as the independent variable, post-experimental trust as the dependent variable and pre-experimental trust as the covariate variable. The one-way ANCOVA results revealed that under each scenario, after controlling the pre-experimental trust, different manual driving styles did not significantly affected the post-experimental trust for all drivers with different manual driving style (Scenario I: $F(1, 29) = 1.083, p = 0.307$, partial $\eta^2 = 0.036$; Scenario II: $F(1, 29) = 0.276, p = 0.603$, partial $\eta^2 = 0.009$). Under Scenario I, pre-experimental trust of drivers with different manual driving styles has significantly affected their post-experimental trust in AVs (Scenario I: $F(1, 29) = 6.172, p = 0.019$, partial $\eta^2 = 0.175$; Scenario II: $F(1, 29) = 2.337, p = 0.137$, partial $\eta^2 = 0.075$). For more details, please see Table 5.

The ANCOVA among different manual driving styles was conducted with the autonomous driving scenario as the independent variable, post-experimental trust as the dependent variable and pre-experimental trust as the covariate variable. For drivers with all driving styles, after controlling the pre-experimental trust, different driving scenarios did not significantly affect their trust in the AV after the experiment (Anxious: $F(1, 13) = 1.957, p = 0.185$, partial $\eta^2 = 0.131$; Angry: $F(1, 13) = 3.568, p = 0.081$, partial $\eta^2 = 0.215$; Risky: $F(1, 13) = 0.775, p = 0.395$, partial $\eta^2 = 0.056$; Careful: $F(1, 13) = 0.095, p = 0.763$, partial $\eta^2 = 0.007$), while their pre-experimental trust also did not significantly affect their trust in AVs after the experiment (Anxious: $F(1, 13) = 0.046, p = 0.833$, partial $\eta^2 = 0.004$; Angry: $F(1, 13) = 3.823, p = 0.072$, partial $\eta^2 = 0.217$; Risky: $F(1, 13) = 0.681, p = 0.424$, partial $\eta^2 = 0.050$; Careful: $F(1, 13) = 0.515, p = 0.486$, partial $\eta^2 = 0.038$). For more details, please see Table 6.

Table 5. One-way ANCOVA under different scenarios for trust

	Manual driving styles		Pre-trust	
	F	η^2	F	η^2
Scenario I	1.083	0.036	6.172*	0.175
Scenario II	0.276	0.009	2.337	0.075

Note: Different superscripts within rows are statistically different at p < 0.05 or better, * p < 0.05; ** p < 0.01; *** p < 0.001

4.5 Post-Experimental Acceptance of AVs

The ANCOVA under different scenarios was conducted with the manual driving style as the independent variable, post-experimental acceptance as the dependent variable and pre-experimental acceptance as the covariate variable. According to the results, under each scenario, after controlling the pre-experimental acceptance of the scenario, different manual driving styles did not significantly affect the post-experimental acceptance (Scenario I: $F(1, 29) = 0.309, p = 0.583$, partial $\eta^2 = 0.011$; Scenario II: $F(1, 29) =$

Table 6. One-way ANCOVA under different driving styles for trust

	Scenarios		Pre-trust	
	F	η^2	F	η^2
Anxious	1.957	0.131	0.046	0.004
Angry	3.568	0.215	3.823	0.217
Risky	0.775	0.056	0.681	0.050
Careful	0.095	0.007	0.515	0.038

Note: Different superscripts within rows are statistically different at p < 0.05 or better, * p < 0.05; ** p < 0.01; *** p < 0.001

0.148, $p = 0.704$, partial $\eta^2 = 0.005$). However, under both scenarios, different types of drivers' pre-experimental acceptance under this scenario significantly affected their post-experimental acceptance of AVs (Scenario I: $F(1, 29) = 29.117, p < 0.001$, partial $\eta^2 = 0.501$; Scenario II: $F(1, 29) = 5.056, p = 0.032$, partial $\eta^2 = 0.149$). For more details, please see Table 7.

The ANCOVA among different manual driving styles was conducted with the autonomous driving scenario as the independent variable, post-experimental acceptance as the dependent variable and pre-experimental acceptance as the covariate variable. For drivers with all driving styles, after controlling the pre-experimental acceptance, different driving scenarios did not significantly affect their acceptance in the AV after the experiment (Anxious: $F(1, 13) = 0.021, p = 0.886$, partial $\eta^2 = 0.007$; Angry: $F(1, 13) = 2.968, p = 0.109$, partial $\eta^2 = 0.186$; Risky: $F(1, 13) = 0.595, p = 0.454$, partial $\eta^2 = 0.044$; Careful: $F(1, 13) = 0.124, p = 0.731$, partial $\eta^2 = 0.009$). For angry manual driving style drivers, after controlling the pre-experimental acceptance of the scenario, the pre-experimental acceptance significantly affected their post-experimental acceptance ($F(1, 13) = 5.795, p = 0.032$, partial $\eta^2 = 0.308$), but the effect was not obvious for other types of drivers (Anxious: $F(1, 13) = 0.337, p = 0.572$, partial $\eta^2 = 0.025$; Risky: $F(1, 13) = 1.702, p = 0.215$, partial $\eta^2 = 0.116$; Careful: $F(1, 13) = 0.236, p = 0.636$, partial $\eta^2 = 0.018$). For more details, please see Table 8.

Table 7. One-way ANCOVA under different scenarios for acceptance

	Manual driving style		Pre-acceptance	
	F	η^2	F	η^2
Scenario I	0.309	0.011	29.117***	0.501
Scenario II	0.148	0.005	5.065*	0.149

Note: Different superscripts within rows are statistically different at p < 0.05 or better, * p < 0.05; ** p < 0.01; *** p < 0.001

Table 8. One-way ANCOVA under different driving styles for acceptance

	Scenarios		Pre-acceptance	
	F	η^2	F	η^2
Anxious	0.021	0.007	0.337	0.025
Angry	2.968	0.186	5.795*	0.308
Risky	0.595	0.044	1.702	0.116
Careful	0.124	0.009	0.236	0.018

Note: Different superscripts within rows are statistically different at $p < 0.05$ or better, * $p < 0.05$; ** $p < 0.01$; *** $p < 0.001$

5 Discussion

5.1 Driver's Preferences of Automated Driving Style

The experimental results showed that there was no significant difference in the preferences of drivers with different manual driving styles for the automated driving style in both Scenario I (car following) and Scenario II (lane changing). The finding was consistent with Lee et al.'s finding that despite the fact that drivers had clear preferences for the manner in which AVs should drive, drivers' self-reported driving style had minimal impact on their preferences for automated driving styles (Lee et al., 2021). However, although the statistical results were not significant, we could see some differences from the data, that was, Risky drivers were more likely to accept a more aggressive automated driving style, while Careful drivers were more likely to accept a more defensive automated driving style. This was in line with Sagberg et al. 's views to some extent (Sagberg et al., 2015).

For drivers with different manual driving styles, there was no statistically significant difference in their preferences for automated driving styles in Scenario I and Scenario II. However, through the observation of statistical results, we could find that drivers prefer Personal automated driving style in all scenarios. Meanwhile, in Scenario I, more drivers chose More Aggressive automated driving style; while in Scenario II, more drivers chose More Defensive automated driving style. Further tests were conducted and the results showed a significant difference in their preference for automated driving style between Scenario I and Scenario II, without distinguishing the type of driver. One possible explanation is that in Scenario I, the AV was reacting passively to an unexpected event (sharp braking). Whereas in Scenario II, the AV performed lane change as an active behavior. In passive events, drivers were more likely to accept a more aggressive response, while in active events, drivers were more likely to expect the AV to drive in the safest possible conditions.

5.2 Post-experimental Trust and Acceptance

After controlling for pre-experimental trust and pre-experimental acceptance, the effect of the driver's manual driving style on post-experimental trust and post-experimental acceptance was not significant in each scenario. The results revealed that the driver's trust and acceptance of AVs was independent of the driver's own driving characteristics, which was in line with Lee et al.'s findings (Lee et al., 2021).

At the same time, an interesting finding was that after controlling pre-experimental trust and pre-experimental acceptance, in Scenario I, driver's post-experimental trust and acceptance for AVs significantly depended on pre-experimental trust and acceptance, and this effect is more obvious in acceptance than in trust. While in scenario II, driver's post-experimental acceptance for AVs significantly depended on pre-experimental acceptance. This revealed that when the AV reacted passively to emergency events, the driver's trust could hardly be changed by its operating performance. However, when the AV took the initiative to carry out driving operations, after observing its excellent or poor performance, the driver's trust of the AV was prone to change. Different from trust, drivers' acceptance of AVs seems to be more stable, and it was difficult to be affected by driving scenarios or specific operational performance of the AV. Compared with trust, acceptance was more of a long-term impression and judgment than an instantaneous and task-dependent feeling.

The post-experimental acceptance of Angry drivers significantly depended on their pre-experimental acceptance. Meanwhile, their post-experimental trust also relies more on pre-experimental trust than other types of drivers. To some extent, this indicated that, compared with other types of drivers, Angry drivers' trust in and acceptance of AVs was more difficult to be changed by the driving performance of vehicles in a short time.

6 Conclusion

This study investigated the driver's preference of automated driving styles, as well as the change of trust and acceptance of drivers with different manual driving styles after experiencing different driving scenarios. It was concluded that there was no significant difference between drivers' preferences for automated driving styles, but Risky drivers were more likely to prefer More Aggressive driving style, while Careful drivers were the opposite. At the same time, drivers were more likely to accept aggressive driving styles in passive driving scenarios and more likely to prefer conservative driving styles in active driving scenarios. Drivers' trust in AVs is difficult to be changed in passive driving scenarios, while it can be in active driving scenarios. Acceptance of AVs, on the other hand, is a stable perception that can hardly be influenced by the scenario.

The current initial results revealed the influence of different driving scenarios on drivers' preference, trust and acceptance of automated driving style, as well as the differences in automated driving style preferences of drivers with different manual driving styles. This provides the direction for future scientific and industrial research, pointing out that simply making the AV to imitate driver's own driving style may not be the most satisfying solution.

Acknowledgement. This study was supported by Changan Automobile Co., Ltd and the National Natural Science Foundation of China (72192824, and 71942005). We would like to show our thanks to Changan Automobile for providing the experimental site and equipment, as well as for providing us with great help in the recruitment of participants.

References

Asgari, H., Jin, X.: Incorporating attitudinal factors to examine adoption of and willingness to pay for autonomous vehicles. Transp. Res. Rec. **2673**(8), 418–429 (2019). https://doi.org/10.1177/0361198119839987

Baccarella, C.V., Wagner, T.F., Scheiner, C.W., Maier, L., Voigt, K.I.: Investigating consumer acceptance of autonomous technologies: the case of self-driving automobiles. Eur. J. Innov. Manag. (2020). https://doi.org/10.1108/EJIM-09-2019-0245

Beiker, S.A.: Legal aspects of autonomous driving. Santa Clara L. Rev. **52**(4), 1145 (2012)

Bellem, H., Schönenberg, T., Krems, J.F., Schrauf, M.: Objective metrics of comfort: developing a driving style for highly automated vehicles. Transp. Res. F: Traffic Psychol. Behav. **41**, 45–54 (2016). https://doi.org/10.1016/j.trf.2016.05.005

Bellem, H., Thiel, B., Schrauf, M., Krems, J.F.: Comfort in automated driving: an analysis of preferences for different automated driving styles and their dependence on personality traits. Transp. Res. F: Traffic Psychol. Behav. **55**, 90–100 (2018). https://doi.org/10.1016/j.trf.2018.02.036

Berliner, R.M., Hardman, S., Tal, G.: Uncovering early adopter's perceptions and purchase intentions of automated vehicles: insights from early adopters of electric vehicles in California. Transp. Res. F: Traffic Psychol. Behav. **60**, 712–722 (2019). https://doi.org/10.1016/j.trf.2018.11.010

Gurumurthy, K.M., Kockelman, K.M.: Modeling Americans' autonomous vehicle preferences: A focus on dynamic ride-sharing, privacy & long-distance mode choices. Technol. Forecast. Social Change, **150**(November 2019), 119792 (2020). https://doi.org/10.1016/j.techfore.2019.119792

Haghzare, S., Campos, J.L., Bak, K., Mihailidis, A.: Older adults' acceptance of fully automated vehicles: Effects of exposure, driving style, age, and driving conditions. Accid. Anal. Prev. **150**, 105919 (2021). https://doi.org/10.1016/j.aap.2020.105919

Jian, J.-Y., Bisantz, A.M., Drury, C.G.: Foundations for an empirically determined scale of trust in automated systems. Int. J. Cogn. Ergon. **4**(1), 53–71 (2000). https://doi.org/10.1207/S15327566IJCE0401_04

Lee, J.D., Liu, S.Y., Domeyer, J., DinparastDjadid, A.: Assessing drivers' trust of automated vehicle driving styles with a two-part mixed model of intervention tendency and magnitude. Hum. Factors **63**(2), 197–209 (2021). https://doi.org/10.1177/0018720819880363

Ma, Z., Zhang, Y.: Drivers trust, acceptance, and takeover behaviors in fully automated vehicles: effects of automated driving styles and driver's driving styles. Accid. Anal. Prev. **159**, 106238 (2021). https://doi.org/10.1016/j.aap.2021.106238

Molnar, L.J., Ryan, L.H., Pradhan, A.K., Eby, D.W., St. Louis, R.M., Zakrajsek, J.S.: Understanding trust and acceptance of automated vehicles: an exploratory simulator study of transfer of control between automated and manual driving. Transp. Res. F: Traffic Psychol. Behav. **58**, 319–328 (2018). https://doi.org/10.1016/j.trf.2018.06.004

Poó, F.M., Ledesma, R.D.: A Study on the relationship between personality and driving styles. Traffic Inj. Prev. **14**(4), 346–352 (2013). https://doi.org/10.1080/15389588.2012.717729

Price, M.A., Venkatraman, V., Gibson, M., Lee, J., Mutlu, B.: April 5). Psychophys. Trust Veh. Contr. Alg. (2016). https://doi.org/10.4271/2016-01-0144

Rahimi, A., Azimi, G., Asgari, H., Jin, X.: Adoption and willingness to pay for autonomous vehicles: Attitudes and latent classes. Transp. Res. Part D: Transp. Environ. **89**(November), 102611 (2020). https://doi.org/10.1016/j.trd.2020.102611

Sagberg, F., Selpi, Piccinini, G.F., Engström, J.: A review of research on driving styles and road safety. Human Factors: J. Human Factors Ergonom. Society **57**(7), 1248–1275 (2015). https://doi.org/10.1177/0018720815591313

Van Der Laan, J.D., Heino, A., De Waard, D.: A simple procedure for the assessment of acceptance of advanced transport telematics. Transp. Res. Part C: Emerg. Technol. **5**(1), 1 (1997). https://doi.org/10.1016/S0968-090X(96)00025-3

van Huysduynen, H.H., Terken, J., Martens, J.-B., Eggen, B.: Measuring driving styles. In: Proceedings of the 7th International Conference on Automotive User Interfaces and Interactive Vehicular Applications, pp. 257–264 (2015). https://doi.org/10.1145/2799250.2799266

Wang, X., Wong, Y.D., Li, K.X., Yuen, K.F.: This is not me! technology-identity concerns in consumers' acceptance of autonomous vehicle technology. Transp. Res. F: Traffic Psychol. Behav. **74**, 345–360 (2020). https://doi.org/10.1016/j.trf.2020.06.005

Wang, Y., Qu, W., Ge, Y., Sun, X., Zhang, K.: Effect of personality traits on driving style: psychometric adaption of the multidimensional driving style inventory in a Chinese sample. PLoS ONE **13**(9), e0202126 (2018). https://doi.org/10.1371/journal.pone.0202126

Zhang, T., Tao, D., Qu, X., Zhang, X., Lin, R., Zhang, W.: The roles of initial trust and perceived risk in public's acceptance of automated vehicles. Transp. Res. Part C: Emerg. Technol. **98**(December 2018), 207–220 (2019). https://doi.org/10.1016/j.trc.2018.11.018

Research on the Influence of Vehicle Head-Up Display Warning Design on Driver Experience with Different Driving Styles

Ruiying Zhang[1], Zhizi Liu[2], Zhengyu Tan[1(✉)], Ruifo Zhang[1], and Shiyu Yu[2]

[1] Hunan University, Changsha, Hunan, China
{ruiyingz,tanzhengyu}@hnu.edu.cn
[2] Chongqing Changan Automobile Co., Chongqing, China

Abstract. Compared with manual driving, the advanced driver assistance system reduces the pressure on the driver, and at the same time, the driver does not have to pay too much attention to driving safety issues. However, there are also potential conditions that exceed the ability of Autonomous Vehicles (AVs). The driver needs to quickly understand the driving intention and behavior warning of autonomous vehicles during the driving process, so as to take over the vehicle in time. Autonomous driving scenarios show that In-vehicle Head-Up Display (HUD) is considered to be the promising way of in-vehicle information presentation in the future. In the context of intelligent assisted driving technology, HUDs which are installed in AVs aim to introduce seamless visual information to drivers, while at the same time, allowing drivers to understand the driving intention of the vehicle more intuitively, which significantly helps to build more effective collaboration between drivers and AVs. Based on the fact listed above, current research on display timing focuses on the takeover prompt in non-driving activity scenarios, while there is less research on the timing of vehicle HUD warning feedback. The experiments were analyzed by qualitative and quantitative methods. We use HUD to present information in our experiments. In this study, a HUD design for AVs driving behavior warning is established to evaluate the experience and impact of different timing warning displays for users with various driving styles in autonomous driving scenarios.

Keywords: Advanced Driver Assistance Systems · Driver Experience · Timing of Warning · Head-Up Display · Trust

1 Introduction

With the breakthrough of artificial intelligence and assisted driving technology and the improvement of drivers' demand for driving experience, in recent years, vehicles are developing towards intelligence, networking, electrification, sharing, and automation. The development of intelligent assisted-driving vehicles is receiving widespread attention from all walks of life. Nowadays, Existing vehicle manufacturers are developing more intelligent and convenient interactive functions to improve driving safety and driver's driving experience. When focusing on solving problems, the development of smart vehicles also faces many challenges, such as the driver's acceptance of smart vehicles, driving experience perception, and driving trust issues.

Perceiving usefulness and trust are important factors in determining the intention to use an autonomous vehicle [1, 2]. In terms of whether and how people will use future self-driving vehicles, at least to a large extent, on how people learn, improvise, and feel in the vehicle in the occasional context of everyday life, (e.g. comfort, familiarity) it may not be rational or evaluative trading thinking [3]. In order to help drivers obtain higher vehicle acceptance and driving experience, it is necessary for the vehicle to provide drivers with appropriate information feedback, which can not only improve the transparency of the vehicle system to a certain extent but also help the drivers be fully prepared for the take-over of the vehicle.

1.1 Driving Transparency

The current urban traffic environment is complex and changeable, which makes it worrying and doubtful whether smart vehicles can accurately and efficiently complete driving tasks on urban roads. It requires a high degree of trust when drivers entrust driving tasks to smart vehicles. In the absence of such trust, assigning decision-making power to smart vehicles may make some drivers feel disempowered, and may feel a loss of control and independence [4]. In the process of driving a traditional vehicle, the driver can autonomously obtain timely feedback during the journey through sensory experiences such as visual and auditory experiences. However, compared with manual driving, advanced driver assistance systems alleviate the pressure on the driver. And at the same time, the driver does not have to pay too much attention to driving safety issues but focuses more on the supervision of vehicle driving status information instead. The rapid development of autonomous driving technology has transformed the role of the driver from an active controller to a passive supervisor, which allows the driver to focus more on non-driving-related tasks (NDRT) in the autonomous driving mode. However, there are also several potential conditions that exceed the capabilities of autonomous vehicles, when drivers need to quickly understand the driving intention behavior warnings from autonomous vehicles in order to take over the vehicles in time. At the same time, increasing the transparency of vehicle status information is a means of building trust and credibility, and transparency can be a driver of product adoption by users [4]. Therefore, if relevant information and driving decisions of smart vehicles in the state of intelligent assisted driving can be displayed in a timely and transparent manner, it can help users better evaluate the reliability of the driving system, thereby improving user acceptance and driving experience.

1.2 Related Research on HUD

To address the discrepancy between the driver's line of sight on the road ahead for safety and infotainment in the vehicle, the vehicle's windshield head-up display (HUD) aims to introduce seamless access to visual information for the driver. By providing driving-related information in the required location, the HUD provides instructions and explanations for the behavior of the automated driving system and supports the driver during monitoring tasks [5], and in the meanwhile, helps the driver operate the vehicle safely. With the continuous development of optical imaging technology and vehicle intelligent systems, the market prospect of vehicle HUD is broad. The HUD projected on the windshield can provide the driver with visual navigation and vehicle data information. It does not require the driver to take his eyes off the road to obtain road information and vehicle status, which is convenient for users to perceive the driving status and prediction the behavior of the vehicle, which can ease the user's trust considerations for assisted driving behaviors and give users an immersive driving experience. Research shows that intelligent systems should be transparent and should provide stable performance and an intuitive user interface. As a new type of driving assistance safety system, HUD is considered to be the main way of presenting vehicle information in the future. At present, many studies have confirmed the advantages of automotive HUD design in terms of driving safety and cognitive load [6]. HUD, integrated into the field of transportation, should have various functions including machine learning, virtual shadow casting systems, and metamaterials, which can realize customizable and transformative image projection functions to improve driver's trust and enhance road safety [7]. However, technology-supported HUD information display system lacks specific research on the design of human-computer interface [8]. Due to the complex and changeable vehicle information, the principle of information feedback should be followed during the design of the vehicle interface. Information design needs to consider the sensitivity of information requirements, information integration methods, and reasonable information flow in order to ensure that users can get feedback quickly and promptly. Considering all the factors listed above, research on HUD information is necessary [9]. However, the timing of displaying warnings has been ignored in previous studies. We consider that while taking the timing into account, we also need to pay attention to the differences in driving styles of drivers in order to provide them with a better driving experience. Research shows that compared with experienced drivers, the development of assisted driving technology HUD is more practical for inexperienced drivers [10]. Research shows that, compared to inexperienced drivers, experienced drivers can selectively change their visual search strategies according to road complexity. Therefore, a HUD with advantages such as reducing "display browsing duration" and shorter "response time to information" can benefit more inexperienced drivers. Therefore, HUD designers and developers should pay more attention to inexperienced novice drivers and help improve driving safety for new drivers [11]. Real-life driving experiences lead to improvements in trust calibration. The improvement in the driver's ability to accurately distinguish between situations in which the automation performed well and those in which it did not play a key role in calibrating participants' trust after the experience. During the driving process, the driver needs to locate the vehicle in the environment and plan the route, and estimate the cost and risk [12]. Previous studies have shown that research on driving experience needs to consider the driver's driving style and investigate its needs in various driving scenarios to address the optimal design of the driving interface [13], and further research is necessary for product development to provide extensive vehicle regulation for personalization. In this paper, we propose a driving simulator study. During the experiment, the driving assistance function is set to operate the vehicle, and the navigation information is presented to the driver through the head-up display (HUD), linking the vehicle operating information with the surrounding environment. This study investigates the impact of HUD warnings at different timings on drivers' driving emotional experience and trust in four scenarios.

2 Method

2.1 Experimental Apparatus

The experiments are set in a real driving seat model to play the experimental video and a smart screen is placed in front of it, which includes a real road environment route, and the HUD information is included in the video. The full set of the experiment is seen in Fig. 1. The experimental phase consists of four driving scenarios: turning, lane changing, braking, and starting. It is used to test the influence of different early warning display timings on the driver's experience and trust of the vehicle HUD in the assisted driving state on urban roads.

Fig. 1. Experimental scene

2.2 Experimental Participants

Through social recruitment, 26 novice drivers in total were invited to the study, and the participants are aged 22–30 (M = 23.54, SD = 1.8). All 26 participants had valid driving licenses. The driving age is between 1–3 years. Education level is the bachelor degree or above. All participants reported normal or corrected-to-normal vision and hearing.

2.3 Experimental Design

Driving Task Design. Previous studies have shown that drivers are familiar with HUD arrow visualization system and have a relatively good experience [14]. Therefore, the

HUD icons in this experiment are represented by visual arrows. During the experiment, the participants did not need to do anything, they only needed to fill in the questionnaire after watching the experimental video each time; after all the videos were finished, retrospective interviews were conducted for different warning timings in each scenario.

Questionnaire Design

1) *Driver Questionnaire*

Both characteristics of the driver's driving ability and driving style have an impact on the vehicle accident possibility [15]. The study uses a self-administered questionnaire to collect drivers' subjective driving characteristics. The questionnaire contains 11 questions that can accurately reflect the driving style, which is widely used to examine the driver's driving ability and self-assessment of driving skills. The questionnaire divides the driving style of the test subjects into three types: calm type, normal type, and aggressive type.

2) *User Experience Questionnaire*

The evaluation questionnaire for the key elements of user experience is filled in after each experimental video, the original MeCUE questionnaire includes 4 parts: product perception, user emotion, consequences of using and overall judgment, with a total of 34 items, which can be applied to all types of interactive system experience survey. The questionnaire is also an evaluation of the reliability on the people's emotions or reactions in different situations. In the study, the questions on user emotions and the Likert scale were selected as reference indicators. These problems can be divided into two dimensions, positive and negative.

3) *Retrospective interview.*

The retrospective interview requires the participants to be in the same room immediately after watching all the test videos and completing the questionnaire. Each participant takes about 5–10 min. The content of the interview includes the experience and trust comparison of the HUD display timing in different scenarios, visual needs, etc. The whole process is recorded by video and experimental observation.

Experimental Video. In this experiment, Adobe Photoshop and Adobe After Effects were used to produce HUD early warning displays at three different timings for four driving behaviors: turning, lane changing, braking and starting. A total of 12 videos are shown in Table 1. In the turning scene, the vehicle is 90 m, 70 m, and 50 m away from the right turn when the HUD displays the vehicle turning warning prompt; in the braking scene, the vehicle is 90 m, 70 m, and 50 m away from the front parking place when the vehicle brake warning prompt is displayed through the HUD; in the lane change scene, the vehicle constant speed is displayed 9 s, 7 s, and 5 s before the lane changing operation, and the vehicle lane change warning prompt is displayed through the HUD; in the starting scene, 9 s, 7 s, and 5 s before starting the vehicle, behavior warning prompt is displayed through the HUD. The HUD prompt information appears at different times in each video content, but the road conditions and vehicle-assisted driving behavior are controlled to be the same.

Table 1. Contents of driving scenarios.

Experimental scene	Display content	Early warning timing
Turn	speed, speed limit, right turn icon, Warning text (for example: turn right after 90m)	90 m, 70 m, 50 m
Brake	speed, speed limit, brake icon, Warning text (for example braking after 90m)	90 m, 70 m, 50 m
Lane change	speed, speed limit, lane change icon, Warning text (example: change lane after 9s)	9 s, 7 s, 5 s
Start	speed, speed limit, start icon, Warning text (example: start after 9s)	9 s, 7 s, 5 s

2.4 Experimental Process

Step 1: Before the start of the experiment, 26 participants were screened through the questionnaire.

Step 2: After the participants signed the informed consent form for the experiment, they were shown an in-vehicle image about Navigator-assisted driving and introduced to the specific performance of the vehicle-assisted driving function and the existing vehicle behavior information feedback is also introduced.

Step 3: Inform the subjects of the specific experimental process and precautions. Before the experiment, the order of the videos is randomly grouped, and the experimental videos are played on a large screen. After each video, the participants rest for 5 s and then fill in the relevant questionnaire; after filling out the scale, we continue to play the next video of the experimental material for them, and the participants fill in the relevant questionnaire again after the video, the procedure continues until all the videos are played.

Step 4: After the video part is completed, conduct retrospective interviews with the participants, which can stimulate the subjects to interpret the evaluation scale.

3 Results

The cluster analysis of the driver's driving style questionnaire was carried out by SPSS software, and reliability analysis, Descriptive statistics, and analysis of variance were performed on the experience scale data of the participants who had driving behavior warnings at different timings, so as to obtain different the impact of HUD early warning timing on the driver's emotions. By sorting out the user interview data, the participants' need for early warning timing and trust experience in different scenarios are obtained.

3.1 Participants' Driving Style

Studies have shown that a driver's driving style can determine his driving behavior to a certain extent [16]. In this experiment, the K-means clustering method was used to analyze the total score of the driving style of the subjects in the questionnaire. The higher the score of the questionnaire, the more obvious the aggressiveness the participant shows. Through the analysis in SPSS software, it is concluded that there is 1 person in the aggressive type, and the cluster center is 1.45; there are 12 people in the ordinary type, and the cluster center is 0.33; there are 13 people in the calm type, and the cluster center is −0.41. Because the number of aggressive subjects was too small and the reference value was small, this group of data was excluded.

Table 2. Clustering of driving styles.

	Normal type	Calm type	Aggressive type
Cluster Center Total Score	.33	.41	1.45
Number of people	12	13	1

3.2 Mood Questionnaire

In this study, the correlation between the score of each question of the questionnaire and the total score of the questionnaire is expressed by Cronbach's coefficient, as shown in Table 2, which is 0.858 > 0.70, which indicates that the user experience questionnaire has high reliability and is acceptable.

Table 3. Questionnaire Credibilit

KMO Sampling Suitability Quantity		.858
Bartlett's test for sphericity	Approx. Chi-Square	3447.069
	degrees of freedom	66
	significant	.000

As shown in Table 3, by conducting the descriptive statistical analysis of the emotional experience of the participants in the three HUD early warning timings under different driving scenarios, it can be seen that the experience scores of drivers with normal and calm driving styles are different in each driving scenario.

Figure 2 is the analysis of the emotional experience of the two groups of participants on different HUD early warning timings in each driving scene. From the figure, it can be seen intuitively that the participants have different emotional experience of the three HUD early warning timings in each scene. In order to judge whether the difference between the two groups is statistically significant, we also conducted a statistical test.

Use analysis of variance (one-way analysis of variance) to study the differences in the driving experience emotions of normal and calm participants in turning scenes at different HUD warning timings. It can be seen from Table 6 that the samples of different driving styles are negative emotional experience was significant ($p < 0.05$), which means

Table 4. Descriptive Statistical Analysis

Driving style	Scenario	Timing	Average Score	Driving style	Scenario	Timing	Average Score
Normal type (Negative questions)	Turn	90 m	2.182	Normal type (Positive questions)	Turn	90 m	3.03
		70 m	1.636			70 m	3.424
		50 m	1.894			50 m	3.455
	Brake	90 m	2.303		Brake	90 m	2.97
		70 m	2.015			70 m	3.318
		50 m	2.076			50 m	3.106
	Lane change	9 s	2.083		Lane change	9 s	3.347
		7 s	2.742			7 s	2.485
		5 s	2.8			5 s	2.517
	Start	9 s	1.818		Start	9 s	3.394
		7 s	1.5			7 s	3.833
		5 s	1.561			5 s	3.864
		Total	2.64			Total	3.235
Calm type (Negative questions)	Turn	90 m	1.226	Calm type (Positive questions)	Turn	90 m	3.714
		70 m	1.69			70 m	3.345
		50 m	1.667			50 m	3.25
	Brake	90 m	2.369		Brake	90 m	3.155
		70 m	2.31			70 m	3.464
		50 m	1.714			50 m	3.774
	Lane change	9 s	2.512		Lane change	9 s	3.333
		7 s	3.036			7 s	3.083
		5 s	2.885			5 s	3.013
	Start	9 s	1.857		Start	9 s	3.94
		7 s	2.321			7 s	3.5
		5 s	2.381			5 s	3.345
		Total	2.16			Total	3.412

that in the turning scene, different driving style samples have differences in the negative experience. Specific analysis shows that different driving styles have a significant level of 0.05 for the negative total score (F = 6.080, p = 0.022), and the specific comparison shows that the average value of the ordinary type (13.09) will be significantly higher than the average value of the calm type Value (7.36).

In this study, we analyze the variance to study the differences in driving experience emotions of ordinary and calm participants in the starting scene at different HUD warning timings, it can be seen from the Table 5 listed above that the samples of different driving

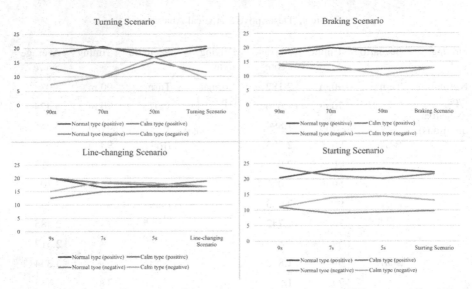

Fig. 2. Analysis of variance of participants' emotional experience of different warning timings in various driving scenarios

Table 5. Variance analysis for turning scenario

Driving style (mean ± standard deviation)	Positive score	Negative score
Normal type	18.18 ± 6.10	13.09 ± 8.48
Calm type	22.29 ± 6.71	7.36 ± 1.91
F	2.495	6.080
p	0.128	0.022*

* p < 0.05 ** p < 0.01

Table 6. Variance analysis for starting scenario

Driving style(mean ± standard deviation)	Positive score	Negative score
Normal type	22.18 ± 8.24	9.76 ± 4.92
Calm type	21.57 ± 8.17	13.12 ± 7.70
F	0.102	4.757
p	0.750	0.032*

* p < 0.05 ** p < 0.01

styles have a significant effect on negative emotional experience (p < 0.05), which means that in the starting scenario, different driving style samples have differences in the negative experience. Specific analysis shows that different driving styles have a significant level of 0.05 for the negative total score (F = 4.757, p = 0.032), and specific comparisons show that the average value of the ordinary type (9.76) will be significantly lower than the average value of the calm type value (13.12).

3.3 Interview

During the interview, we asked the participants about the reasons for their preferences of the warning information timing, the driver's information feedback needs for the HUD while driving, and suggestions for the experimental settings. According to the retrospective interviews, the participants in different task scenarios carried out weight assignment analysis and processing on the interview content of the three warning timing preference sequences. On the whole, in the lane-changing scene, the subjects agreed that the sooner the warning is given, the better. The reason is that the operation of changing lanes is difficult, and the driver needs to pay attention to the driving state of the vehicle in advance.

The ordinary participants in the sample showed a clearer preference awareness in the parking and lane changing scenarios and believed that the earlier (9s) warning time of the two is the best. The calm participants in the sample showed a clear preference consciousness in the turning and lane changing scenarios and believed that the later the turning (50 m), the better, and the earlier the lane changing is better (9 s).

In the interview, 21 participants described that the timing of displaying warning behaviors on the HUD would make the driving experience better and they would feel safer; 3 participants maintained a neutral attitude towards the existence of HUD warning information; 1 participant expressed doubts about the early warning capability of HUD, worrying that in the event of a traffic emergency, HUD would not be able to give feed-back information to the driver in a timely and accurate manner as before.

The participants hope that the HUD can flexibly adapt to multiple driving scenarios, not just those in the experiment, which will greatly improve the driver's monitoring of vehicle behavior. Secondly, for the interface in the experiment, the participants hope that there is more graphic guidance on the central position of the HUD, which can help the driver better understand the expected behavior of the vehicle. There are also 2 users with conservative driving styles saying that the warning information will bring a certain sense of oppression and tension.

Finally, in the interview, we found that, because of the early display of HUD warning information, the participants stated that they would pay more attention to the road instead of NDRT, which confirmed the help of HUD to driving safety to a certain extent.

4 Discussions

Tracing back to the experimental results above, participants with different driving styles will have different negative emotional experiences in the starting scenario. The analysis of the interview data infers that the reason why the emotional perception of the participants is comparably obvious is that the starting scene changes from static to dynamic driving behavior. In the turning scene where the HUD warned 90 m in advance, the negative emotions of the ordinary type were significantly greater than those of the calm type. Some participants reported that they felt anxious and negative because they felt that the warning was too early. There is not much difference in the choice of timing in the experiment, and the situation that the difference in emotional perception of some participants is not obvious enough can also influence the fact.

According to the experiment listed above, it is a research on the influence of the driver watching the driving under the simulated assisted driving state, but the specific vehicle conditions cannot be observed in all directions. Related studies have shown that real-life driving experience will improve the driver's trust calibration for smart cars [17, 18]. Therefore, the novice driver's experience and trust preference for early warning timing may be dynamically adjusted with the future realistic driving experience. Secondly, there is a small gap between the three HUD warning timings set in the experiment, and some subjects have unclear perceptions. Finally, it is also possible to compare the experience and trust of a HUD to an in-vehicle screen when displaying different types of driver assistance information.

5   Conclusions and Prospects

5.1   Conclusion

This paper studies the driver's experience and trust in the timing of the on-board HUD warning display under different operating scenarios in the assisted driving state, and draws the following main conclusions:

(1) For different HUD early warning display timings in the starting and turning scenes, there is a significant difference in the emotion scores of ordinary and calm novice drivers;
(2) Under difficult task scenarios such as lane-changing, the drivers agreed that the earlier the HUD warning time, the better;
(3) Drivers with different driving styles have a clear sense of preference in parking, lane changing, and turning scenarios. Novice drivers with different driving styles have different timings for accepting driving behavior warnings.

Drivers should be allowed to actively adjust the settings of non-critical system functions in the vehicle HMI design in order to customize the system according to personal preferences [19]. Therefore, the HUD early warning timing for assisted driving behavior can be adaptively adjusted according to the driver's preference and driving style, but the timing range of the adjustment needs to be further studied.

5.2   Limitation and Future Work

The study found that the design of the early warning information of the vehicle HUD should be different according to the characteristics of the driver (such as gender) [20]. The information about the participants during the experiment can be further studied according to the clustering, classification, and behavioral characteristics, and the differences in the behavior of different types of drivers for the early warning information of the vehicle-mounted HUD at different times can be deeply excavated.

For this study, first, this study was conducted in a simulated driving environment, unlike real-world driving environments. Future research should further conduct field studies to support the results of this simulation. Secondly, the participants of this study are limited to relatively homogeneous drivers with little driving experience, the participants' class, age, and education level span is small, and the reference value is limited. Therefore, the group of participants in future studies should be more diverse, for example, including different levels of education and ages. Third, future studies should have more diverse driving scenarios to better understand the adaptation effect of driving behavior.

Evaluating and designing specific mechanisms based on various perspectives and factors will also need to be adjusted as the needs of drivers are constantly changing. Finally, physiological measurement tools can be introduced into experiments, and one study showed that EEG-based measurements can be a powerful tool for studying driver behavior [21].

References

1. Investigating the Importance of Trust on Adopting an Autonomous Vehicle: Int. J. Hum.–Comput. Interact. **31**(10). https://www.tandfonline.com/doi/full/, https://doi.org/10.1080/10447318.2015.1070549
2. Aydogdu, S., Seidler, C., Schick, B.: Trust Is good, control is better? – the influence of head-up display on customer experience of automated lateral vehicle control. In: Krömker, H. (ed.) HCII 2019. LNCS, vol. 11596, pp. 190–207. Springer, Cham (2019). https://doi.org/10.1007/978-3-030-22666-4_14
3. Pink, S., Osz, K., Raats, K., Lindgren, T., Fors, V.: Design anthropology for emerging technologies: Trust and sharing in autonomous driving futures. Des. Stud. **69**, 100942 (2020)
4. Michler, O., Decker, R., Stummer, C.: To trust or not to trust smart consumer products: a literature review of trust-building factors. Managem. Rev. Quart. **70**(3), 391–420 (2019). https://doi.org/10.1007/s11301-019-00171-8
5. Augmented Reality Head-Up Display: A visual support during malfunctions in partially automated driving? | IEEE J. Mag. | IEEE Xplore. https://ieeexplore.ieee.org/document/9599523
6. Betancur, J.A., Gómez, N., Castro, M., Merienne, F., Suárez, D.: User experience comparison among touchless, haptic and voice Head-Up Displays interfaces in automobiles. Int. J. Interactive Design Manufact. (IJIDeM) **12**(4), 1469–1479 (2018). https://doi.org/10.1007/s12008-018-0498-0
7. Skirnewskaja, J., Wilkinson, T.D.: Automotive holographic head-up displays. Adv. Mater. 27 (2022)
8. Human-Computer Interaction. Design and User Experience Case Studies: Thematic Area, HCI 2021, Held as Part of the 23rd HCI International Conference, HCII 2021, Virtual Event, July 24–29, 2021, Proceedings, Part III. vol. 12764. Springer International Publishing. (2021). https://doi.org/10.1007/978-3-030-78468-3
9. Li, X., Rong, J., Li, Z., Zhao, X., Zhang, Y.: Modeling drivers' acceptance of augmented reality head-up display in connected environment. Displays **75**, 102307 (2022)
10. Li, R., Chen, Y.V., Sha, C., Lu, Z.: Effects of interface layout on the usability of In-Vehicle Information Systems and driving safety. Displays **49**, 124–132 (2017)

11. Li, R., Chen, Y.V., Zhang, L., Shen, Z., Qian, Z.C.: Effects of perception of head-up display on the driving safety of experienced and inexperienced drivers. Displays **64**, 101962 (2020)
12. Bauerfeind, K., et al.: Navigating with Augmented Reality – How does it affect drivers' mental load? Appl. Ergon. **94**, 103398 (2021)
13. Ma, Z., Zhang, Y.: Drivers trust, acceptance, and takeover behaviors in fully automated vehicles: Effects of automated driving styles and driver's driving styles. Accid. Anal. Prev. **159**, 106238 (2021)
14. von Sawitzky, T., Wintersberger, P., Riener, A., Gabbard, J.L.: Increasing trust in fully automated driving: route indication on an augmented reality head-up display. In: Proceedings of the 8th ACM International Symposium on Pervasive Displays 1–7. ACM (2019). doi:https://doi.org/10.1145/3321335.3324947
15. Zhang, L., Wang, J., Yang, F., Li, K.: Factor analysis and fuzzy clustering of driver behavior patterns. J. Traffic Trans. Eng. **9**, 121–126 (2009)
16. Ji, S., et al.: Driving style classification method based on high-frequency data of pure electric vehicles. J. Tongji Univ. (Nat. Sci. Edn.) **50**, 273–282 (2022)
17. Walker, F., Boelhouwer, A., Alkim, T., Verwey, W.B., Martens, M.H.: Changes in Trust after Driving Level 2 Automated Cars. J. Adv. Transp. **2018**, e1045186 (2018)
18. Walker, F., Boelhouwer, A., Alkim, T., Verwey, W.B., Martens, M.H.: Changes in trust after driving level 2 automated cars. J. Adv. Transp. **2018**, 1–9 (2018)
19. Ekman, F., Johansson, M., Sochor, J.: Creating appropriate trust in automated vehicle systems: a framework for HMI Design. IEEE Trans. Human-Mach. Syst. **48**, 95–101 (2018)
20. Park, J., Abdel-, M., Wu, Y., Mattei, I.: enhancing in-vehicle driving assistance information under connected vehicle environment. IEEE Trans. Intell. Transport. Syst. **20**, 3558–3567 (2019)
21. Di Flumeri, G., et al.: EEG-based mental workload neurometric to evaluate the impact of different traffic and road conditions in real driving settings. Front. Human Neurosc. **12** (2018)

Designing Driver and Passenger User Experience

Objective Metrics for Assessing Visual Complexity of Vehicle Dashboards: A Machine-Learning Based Study

Huizhi Bai[1], Zhizi Liu[2], Ziqi Fu[3], Zihao Liu[4], Huihui Zhang[5], Honghai Zhu[2], and Liang Zhang[1(✉)]

[1] Institute of Psychology, Chinese Academy of Sciences, Beijing, China
zhangl@psych.ac.cn
[2] Chongqing Changan Automobile Co., Chongqing, China
[3] Department of Cognitive Science, McGill University, Montreal, Canada
[4] Research Center of Adolescent Psychology and Behavior, School of Education, Guangzhou University, Guangzhou, China
[5] School of Psychology, Nanjing Normal University, Nanjing, China

Abstract. Dashboard is a central component of an in-vehicle information system (IVIS), and plays a crucial role in providing drivers with key information related but not limited to driving. With the expansion of the IVIS features, modern dashboards additionally integrate various new elements, which often leads to an increase in their visual complexity. Since high visual complexity of the dashboards threatens driving safety and performance, it is essential for researchers and designers to understand what objective features of the dashboards are related to their perceived visual complexity (PVC) so as to establish more cognitively efficient dashboards. In the present study, we refined the objective metrics of assessing visual complexity proposed in previous research and added two new dimensions, colors and animation, to better characterize recent development in the dashboard displays. We then utilized the indicators in the metrics to predict the dashboard PVC. Machine learning was innovatively applied, and the models were found to have stable performance. The study contributes reliable metrics and novel methodology to evaluate the visual complexity of the dashboards for the reference of future studies.

Keywords: Perceived visual complexity · Vehicle dashboard · Machine learning models · Human-vehicle interface design

1 Introduction

1.1 Study Background

Vehicle dashboard, also called instrument panel, is one of the main components of the in-vehicle information system (IVIS). Functioning as a central human-vehicle interface, dashboards not only provide important information that promotes driving safety and performance, but also assist drivers to engage in some non-driving related tasks. Traditional dashboards often contain the speedometer, tachometer, odometer, gauges, and warning for vehicle conditions, which are displayed as texts, icons, and charts. Many studies have investigated how individual dashboard elements can be designed in a cognitively optimal manner. Based on the results, an international standard for the visual presentation of dashboards has been established to guide and regulate vehicle dashboard visual designs [1].

With the development of smart features, various novel elements have been integrated into the modern design of dashboards, which in turn increase the overall visual complexity of the dashboards. Visual complexity generally refers to the level of intricacy of stimulus appearance [2]. In the context of human-vehicle interface design, too much visual complexity hampers humans' limited information-processing capacities, and forms a bottleneck to restrict the usefulness of the system [3]. Specifically, high visual complexity of dashboard displays exerts negative effects on drivers' visual search performance, driving performance, and subjective workload [4, 5], which in turn poses a serious threat to driving safety. Therefore, it is of importance to determine what factors are associated with the visual complexity of the dashboards and construct an evaluation framework.

Visual complexity depends on both the intrinsic visual characteristics of the dashboards as well as mental processing of the perceivers. In previous studies on dashboard visual complexity, three dimensions were acknowledged to have a significant contribution to the overall visual complexity of dashboards, which are quantity (numeric size) – number of basic visual information units, variety – discrete characteristics of the units, and relation (interconnections) – the structural connections between each component [3, 6–8]. They were generally validated using questionnaires [3, 6], where participants reported their perception of each dimension, as well as rated the overall visual complexity of the dashboards. As perceiver-dependent cognitive characteristics played a key role in the evaluation, the related psychological construct is termed perceived visual complexity (PVC) [9]. On the other hand, there is a growing interest to design metrics to quantify the visual complexity of dashboards purely based on their objective visual features, often referred to as objective visual complexity (OVC) [7, 8]. Despite not having been widely validated, many components of the objective metrics showed good correlations with PVC reported in questionnaires and glance behaviors in a visual search task [7].

1.2 The Present Study

The present study first aims to refine the metrics of characterizing the OVC of the dashboards. On the basis of previous research [7, 8], we innovatively integrated two dimensions – colors and animation. While the variety dimension includes some measures of color types and colorfulness, the roles of individual colors, and the degree to which colors are present on the dashboards may also be relevant as they do in website design [10]. Moreover, with the development of presenting methods and visual effects, dashboards are more likely to incorporate animation, i.e., motion effects. We hypothesized the two new dimensions would account for unique aspects of the dashboards' OVC. Together, a total of 45 predictors formed the novel metrics. Moreover, using items in the metrics as inputs, we attempted to construct machine learning models to predict and classify the dashboards into 3 PVC levels determined by questionnaire data.

2   Methods

2.1   Study 1: PVC Rating

Participants. 160 participants aged from 18 to 44 years old (Mean = 24, SD = 4.21) were recruited for a subjective visual complexity rating experiment through advertisement in group chats and posts on social media.

Materials. Initially, 1400 images of vehicle dashboards from 170 different brands were collected from a major online vehicle forum in China [11] (see Fig. 1 for an example). Most of them were pictures taken and uploaded by car enthusiasts of their own cars or model cars at automobile sales service shops, while some were advertisement pictures taken from the official websites of vehicle manufacturers. Although a large number of raw stimulus materials were gathered in the first place, many of them display 3 common issues that could impede the subjective rating of dashboard visual complexity: insufficient image resolution, too much reflection that blocked parts of the dashboard from being seen or made dashboard elements indistinguishable, and highly distinct dashboard size or shape.

Fig. 1. An example of a raw vehicle dashboard image.

In order to enhance the quality of the stimuli, a preliminary screening of the original images was conducted. A raw image would be excluded if its resolution was too low (below 480*360 px), had a large reflection or shadow, or if the dashboard was obscured by the steering wheel (Fig. 2). 6 trained research assistants worked in groups of two for the evaluation. If a group disagreed on the eligibility of an image, a third research assistant would make the decision. A total of 200 images were determined to be eligible and included in the stimuli set, which were then corrected by cutting off areas irrelevant to the dashboard (Fig. 3).

Fig. 2. Major issues of the raw dashboard images. Left: Too much reflection. Middle: Distinct dashboard shape. Right: Some parts of the dashboard were blocked from sight by the steering wheel.

Fig. 3. An example of an image after the preliminary editing. The steering wheel and the irrelevant background were cut off from the image so that only the dashboard area was left.

Furthermore, the research assistants worked in groups to label the core dashboard area and dashboard elements, as well as calculate their sizes using LabelMe [10]. In order to normalize general visual features, a self-made program in Python [13] was utilized to add a black rectangular background to each image, which therefore ensures the unified appearance of the stimuli (Fig. 4).

Fig. 4. An example of a dashboard image after further labeling and adding the black background

After the abovementioned pre-processing, a super-resolution reconstruction was performed using a validated Real-ESRGAN algorithm [14] to improve the image resolution of the stimuli. Based on the classic Real-ESRGAN algorithm, Wang and colleagues employed the U-Net discriminator to replace the original VGG discriminator and introduced spectral normalization to make training more stable and reduce artifacts. Moreover, they used purely synthesized data and a "second-order" degradation model for training so that the algorithm could be applied to image repair in real-world scenarios. A self-developed Python program was then used to convert all stimuli to 2560*1080 px resolution. Images that were still blurry after the reconstruction were excluded, and eventually, a total of 100 images formed the stimuli set (Fig. 5).

Fig. 5. An example of a dashboard image included in the final dataset

Measures. Dashboard PVC was measured by a questionnaire adapted from previous research [6, 7]. The questionnaire asks participants about the perceived complexity of a dashboard from 4 aspects: quantity, variety, and layout of the displayed visual elements, and overall complexity. The questionnaire contains 13 items: 3 items for each aspect, and an additional item that concerns participants' general preference for a dashboard. Participants rate how much they agree to a description on a 7-point Likert scale. The questionnaire has been translated into Chinese as all participants are Chinese.

Procedure. Eligible participants were first invited to the study group chat to read the instruction and fill out the consent form. Then, they were assigned to a timeslot for the actual experiment according to their availability. In the experiment, participants completed the rating task on an online meeting platform under the instruction of a trained research assistant. A self-designed program in Python was utilized to play stimuli on participants' computers. The program would start the display only when the window resolution has been set to 1920*1080 px and layout zooming has been set to 125%, since the size of the window where an image is presented may interfere with the judgment of PVC [15]. While the stimuli were being exhibited on their computer screen, participants were asked to rate the visual complexity of the corresponding dashboard on their computer based on their first impression. Throughout the experiment, participants were also asked to keep their cameras and microphone on so that the research assistant was able to monitor their status and ensure their careful completion. The whole experiment takes 50 min on average to complete.

2.2 Study 2: OVC Rating

Materials. Same as in Study 1.

Metrics. 14 objective indicators proposed in previous studies [7, 8] were referred to as the measurement of the dashboards' objective visual complexity in the current study. In addition, 31 indicators related to animation effects and colors were innovatively included. 2 indicators of animation effects are component animation and background animation. Color-related indicators include 24 different colors, background color (black), color types, large color block types, size of the large color blocks, and size of the area other than the background color.

Procedure. 6 research assistants were trained to evaluate the objective visual complexity of the dashboard based on the original 14 indicators and animation-related indicators. They scored each indicator manually in pairs and reached a consensus on their evaluation. With regard to the color dimension, a self-developed program in Python was used to convert the color format in the image from RGB to HSV, and then judged the hue value of each pixel in the image. Data of the color-related indicators were yielded automatically.

2.3 Data Analysis

Overview of Models and environment. 3 types of machine learning models: support vector machine (SVM), K-proximity classification (KNN), and linear discriminant analysis (LDA) were constructed to predict the subjective visual complexity of dashboards. The models were built in the environment of Python 3.7.8 on a server with an operating system of 64-bit Windows 10, a CPU of AMD 3950X, a memory of 32G, and a related machine learning library of Scikit-learn 0.24.2. The average score of the overall PVC dimension in the questionnaire was averaged, based on which the images were divided into 3 groups: low PVC (26 images), medium PVC (47 images), and high PVC (27 images), which formed the calibration labels used in all models.

Firstly, Support Vector Machine (SVM) is a generalized linear classifier based on supervised learning [16]. As one of the common kernel learning methods, the SVM works by looking for the best separation hyperplane in the feature space to maximize the interval between positive and negative samples in the training set, which is commonly used to solve binary classification as well as nonlinear issues. The SVM has advantages in working with data of limited sample size or high dimensionality, and is also able to eliminate problems such as neural network structure selection and local minimum points [17].

K-nearest neighbor (KNN) algorithm is one of the classical machine learning algorithms. The idea is that in an eigenspace, if the majority of k nearest training samples of a test sample belong to a certain category, the test sample should be assigned the label of this category [18]. The KNN is a simple but powerful method that has good performance without the need to make any assumptions about data distribution.

Linear discriminant analysis (LDA), also known as Fisher linear discriminant (FLD), is a classical supervised learning algorithm often used for linear classification and dimensionality reduction. The LDA works by projecting data from a high-dimensional space into a lower-dimensional space and ensuring that the intra-class variance is small and the difference between the mean of each class is large after the projection. In this way, data points which belong to the same category are clustered together and can easily be distinguished from other categories due to the large intra-class distance [19].

Type-1 Models: Models Based on Traditional Feature Selection. *Input Screening.* In the first attempt, 4 feature selection approaches, namely, variance selection, correlation analysis, chi-square test, and Fisher Score, were applied to screen the 45 features (objective indicators). Specifically, the variance selection method calculates the variance in each feature, and filters out the features with variance below the threshold (threshold = 2 in the present study); The correlation analysis filters out the uncorrelated features based on the correlation coefficient between each feature and the average PVC scores; The chi-square test, often used for classification, automatically filters out k features according to the setting ($k = 10$ in the present study); Lastly, the Fisher Score ranks the classification power of indicators by calculating the ratio of within-group variance to between-group variance for each feature and manually filters them.

After the screening, 13 features were retained for modeling based on a comprehensive evaluation of all approaches' results, and they were: the total number of characters, total number of icons, chart percentage, image percentage, font type, speedometer components, Variability in section sizes, text-to-image ratio, number of information, the total size of the dashboard, component animation, color types occupying more than 0.01% of the dashboard size, and large color block types occupying more than 5% of the dashboard size.

Models Training and Optimization. The final ratio of the training set to the test set is 8:2. In order to enhance model performance, hyperparameter optimization was performed, which refers to the process of finding the best combination of model hyperparameter values to achieve the maximum performance in a reasonable time. Considering the difficulty of hyperparameter tuning, we adjusted the model parameters using the GridScarchCV algorithm [20]. The algorithm first lists the parameter space to be searched, and then determines the optimal value by comparing all the points in the search range in a brute-force fashion. In practice, grid search generally applies a large search range and step size to find the possible site of the global optimal value. Then, it will gradually narrow down the search range and step size to find more accurate optimal values.

Models Comparison. In order to compare different machine learning models in dashboard complexity evaluation and avoid chance bias, k-fold cross-validation was chosen as the cross-validation method. Specifically, the dataset was divided into k equally sized and mutually exclusive subsets, each of which was as consistent as possible with the frequency distribution of the entire sample. Each time, the union of k - 1 subsets was used as the training set of the machine learning model to be compared, and the remaining subset was used as the test set so that k pairs of training sets and test sets were yielded. The average value of k results was returned to compare the prediction accuracy of each model. Due to the limited sample size, $k = 5$ was selected to evaluate model stability.

Type-2 Models: Models Based on Principal Component Analysis (PCA). *Dimensionality Reduction.* Given that a large number of features per stimulus may prevent the models from converging, the PCA was used to perform dimensionality reduction. Invented by Pearson [21], PCA is a pre-processing technique that is widely used to remove redundant information from the data and make the model more parsimonious [22]. With the PCA, all objective indicators were used as the inputs of the models.

Models Training and Optimization. Same as in Type-1 Models.

Models Comparison. Same as in Type-1 Models.

3   Results

Machine learning models are often evaluated on accuracy, error rate, and sensitivity (i.e., recall) [23]. In the present study, accuracy was selected as the performance indicator. The accuracy of machine learning models refers to the proportion of the cases correctly classified in the samples and is the most commonly used and most intuitive.

Due to the skewed distribution of PVC, the cross-validation parameter "Stratify" was set to "True". Inputs of this type of model were the 13 features kept after feature selection. Results of the Type-1 Models are shown in Table 1. Accuracy: SVM > LDA > KNN. The average accuracy of the models is 53%, and varies relatively much across models.

Table 1. The results of the models based on traditional feature selection. SVM: Support Vector Machine. LDA: Linear Determinant Analysis. KNN: K-Nearest Neighbors

Model	Pre-processing	C	Gamma	n-conpoments	K	Stratify	Accuracy
SVM	MinMaxScaler	9	1	-	-	True	0.61
LDA	MinMaxScaler	-	-	2	-	True	0.53
KNN	MinMaxScaler	-	-	-	15	True	0.45

The results of the Type-2 models are shown in Table 2. Inputs of this type of model were all 45 objective indicators after being processed by the PCA. Overall, SVM still performs the best, and the average accuracy (56%) is higher than type-1 models. The accuracy is also relatively stable across models. Comparisons between all models in terms of their accuracy are shown in Fig. 6.

Table 2. The results of the models based on principal discriminant analysis (PCS). SVM: Support Vector Machine. LDA: Linear Determinant Analysis. KNN: K-Nearest Neighbors.

Model	Pre-procesing	PCA	C	Gamma	n-conpoments	K	Stratify	Accuracy
SVM	StandardScaler	1	6	50	-	-	True	0.58
LDA	StandardScaler	20	-	-	1	-	True	0.55
KNN	StandardScaler	1	-	-	-	17	True	0.55

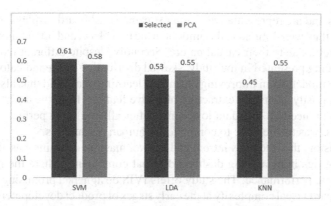

Fig. 6. Models Comparison. Selected: Models based on feature selection. PCA: Models based on the principal factor analysis (PCA). SVM: Support Vector Machine. LDA: Linear Discriminant Analysis. KNN: K-Nearest Neighbors. X-axis: Model Type. Y-axis: Accuracy.

4 Discussion

The objective of the current study is to construct objective metrics for predicting the visual complexity of vehicle dashboards. Innovatively, the study integrated color and animation on top of the three dimensions proposed in previous research: quantity, variety, and relations, and resulted in a total of 45 objective indicators. Furthermore, the study inventively connected quantifiable visual features to PVC of the dashboard perceivers using machine learning modeling. The new objective metrics were found to have 56% accuracy in predicting the level of the PVC of the dashboards, as measured in a subjective questionnaire. The introduction of machine learning models was shown to be successful, marking them a promising method in future user interface design and research.

Two types of machine learning models were applied to predict the PVC of dashboards using the objective indicators in the metrics. The first type of model was built based on feature selection. Eventually, 13 features were kept in the models. The average accuracy is 53%, which may be due to the data loss in the selection process. Although the objective indicators abandoned were determined to be least relevant to the PVC by all approaches comprehensively, the screening would inevitably lead to the loss of underlying information about the indicators, and thus the performance of the models contrasted based on it was limited. To reduce the data loss associated with feature selection, dimensionality reduction was applied alternatively using the PCA. Models whose inputs were pre-porcessed by the PCA are generally more accurate and stable than those built with feature selection, which suggests that the objective metrics as a whole are able to provide more information with judging the dashboard visual complexity.

As a preliminary attempt to utilize machine learning models in dashboard visual complexity research, several limitations exist in the current study. Firstly, small training set size constrained model performance. Since clear and complete pictures of dashboards are seldom exhibited on the official websites or in the advertisement materials of automobile companies, study materials were gathered from amateur-uploaded pictures on vehicle forums. While in a positive sense, these pictures are from the most up-to-date vehicle series and are representative of mainstreamed dashboard displays in the current marketplace, they were limited in the amount in the first place, and often failed to meet the inclusion criteria due to their casual nature. Secondly, despite different machine learning models being presented in the study, skewed distribution of the indicators may have prevented the models from improving during the learning phase and thus also resulted in under-fitting. Lastly, manually extracting objective features from the dashboard images was likely to induce bias and data loss, since, after all, circling a perfectly-fitting area using a mouse was too delicate to complete for human researchers.

In conclusion, the study has introduced a novel machine-learning-based method for validating metrics in measuring dashboard visual complexity, where the models have good and stable performance. The study offers IVIS designers a promising approach to assess dashboard visual complexity at the early stages of product development, which can also be generalized to other human-computer interfaces visual design research. Future studies should work on expanding the modeling dataset, as well as using original images as learning materials through image enhancement techniques. Furthermore, by combining fundamental perceptual theories in psychology with deep learning, the performance of the evaluation models will increase largely to reach and exceed humans.

Acknowledgement. The study was supported by the National Natural Science Foundation of China (Grant No. T2192932) and the Scientific Foundation of Institute of Psychology, Chinese Academy of Sciences (No. E2CX4535CX). The authors would like to thank Chongqing Changan Automobile Co., Ltd. For project support.

References

1. ISO 15008: 2017.: Road vehicles–ergonomic aspects of transport information and control systems–specifications and test procedures for in-vehicle visual presentation. (2017)
2. Snodgrass, J.G., Vanderwart, M.: A standardized set of 260 pictures: norms for name agreement, image agreement, familiarity, and visual complexity. J. Exp. Psychol. Hum. Learn. Memory **6**, 174 (1980)
3. Xing, J.: Designing questionnaires for controlling and managing information complexity in visual displays. FEDERAL AVIATION ADMINISTRATION OKLAHOMA CITY OK CIVIL AEROSPACE MEDICAL INST (2008)
4. Lee, S.C., Kim, Y.W., Ji, Y.G.: Effects of visual complexity of in-vehicle information display: age-related differences in visual search task in the driving context. Appl. Ergon. **81**, 102888 (2019). https://doi.org/10.1016/j.apergo.2019.102888
5. Lobjois, R., Faure, V., Désiré, L., Benguigui, N.: Behavioral and workload measures in real and simulated driving: do they tell us the same thing about the validity of driving simulation? Saf. Sci. **134**, 105046 (2021). https://doi.org/10.1016/j.ssci.2020.105046
6. Lee, S.C., Hwangbo, H., Ji, Y.G.: Perceived visual complexity of in-vehicle information display and its effects on glance behavior and preferences. Int. J. Human-Comput. Interact.. **32**, 654–664 (2016). https://doi.org/10.1080/10447318.2016.1184546
7. Yoon, S.H., Lim, J., Ji, Y.G.: Assessment model for perceived visual complexity of automotive instrument cluster. Appl. Ergon. **46**, 76–83 (2015). https://doi.org/10.1016/j.apergo.2014.07.005
8. Kim, J.M., Hwangbo, H., Ji, Y.G.: Developing visual complexity metrics for automotive human-machine interfaces. J. Ergonom. Society Korea. **34**, 235–245 (2015). https://doi.org/10.5143/JESK.2015.34.3.235

9. Cummings, M., Sasangohar, F., Thornburg, K., Xing, J., D'Agostino, A.: Human-system interface complexity and opacity part i: literature review. Massachusettes Institute of Technology, Cambridge, MA (2010)

10. Reinecke, K., et al.: Predicting users' first impressions of website aesthetics with a quantification of perceived visual complexity and colorfulness. In: Presented at the Proceedings of the SIGCHI Conference on Human Factors in Computing Systems (2013)

11. Pacifics Vehicle Forum, https://bbs.pcauto.com.cn/, Accessed 6 Feb 2023

12. Wada, K., et al.: wkentaro/labelme: v4.6.0 (2021). https://doi.org/10.5281/zenodo.5711226

13. Van Rossum, G., Drake, F.L.: Python 3 Reference Manual. CreateSpace, Scotts Valley, CA (2009)

14. Wang, X., Xie, L., Dong, C., Shan, Y.: Real-ESRGAN: Training Real-World Blind Super-Resolution with Pure Synthetic Data (2021). https://arxiv.org/abs/2107.10833

15. Yared, T., Patterson, P.: The impact of navigation system display size and environmental illumination on young driver mental workload. Transport. Res. F: Traffic Psychol. Behav. **74**, 330–344 (2020). https://doi.org/10.1016/j.trf.2020.08.027

16. Boser, B.E., Guyon, I.M., Vapnik, V.N.: A training algorithm for optimal margin classifiers. In: Proceedings of the Fifth Annual Workshop on Computational Learning Theory, pp. 144–152 (1992)

17. Müller, B., Reinhardt, J., Strickland, M.T.: Neural networks: an introduction. Springer Science & Business Media (1995)

18. Fix, E., Hodges, J.L.: Discriminatory analysis. Nonparametric discrimination: Consistency properties. Int. Statist. Rev./Revue Internationale de Statistique. **57**, 238–247 (1989)

19. Boedeker, P., Kearns, N.T.: Linear discriminant analysis for prediction of group membership: a user-friendly primer. Adv. Methods Pract. Psychol. Sci. **2**, 250–263 (2019). https://doi.org/10.1177/2515245919849378

20. LaValle, S.M., Branicky, M.S., Lindemann, S.R.: On the relationship between classical grid search and probabilistic roadmaps. Int. J. Robot. Res. **23**, 673–692 (2004)

21. Pearson, K.: LIII. On lines and planes of closest fit to systems of points in space. The London, Edinburgh, and Dublin Philosophical Magazine and Journal of Science. **2**, 559–572 (1901). https://doi.org/10.1080/14786440109462720

22. Shlens, J.: A tutorial on principal component analysis. arXiv preprint arXiv:1404.1100. (2014)

23. Raschka, S.: Model Evaluation, Model Selection, and Algorithm Selection in Machine Learning, https://arxiv.org/abs/1811.12808 (2018)

A Systematic Analysis for Multisensory Virtual Artifacts Design in Immersive E-Sport Applications and Sim-Racing

Alberto Calleo[1]([⊠]) [iD], Giorgio Dall'Osso[2] [iD], and Michele Zannoni[1] [iD]

[1] Department of Architecture, University of Bologna, Bologna, Italy
{alberto.calleo,michele.zannoni}@unibo.it
[2] University of the Republic of San Marino, San Marino, San Marino
giorgio.dallosso@unirsm.sm

Abstract. The mechanics of human-machine interaction have been subjected to numerous influences from the video game industry. With respect to the relationship between transportation and vehicle design and game tech, the area of e-sports and, in particular, that of sim-racing appears to be of particular interest. The perceptual dimension, in its potential as a vehicle for distributed multilevel feedback, becomes a tool that blurs the boundaries of the real-virtual experience due to the potential for spatial and emotional connection that have already been verified in the mixed reality context. The contribution aims to establish and develop a systemic analysis of technologies and approaches used to recreate multimodal perceptual-sensory mechanisms in the context of virtual simulations and in particular in sim-racing devices and e-sports. The analysis and classification of the state of the art is meant to represent a design map to support the designer of digital interfaces in the simulation domain. For the designer of human-computer interaction interfaces, the opportunity emerges to develop innovative interaction patterns and dynamics that exploit multimodality as an input and feedback channel. In this scenario, the role of the interaction designer emerges as a mediator between technological development and the design of the user experience of human-machine interfaces.

Keywords: E-Sport · Simulation · HCI · Sense · Human Body Interaction

1 Introduction

The use of tools and practices derived from the video game industry in productive sectors different from that of digital entertainment is a topic of growing interest in both academia and industrial research, development and production. Whether we are talking about gamification, applied games, or simulations, the mechanics of human-machine interaction have been subjected to numerous influences from the video game industry. Indeed, it is possible to observe a growing integration between specialized technologies developed and used specifically for different production areas the technologies used in video game production (game-tech). An illustrative case is the use of real-time rendering engines

H. Krömker (Ed.): HCII 2023, LNCS 14049, pp. 114–124, 2023.
https://doi.org/10.1007/978-3-031-35908-8_9

for the construction of digital twins for applications ranging from architectural planning to logistics management of production plants. Such development platforms enable the construction of visualization environments in which topographical and architectural surveys, instrumental data of various types collected through sensors, and processing performed by specialized software converge.

Digital technologies related to Industry 4.0 are bringing a wide variety of innovations in the automotive industries pushing companies to develop new strategies of innovation and fostering the transformation of the relationships between the industry players [1]. With respect to the relationship between transportation and vehicle design and game tech, the area of e-sports and, in particular, that of sim-racing appears to be of particular interest. Competitive driving simulators that take advantage of high-fidelity driving simulation software and real-world driving control hardware replicas, around which there is growing commercial and research interest. During the pandemic crisis, for example, both Formula 1 and NASCAR used virtual sim-racing to continue their activities on live streaming platforms helping to increase interest in this form of entertainment [2]. The transposition of car racing into virtual competitions raises interesting considerations with respect to new dynamics of interaction between athletes' bodies and simulation interfaces. The athlete's body in eMotorSport is "in negotiation with algorithms and processes to quantify its movement. An assemblage that links human and nonhuman, matter and information" [3].

Growing in parallel is the interest in the video game as a domain for the development of virtual economies in which new modes of consumption and value creation of the commodity form are redefined. This represents a dimension of design and speculation that evolves from the established habit of car companies to design concept cars. In the videogame dimension, players have the opportunity to experience the driving of a concept car whose performances are simulated by the game engine with high level of accuracy. One example is the Ferrari Vision GT, a concept car developed by Ferrari and unveiled in November 2022 and designed specifically for the video game Gran Turismo 7[1].

Within the redefinition of the human-machine interaction paradigm, research poses as an additional element of debate in this problematic field: the assemblage [3], the cyborg [4] and prosthetic [5, 6] connection between human and multisensory interfaces.

The interpretation of perceptual stimuli coming from multiple sources and experienced through different interfaces is a field of the broad project in which multidisciplinary expertise converges. From a design perspective, it is of particular interest to understand the potential of particular sensory feedbacks in relation to human behavior.

Academic research shows that scientific investigation on the relationship between digital technologies and human senses is exponentially increasing with a clear predominance over the sense of sight and proprioception. Less significantly, research is moving on hearing, touch, and forms of multisensory perception [7]. The sense of sight and proprioception appear to be central to the market for immersive digital experiences. Hence, the opportunity for multimodal integration on the other senses within simulation, e-sports and even mixed reality experiences to control real-world parameters is highlighted [8]. With respect to this integration, literature research highlights the potential related to the

---

[1] https://www.ferrari.com/it-IT/corporate/articles/ferrari-vision-gran-turismo.

implementation of feedback through haptic channels. This sensory channel is characterized by multiple properties that can be related to digital applications that exploit the possibilities of transferring sensations of temperature [9], pressure [10], vibration[11], proprioceptive stimuli [12, 13].

The haptic channel turns out to be a guide for the qualitative interpretation of information gathered through the other sensory channels [14], this means that its conscious use can direct the perception of the other senses as well. Through the haptic channel also information can be sent silently and personally, capable in each case of capturing the user's attention [15] with different levels of urgency. It is also evident how in human-machine interfaces the haptic channel is underutilized with respect to the potential related to the widely analyzed body application surface with respect to individual implementable functionalities [16].

Touch turns out to be a suitable channel for delivering information and establishing communication languages for stimulating and interacting with user's peripheral attention [17, 18]. Finally, the tactile channel turns out, as confirmed by experimental validations, to be effective from the point of view of emotional communication [19]. In this direction, haptic sensations are used both parallel to text messages [20] and in synchrony with visual information experienced from the surrounding space [21]. In the former case, haptic cues support the meaning of textual language by increasing its communicative effectiveness; in the latter, they increase engagement with respect to the situation observed by sight.

In describing the driving experience, the vehicle is often described by drivers as an extension of their own body. This description can be ascribed to the debated concept of homuncular flexibility (HF), or the ability of a part of the cerebral cortex to map the body's movements and perceived sensibility onto extraneous physical appendages. The phenomenon of HF has been shown to extend also to virtual appendices [22] opening to a wide range of opportunities and challenges within the human-machine interface research field.

As frequently happens in technological innovation processes, in the area of driving simulators (sim-racing) there is an increasing consumer-level diffusion of instrumentation and technologies that were previously only used in research and industrial development settings. Indeed, it is possible to identify a growing number of commercial applications available on the market that integrate software and hardware to enrich the immersive component of virtual experiences. Through physical interfaces and haptic feedback, it is possible to increase the level of embodiment within the virtual simulation by making it verisimilar to the physical experience [23]. Moreover, professional racing suits and boots, helmets, and gloves are often used in combination with the simulator to further increase the immersion level acting both as diegetic components and physical body constrain/support systems. The perceptual dimension, in its potential as a vehicle for distributed multilevel feedback, becomes a tool that blurs the boundaries of the real-virtual experience due to the potential for spatial and emotional connection that have already been verified in the mixed reality context.

In this contribution we aim to conduct an analysis of the technologies and ways in which the body is related to virtual simulation by providing a useful design map in the development of immersive interaction mechanics.

2 Methodology

The combined use of instrumentation that merges different perceptual stimuli to recreate the illusion of presence and verisimilitude of experience represents a broad project space that crosscuts different disciplines ranging from neuroscience, computer science, game studies, mechanical engineering, and design. The contribution aims to establish and develop a systemic analysis of technologies and approaches used to recreate multimodal perceptual-sensory mechanisms in the context of virtual simulations and in particular in sim-racing devices and e-sports.

Classification criteria were adopted in the collection of case studies in order to better describe the design approaches and the ways in which different technologies were employed. The study compared across multiple aspects forty case studies expressing interfaces and experiences dedicated to simulation and e-sports. The number of cases has been defined in order to develop a flexible, yet detailed classification that will be used to map new case studies during the research activity.

An initial classification of the interface systems between the body and the simulation that is proposed is between diegetic interfaces and mimetic interfaces:

- Diegetic interfaces: the input interface reproduces the morphological, material, and usage characteristics of the virtual counterpart. It is a homologous interface.
- Mimetic interfaces: the input interface has no direct correspondence with the motion or action reproduced in the simulation.

This classification allows interfaces to be categorized on the basis of different modes of interaction. Diegetic interfaces place themselves in a relationship of continuity with the simulated environment by articulating the dialogue between physical space, perception, real environment and simulated environment. An example of this is the steering wheels used in sim-racing whose finishes, geometric and functional characteristics reproduce, in the most advanced and professional models, the features and details of their real-life counterparts. Mimetic interfaces do not preserve a homology relationship between the body movement exerted on the input interface and the action reproduced in the simulation and, therefore, the morphological and material characteristics are also independent of the simulated medium. One example is game controllers, which, through different forms of input (button presses, trigger presses, lever rotations) allow the control of different avatars (whether these are anthropomorphic or vehicles).

The availability of different technologies to return interaction and feedback mechanisms articulates the possibility of combining different technical solutions to be implemented in the design. In defining the combination of components, the designer has the opportunity to develop customized solutions on multiple parameters: cost, complexity of deployment, sensory multichanneling, special needs. In addition, prototyping systems structured according to different combinations of components to return a particular interaction or feedback allows for evaluation of perceptual differences and spillovers in terms of affordance, embodiment, and immersiveness. Technological components can be used to give control inputs or to return response feedback. In the analysis performed on the case studies, information was collected about the presence of visual, auditory and tactile feedback. For completeness, tactile feedbacks were divided into proprioceptive, vibratory, pressure, temperature and electromuscular.

A further criterion for cataloging the case studies emerged with respect to the role that the interfaces play in relation to the human body. To this end, four categories were chosen: prostheses, braces, protectors, and substitutes:

- Prosthetics are understood to be those design elements that intervene in the relationship with the human body in augmenting human motor, perceptual and intellectual capabilities [5] using both passive and interactive approaches [6]. This category has very broad disciplinary boundaries that link the e-sports and simulation industry to artistic experiments related to cyborg culture represented by the works of Sterlac, Harbisson, and Ribas[2], among others;
- Braces are understood to be the category of objects that, through an action of resistance to bodily dynamics [24], provide the virtual reality experience with proprioceptive sensations that contribute to the perception of the shape and size of the elements;
- Protections are the designs that included in the experiences components that have the task of protecting the body immersed in a simulative dimension estranged from the dangers of the surrounding space;
- Substitutes are understood as those tools that replace a human body within interactions.

Therefore, to describe the different types of motion, a classification is proposed that interprets in spatial terms the action performed by the user in interacting with the interface or the feedback of the system on the user's body. A classification is adopted that distinguishes signals (output feedback or input actions) into:

- Punctual: when the signal can be described by locating it at a point (e.g., pulling a trigger);
- Linear: when the signal develops along one direction only or very prevalent over the others (e.g., the sliding of a slider along a rail);
- Planar or superficial: when the signal develops along a plane or surface (e.g., electromuscular stimulation distributed over a supporting surface);
- Volumetric: when the signal is collected or distributed in three-dimensional space (e.g., spatial tracking systems).

The choice of sensing system may depend on reasons related to technical/constructive needs or, in the case of diegetic interfaces, on the ergonomic and morphological characteristics of the interface itself. Analysis of this aspect makes it possible to identify parameters of accessibility of the interface with respect to the variety of possible physical limitations or disabilities.

A systematic analysis of the hardware technologies employed in the field of e-sports and the ways in which these can be integrated with each other and with simulation software aims to return design tools based on the perceptual transfer of virtual experience onto the body.

---

[2] https://www.cyborgarts.com/.

3 Discussion

The analysis of the case studies revealed some design, methodological and technological trends of interest. Among the interfaces investigated, the category of prosthetics clearly prevails (Fig. 1). Within the design field related to simulative interfaces there appears to be a greater tendency to develop elements that are intended to implement human capabilities within the virtual environment. Cases studies in which the design limit, protect or replace the body have been collected in much smaller numbers. Braces are used in more advanced designs to stimulate proprioceptive sensations within the VR experience. Braces are used to ensure the preservation of the integrity of the physical body when the virtual experience involves fast and large movements in space. Finally, only one case of an object substituting itself for an opponent's body has been detected; this is the case of BotBoxer[3] in which the physical punching bag comes alive and dodges punches by real-time control of the human's movement.

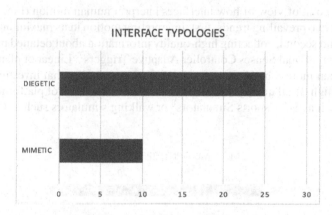

Fig. 1. Interface typologies

Among the collected case studies, the typology of diegetic interfaces prevails (Fig. 2), that is, capable of activating, thanks to symbolic forms, the narrative related to the experience in which users will be immersed. Examples of these interfaces are the multiple immersive experiences related to sports such as golf (e.g., Golfzon[4]) or baseball (e.g., Win Reality[5]). Mimetic interfaces are predominantly adopted when attempting to build multi-experience elements or when the interface becomes the movement of the body and thus the devices invade the entire body to track its movement as in the case of climbing of the Red Bull project The Edge[6].

---

[3] https://www.skytechsport.com/botboxer-home.

[4] https://golfzongolf.com/.

[5] https://winreality.com/.

[6] https://www.redbull.com/int-en/projects/the-edge-matterhorn-vr.

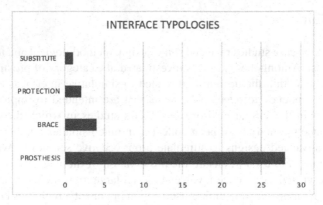

Fig. 2. Diegetic/mimetic interfaces

From the point of view of how interfaces interpret human motion (Fig. 3), the case studies detect two prevailing trends: first, interpreting motion in its maximum volumetric expression, and second, collecting high-quality information about detailed motion, as in the case of Sony's Dual Senses Controller Adaptive Triggers[7]. Linear or planar interpretation of human motion by interfaces is present in case studies that investigate specific motion with high detail quality. Prominent case studies are those of platforms for skiing simulation such as Snowsports Simulators[8] or walking simulators such as Cyberith[9].

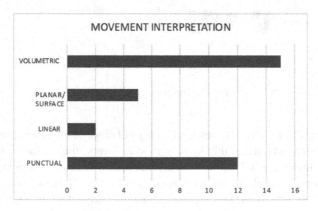

Fig. 3. Movement interpretation

The analysis of the material collected from commercial case studies shows heterogeneous technologies used to stimulate different sensory channels. Detailed data on the technologies used are also often not available. However, the results related to reading the

---

[7] https://www.theverge.com/21562206/ps5-dualsense-controller-review-games-features-vibrat ions.

[8] https://www.skytechsport.com/ski-simulators-home.

[9] https://www.cyberith.com/virtualizer-elite/.

type of sensation sent to users during the experiences appear interesting (Fig. 4). Since these are mostly immersive experiences, sight and hearing stimuli are always present. In contrast, stimuli related to touch are not yet widespread in market solutions. The observation related to the greater presence of pressure stimuli than vibration is of interest. To a slightly greater extent there are case studies that use systems to give proprioceptive feedback through mechanical actions on the user's body, as in the case of the Blackbox[10]. No designs related to temperature feedback were found in the case studies analyzed, and only Teslasuit[11] uses electromuscular feedback.

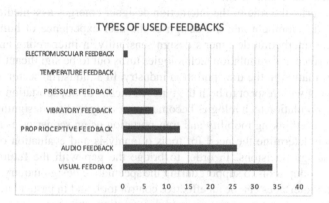

Fig. 4. Types of Feedback

Finally, from the analysis, it is relevant to compare the number of stimuli that are delivered within the individual experiences. In Fig. 5, the columns represent the number of identified stimuli. Removing the sight and hearing stimuli, the lack of multimodal stimuli is evident for most of the case studies.

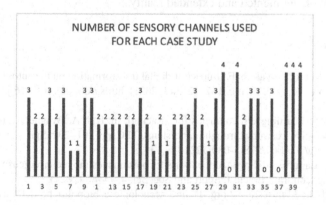

Fig. 5. Number of involved sensory channels

---

[10] https://www.blackbox-vr.com/the-hardware/.
[11] https://teslasuit.io/.

4 Conclusions

The analysis of the case studies confirms the interest in the development of multimodal input/output platforms already investigated in research. For the designer of human-computer interaction interfaces, the possibility of developing innovative interaction models and dynamics that exploit multimodality as an input and feedback channel emerges. The analysis confirms the growing interest in haptic and auditory feedback interfaces. Further research could be directed at the use of feedback acting through alternative channels such as taste and smell.

In this scenario, the role of the interaction designer emerges as a mediator between technological development and the design of the user experience of human-machine interfaces. The interaction designer's design sensibility in interpreting and mediating the use of interface and simulation technologies turns out to be significant in a phase of strong transformation in the transportation industry. In a industrial sector undergoing a major rethinking with respect to both the types of means of transportation and their use during travel, simulation technologies become a necessary tool for designing and testing new scenarios. Rethinking mobility and transportation using scenarios as a speculative design practice determine the need for tools of analysis and evaluation of the impact of strategic design decisions. In order to bridge the gap with the future, it appears necessary to develop tools to support during the speculative design inquiry. Simulations and interaction interfaces can be an effective control tool, and in particular, it is deemed important to leverage on sensory multichannel interaction whit more accuracy to evaluate choices and directions of innovation in greater detail.

The analysis and classification of the state of the art is meant to represent a design map to support the designer of digital interfaces in the simulation domain. Although the scope of the investigation focuses on human-vehicle interaction, the results may be extended to other areas of human-computer interaction in general to investigate what are possible corridors for future development and innovation on different domains in the context of virtual, augmented and extended reality.

References

1. Llopis, C., Rubio, F., Valero, F.: Impact of digital transformation on the automotive industry. Technol. Forecast. Soc. Chang. **162**, 120343 (2021). https://doi.org/10.1016/j.techfore.2020.120343
2. Scacchi, W.: [Learning Environments for , Racing]. In: Hui, A., Wagner, C. (eds.) Creative and Collaborative Learning through Immersion. CTFC, pp. 201–232. Springer, Cham (2021). https://doi.org/10.1007/978-3-030-72216-6_13
3. Ruffino, P.: La mossa di Pagenaud: eSports tra postumanesimo e convergenze tecnologiche, economiche, culturali. Eracle J. Sport and Soc. Sci. **4**, 74–88 (2021)
4. Haraway, D.J.: Manifesto cyborg. Donne, tecnologie e biopolitiche del corpo. Feltrinelli (1995)
5. Maldonado, T.: Critica della ragione informatica. Feltrinelli, Milano, Italia (1997)
6. Zannoni, M.: Progetto e interazione. Il design degli ecosistemi interattivi. Quodlibet, Macerata (2018)

7. Berry, J.A., Valero-Cuevas, F.J.: Sensory-motor gestalt: sensation and action as the foundations of identity, agency, and self. presented at the ALIFE 2020: The 2020 Conference on
   Artificial Life July 1 (2020). https://doi.org/10.1162/isal_a_00340
8. Reynal, M., et al.: Investigating multimodal augmentations contribution to remote control
   tower contexts for air traffic management: In: Proceedings of the 14th International Joint
   Conference on Computer Vision, Imaging and Computer Graphics Theory and Applications,
   pp. 50–61. SCITEPRESS - Science and Technology Publications, Prague, Czech Republic
   (2019). https://doi.org/10.5220/0007400300500061
9. Matthies, D.J.C., Müller, F., Anthes, C., Kranzlmüller, D.: ShoeSoleSense: proof of concept
   for a wearable foot interface for virtual and real environments. In: Proceedings of the 19th
   ACM Symposium on Virtual Reality Software and Technology, pp. 93–96. Association for
   Computing Machinery, New York (2013). https://doi.org/10.1145/2503713.2503740
10. Kettner, R., Bader, P., Kosch, T., Schneegass, S., Schmidt, A.: Towards pressure-based feed-
    back for non-stressful tactile notifications. In: Proceedings of the 19th International Confer-
    ence on Human-Computer Interaction with Mobile Devices and Services, pp. 1–8. ACM,
    Vienna Austria (2017). https://doi.org/10.1145/3098279.3122132
11. Karuei, I., MacLean, K.E., Foley-Fisher, Z., MacKenzie, R., Koch, S., El-Zohairy, M.: Detect-
    ing vibrations across the body in mobile contexts. In: Proceedings of the SIGCHI Confer-
    ence on Human Factors in Computing Systems, pp. 3267–3276. Association for Computing
    Machinery, New York, NY, USA (2011). https://doi.org/10.1145/1978942.1979426
12. Je, S., Kim, M.J., Lee, W., Lee, B., Yang, X.-D., Lopes, P., Bianchi, A.: Aero-plane: a handheld
    force-feedback device that renders weight motion illusion on a virtual 2D Plane. Presented
    at the October 17 (2019). https://doi.org/10.1145/3332165.3347926
13. Zhang, Z.-Y., Chen, H.-X., Wang, S.-H., Tsai, H.-R.: ELAXO : Rendering Versatile Resistive
    force feedback for fingers grasping and twisting. In: Proceedings of the 35th Annual ACM
    Symposium on User Interface Software and Technology, pp. 1–14. Association for Computing
    Machinery, New York (2022). https://doi.org/10.1145/3526113.3545677
14. Buiatti, E.: Forma Mentis. Neuroergonomia sensoriale applicata alla progettazione. Franco
    Angeli, Milano, Italia (2016)
15. Jones, L.A., Sarter, N.B.: Tactile displays: guidance for their design and application. Hum/
    Factors 50, 90–111 (2008). https://doi.org/10.1518/001872008X250638
16. Zeagler, C.: Where to wear it: functional, technical, and social considerations in on-body
    location for wearable technology 20 years of designing for wearability. In: Proceedings of
    the 2017 ACM International Symposium on Wearable Computers, pp. 150–157. Association
    for Computing Machinery, New York (2017). https://doi.org/10.1145/3123021.3123042
17. Baumann, M.A., MacLean, K.E., Hazelton, T.W., McKay, A.: Emulating human attention-
    getting practices with wearable haptics. In: 2010 IEEE Haptics Symposium, pp. 149–156
    (2010). https://doi.org/10.1109/HAPTIC.2010.5444662
18. Enriquez, M., Afonin, O., Yager, B., Maclean, K.: A pneumatic tactile alerting system for
    the driving environment. In: Proceedings of the 2001 workshop on Perceptive user interfaces,
    pp. 1–7. Association for Computing Machinery, New York (2001). https://doi.org/10.1145/
    971478.971506
19. Hertenstein, M.J., Keltner, D., App, B., Bulleit, B.A., Jaskolka, A.R.: Touch communicates
    distinct emotions. Emotion 6, 528–533 (2006). https://doi.org/10.1037/1528-3542.6.3.528
20. Rovers, A.F., van Essen, H.A.: HIM: a framework for haptic instant messaging. In: CHI '04
    Extended Abstracts on Human Factors in Computing Systems, pp. 1313–1316. Association
    for Computing Machinery, New York (2004). https://doi.org/10.1145/985921.986052
21. McCormick, J., et al.: Feels like dancing: motion capture-driven haptic interface as an added
    sensory experience for dance viewing. Leonardo 53, 45–49 (2020). https://doi.org/10.1162/
    leon_a_01689

22. Won, A.S., Bailenson, J.N., Lanier, J.: Homuncular Flexibility: The Human Ability to Inhabit Nonhuman Avatars. In: Emerging Trends in the Social and Behavioral Sciences, pp. 1–16. John Wiley & Sons, Ltd. (2015). https://doi.org/10.1002/9781118900772.etrds0165
23. Scacchi, W.: Autonomous eMotorsports racing games: emerging practices as speculative fictions. J. Gaming Virtual Worlds **10**, 261–285 (2018). https://doi.org/10.1386/jgvw.10.3.261_1
24. Tsetserukou, D., Sato, K., Tachi, S.: ExoInterfaces: novel exoskeleton haptic interfaces for virtual reality, augmented sport and rehabilitation. In: Proceedings of the 1st Augmented Human International Conference, pp. 1–6. Association for Computing Machinery, New York (2010). https://doi.org/10.1145/1785455.1785456

Digitalization and Virtual Assistive Systems in Tourist Mobility: Evolution, an Experience (with Observed Mistakes), Appropriate Orientations and Recommendations

Bertrand David[(⊠)] 🆔 and René Chalon 🆔

Université de Lyon, CNRS, Ecole Centrale de Lyon, LIRIS, UMR5205, 69134 Lyon, France
{Bertrand.David,Rene.Chalon}@ec-lyon.fr

Abstract. Digitalization and virtualization are extremely active and important approaches in a large scope of activities (marketing, selling, enterprise management, logistics). Tourism management is also highly concerned by this evolution. In this paper we try to present today's situation based on a 7-week trip showing appropriate and shame situations. After this case study, we give a list of appropriate practices and orientations and confirm the fundamental role of User Experience in validating the proposed assistive system and the User Interfaces needed for client/user satisfaction. We also outline the expected role of Metaverse in the future of the evolution of this domain.

Keywords: Digitalization · virtualization · assistive systems · tourism · tourist mobility · User Interface · User Experience · Metaverse

1 Introduction

Tourist trips have always needed supportive assistance [1], which was initially up to 10–15 years ago only physical and paper-based and mainly prior to the trip. The evolution of assistive systems is permanent and will continue in the future. We can mention the evolution of physical agencies to virtual technologies through the emergence of Internet [2]. The evolution of one-shot contact to continuous access to appropriate information, contextualization by geolocalization, and the internet of things must be emphasized. In this new context, User Interfaces and human behaviors have become the main factors of success of this evolution, characterized by utility, usability, and user satisfaction. Due to continuous complexification, these factors are not easy to guarantee.

Digitalization and virtualization are extremely active and important approaches in a large scope of activities (marketing, selling, enterprise management, logistics) [3]. The objective is to replace physical documents and face-to-face contact by digitalized documents shared worldwide and by remote virtual access to their exchange and manipulation. Human face-to-face is replaced by independent manipulation of these documents (data supports). Tourism is also significantly concerned by this evolution, not only in the trip preparation stages, but also and mainly in mobile situations during trips.

H. Krömker (Ed.): HCII 2023, LNCS 14049, pp. 125–141, 2023.
https://doi.org/10.1007/978-3-031-35908-8_10

For this paper we had two possibilities of presentation, either a general (abstract approach) followed by case studies or an opposite one, a case study showing the most frequently observed misbehaviors and bad situations followed by local correction suggestions and only, at the end, a trial to synthesize and give an open list of recommendations. We decided to apply this second approach, after a brief presentation of the state-of-the-art of digitalization in general and its rapid synthesis in the tourism industry.

Our objective is to present first a case study based on a 7-week trip in the preparation, execution, and consolidation stages. The objective is to show appropriate as well as inappropriate behaviors and management situations. Based on this experience and a general appreciation, we draw up a digitalization schema for the tourism industry and a list of recommendations and suggestions to consider.

2 State-of-the-Art

To address this issue, it seems interesting to describe the state-of-the art in three fields that structure our approach: An informal view of digitalization of tourist activities, digitalization approaches and principles in various domains of use, and User Experience principles as the approach to observe and evaluate user behaviors in various situations.

2.1 Digitalization of Tourist Activities

We are not able to define once and for all the entire scope of tourist activity digitalization. As such, we propose to express it in a more informal manner.

The Influence of Digitalization on Tourism.   Digital technology has a profound influence on tourist economy everywhere on the planet. Trends are digital for cities, tourist offices, and travel professionals. To offer an ever more unique and impactful experience, there is no shortage of technological means. Cities are investing in tactile and interactive media [4]. Cities and communities are modernizing and investing in interactive technologies to respond to the digital transformation of uses. In tourist offices it is no longer rare to take advantage of dynamic displays and other tactile supports that enrich and facilitate the presentation of expected vacancy locations to potential clients [5]. The clever combination of broadcast screens, touch terminals, and interactive media makes it possible to create an augmented reception area: a modernized place at the cutting edge of digital technology with real added value for tourists seeking information. The positive consequences are numerous: in addition to improved communication within tourist offices equipped with digital displays, visitors benefit from a better quality of service and information than before. Multiple language facilities are also available. Large amounts of data can also be collected, analyzed, and then used with the aim of improving visitor satisfaction: performance of advisers, attendance, visitors' impressions, and their recurring requests, etc. However, in cities it is not just tourist offices that are equipped with the latest interactive technologies. Museums, shopping malls, fairs, festivals, and other tourist sites are also equipped with them in order to offer travelers an unforgettable and rich experience [6, 7].

The Tourism Industry is Reinventing Itself.   The tourism industry is reinventing itself. Technological developments have caused new forms of tourism to emerge:

- **m-tourism**, also called mobile tourism, is the action of booking holidays directly via a mobile or touch pad [8];
- **Social tourism** consists in choosing your stay based on social networks [9];
- **E-tourism** is the association of web and tourism giving rise to all kinds of digital experiences [10].

To meet the growing expectations of increasingly demanding tourists, comparative sites and tools have multiplied and make it possible to create ever more personalized trips.

The Travel Agency Model is Changing. Another proof that tourism is directly impacted by the digital transition is that traditional travel agencies, where customers used to go to book their stays, are threatened. The Internet has seen countless tourism stakeholders flourish in just over a decade, whether they are online agencies, flight comparators, hotels, or destinations. However, all is not lost for physical agencies. Various studies have shown that 8 out of 10 tourists go on the internet to search for their holidays, but only 1 out of 3 actually book from a screen [11]. Many tourists prefer to speak with an adviser, for example to be reassured. This is where traditional agencies can make their mark by installing interactive screens, touch screens, and digital signage technologies on their premises to offer visitors a more realistic view than ever before. The immersive and interactive travel agency of the 21st century is already here and continues to grow via the Metaverse approach [12].

2.2   Digitalization Principles

Digitalization is the transformation of current working principles in the physical world to a new world that is more or less digital and virtual [13]. This evolution started at least 15 years ago and has progressed constantly rather than linearly. It can be applied to marketing, to CRM (Customer Relationship Management), to production, to public services, to logistics and so on [14]. While information-based services can be totally digitized, physical object-based activities, production, and manipulation mix physical and virtual activities with the Internet of Things (IoT) approach [15]. Information digitalization can ensure substantial improvements to data management, thanks to the quality obtained, process time reduction, and associated costs.

Commonly, digitalization is presented as a 3-step process starting with Digitization; transformation of physical documents (texts, drawings, etc.) to digital ones, which can be more easily transported and modified. The next step is known as Digitalization, which is characterized by the presence of manipulation tools (for these digital documents) allowing these data to be shared and presented on different websites. Their transportation between different locations of use, mainly by computer networks and commonly accessible stores, either in specific servers or increasingly more in no-located cloud infrastructures, are thereby facilitated. The last and main step is known as Digital Transformation, which is concerned by the proposals of new organizations considering previously explained elements, leading to what is generally called Informatization [16].

In this way, economic, social, and cultural relationships and barriers are reconsidered and profoundly modified. As stated earlier, methodological processes are not the same as the scope of actions is different. Nevertheless, generic digitalization processes exist and can be used either directly or after adaptation to the field concerned.

Our paper does not focus on the design process of digitalization but rather adopts the opposite approach. We observe in it an existing digitalized environment (system) from the user/customer point of view and then suggest appropriate evolutions and improvements. This is why we give only a brief survey of possible approaches.

Globally, we can define digitalization as a transformation process, the goal of which is to propose: new practices, a new ecosystem, new stakeholders, a new business model, new management in order to provide better customer services.

For this purpose, more or less elaborated processes are proposed. We shall mention only two: the McKinsey 7S model (Fig. 1) [17] giving the main components of an appropriate digitalization approach with a Strategy, a Structure, Supportive Systems and managing Styles, working Staff, and members' main Skills manipulating Shared Values.

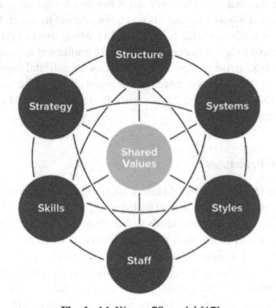

Fig. 1. McKinsey 7S model [17]

More fine modeling is proposed on The Big Picture of Digital Transformation (Fig. 2) [18], synthesizing all aspects, elements, and methods. These approaches organize design and development stages, which are not our objective in our approach, the goal of which is to recall an experience and formulate appropriate evolutions.

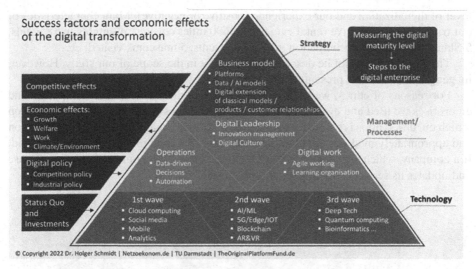

Fig. 2. Digital Transformation - The Big Picture (Edition 2022) [18].

2.3 User Experience (UX)

The term UX (User eXperience) refers to the quality of user experience in any interaction situation. UX describes the overall experience felt by the user when using an interface, a digital device or, more broadly, when interacting with any device or service. UX is therefore to be differentiated from ergonomics and usability.

It was Donald Norman, in the 90 s [19], who first used this term "user experience" that would then go on to enjoy the success we know. Donald Norman recalls in a very short and instructive video [20], the origin of the term "UX" and the vision he has of it today, an open vision that makes us challenge certain ideas. Donald Norman and Jacob Nielsen in their NNGroup [21] continue to promote and evangelize this concept with a series of conferences and videos. User Experience can be seen as design and test methods conducted by professional designers, but also as final user tests in working conditions of the application or system [22].

Concerning the digitalization process, we choose to use User Experience as final user of our case study situations and collect positive and negative observations. However, our real position as computer and UI specialists allows us to formulate later some observations and suggestions for evolution of different elements (services, applications) that we used during our case study.

3 Long-Term Case Study

Our case study is related to a relatively long tourist experience (7 weeks) organized initially from France and taking us to the United States for a trip covering California, Nevada, Arizona, New Mexico, Utah, and Colorado. The objective was to visit several National Parks and museums and to participate in local activities organized during our stay in different locations. We relate the main tourist services used and describe their level of digitalization and our experience (positive or negative). Main characteristics of our trip were: 2 successive rental cars used, 3600 miles covered, 43 nights in 18 hotels, 5 State Parks visited, participation in 2 major events, 5 museums visited, etc.

The trip itself will not be described, as it is not in the scope of our study. However, its geographical print is presented in Fig. 3.

For reasons of ethics, we decided not to publish the names of the companies and enterprises we used and evaluated. There are two reasons for this: first, a company, on which our observation is based, recognizes the described behavior without stigmatization and appropriately modifies the identified services. That is an interesting result. Second, if a company, which was not observed by us, is concerned it could be in the same case and updates its services, the impact will be greater.

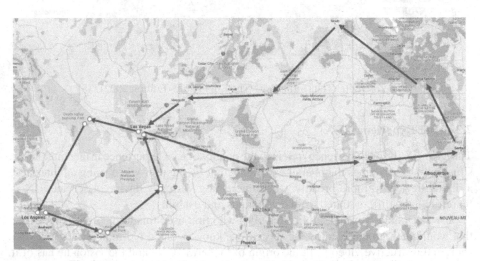

Fig. 3. Two-loop trip (first in red, second in blue). (Color figure online)

3.1 Trip Preparation

The first major activity carried out well before the trip is its preparation for which the physical and digital worlds are appropriately explored. In the physical world, discussions with friends, colleagues, and tourism professionals, consultation of books, flyers, etc., are the practices most commonly deployed. In the digital world, multiple websites are available with search engines, while social networks, such as WhatsApp, Messenger, Twitter, Facebook, and so on, make it easy to find the information you need. In this step, exchanges are open, without commitment, and tools work appropriately. Visa-Platinum and other high-level credit cards also provide a source of inspiration, making it easy to organize quality trip programs tailored to your needs.

3.2 Flight Finalization

The next step is transportation finalization with precise timing decisions and financial agreements. In-city travel agencies have ceased to be the only solution today, as the vast majority of travelers use Internet. A variety of formulas are possible: fly only or fly & hotel packages, and so on.

In this context, appropriate digitalization is mandatory, with integrated support during and after the trip. It is important to share the appropriate data model, managing all necessary data, as well as a User Interface between provider and client as far as possible in real-time.

We shall now describe various problems that may occur. When booking a flight, the website may crash for a variety of reasons, such as network disconnection, evolution of availability of chosen option or flight fare. The wrong solution is to ask the user to restart the process. Usually, websites are able to store the data collected for this partially processed booking and continue via an alternative solution as soon as possible (using a mobile application, phone call with help by an assistant, etc.) based on a multichannel approach, in order to finalize this sales action securely and as quickly as possible.

We had the opportunity to benefit from the possibility of finishing our purchase after an initial website crash with a flight booking operator.

After this initial action, it is important to remain in contact with the operator for the following official procedures (registration, complementary baggage payment, options, etc.), as well as for more personal ones (trip modification, cancelation, etc.). These contacts must be as reactive as possible. Unfortunately, it is not always easy to obtain an answer quickly. As an example, we can mention an answer received a month and a half after our trip concerning additional baggage costs. Some messages related to transportation reservation codes are managed appropriately and in accordance with time constraints, while others are unfortunately treated more randomly.

We observed several difficulties during check-in, for example visa verification, COVID certificates, and vaccinations were impossible to be done on-line, making it necessary for them to be checked at the airport.

For post-trip actions an appropriate behavior is expected: reactivity and a rapid answer, also for complaints, with a two-step approach: confirmation of complaint reception, then later, but not too late, the final answer. Reactivity to emails is a huge problem as answers are returned far too late for the problem in question. Complaints are all too often processed after a long wait, if at all.

Aviation transport introduced at an early stage digitalized services to replace the tourist agency approach. The search for trips and flights are appropriately supported, as well as payment.

Unfortunately, this is not the case for more specific issues, where it is possible to contact companies only by phone (with cost and communication problems). The majority of emails sent by companies are no-respondable, i.e. it is impossible to use them to ask a question, point out a problem or send a complaint. One solution is to use pre-established formulas, but these are often limited to a closed list of situations (FAQ), unfortunately, without open-ended ones.

3.3 Car Rental Experience

Car rental is an important aspect, particularly for trips to other countries or continents. Tourist agency physical reservation is progressively being replaced by digital reservation, making it necessary to contact directly via internet specific car rental servers or a generic server working for multiple car rental companies or to use related services proposed by airlines, hotel reservation servers or high-level credit card bank services.

In most cases the process is easy: choice of location, rental duration, car category, and complementary services. On arrival in the car rental agency, the document (voucher) is presented, together with an ID, a credit card, and a driving license. The employee draws up the rental contract, charges the credit card, and asks the client to sign it. The client can then go to the rental carpark to choose the car in the chosen rented category, check it with the employee, and indicate any problems observed (often car-body impact).

Unfortunately, we did not go through this generic process for our first car rental. Our reservation process started with the bank and its high-level credit card service. This service offered us a rental company and a rental price. On arrival at the rental agency, the employee verified our ID, driving license, and charged our credit card, without any contract to sign, before giving us the car key and indicating the carpark. We left the carpark without any car state checking or document, but with an important statement: we are "PAPERLESS", you will receive an email with all appropriate information. Several days later, we received the promised email in which we could read all the car-body impacts declared by previous rental users. However, we also observed that the rental period was not ours (as indicated in the voucher). It was modified (3 days shorter) at the same price. We tried to find out what had happened but the car rental company's email was non-answerable. We then contacted our bank that had proposed us this rental and their answer was "go back to the rental agency to modify this problem". However, we were 200 miles from the initial location. The second email stated: "the rental company is not able to establish more than a one-month rental (our rental was 32 days), they will send you on the last day of your current rental period a new contract covering the missing period of 3 days". We went along with this, but what actually happened was totally different. We never received the additional email with the new contract and on the last day of the first period, our bank account was charged for the entire amount. Three days later, when we returned the car we went through the same process. The employee asked for our car key, checked the car, and told us everything is OK, you will receive an email, as we are "PAPERLESS". We never received this email, but our card was charged for an extra $900. We tried to inform our bank and both the US and the French car rental agencies. For two weeks we received no answer. Two weeks later our bank account was credited for two sums adjusting the cost of the first and second periods, and we received a partial explanation from the car rental agency stating the impossibility to rent a car for more than one month.

To wind up this case, two weeks later we received from a higher-level manager of the US car rental agency the following and really surprising email (Fig. 4).

Of course, this observation about "special cleaning due to it being dirty" was unfounded as we were two in the car (without kids), never ate inside and never drove off official roads. The check-out employee did not indicate any problem.

Fortunately, our second car rental experience went smoothly as per the normal procedure before, during and after rental, allowing us to verify that our first experience was a counterexample, as well as helping us restore our confidence in this kind of service. In the last section of this paper, we will summarize our suggestions for appropriate behaviors of different digitalization stakeholders.

Please note that all our vehicles are carefully cleaned after every rental. This is already factored into our rental prices. After verifying to one of our associates at our Las Vegas Int Airport location, it has been advised that the vehicle you returned required a special cleaning due to it being dirty. Therefore, cleaning the vehicle took considerable time and expense and for this reason you have incurred the additional charge for cleaning. based on your last contract XXXXXXXXXX, vehicle was returned on AUGUST 30 2022. The charge on our end is $916.09 on AUGUST 31 2022.

But it was removed and the only charge appearing in this contract is for $245.96 which is for your remaining rental days, from AUGUST 26-30 2022. I hope this information clarifies your inquiry.

Fig. 4. Last e-mail from US car rental agency.

We finish with an open question: How would you react if you saw the following message on your dashboard, see Fig. 5.

Fig. 5. Change Engine Oil Soon.

3.4 Hotel Reservation Management

Hotel reservation is another important aspect in tourist trips. Assistive Systems for hotel reservation are progressing accordingly. After physical reservation via a tourist agency, phone calls, and direct snail mail contact with the required locations, websites now increasingly propose appropriate services via specialized, individual websites, via more or less aggregated solutions by hotel chains (homogeneous Best Western, Hilton, etc.), heterogeneous solutions (Choice Privileges) or via totally generic solutions such as Hotels.com or Booking.com. The level of services seems appropriate, with in-depth selection tools, appropriate reservation services, and the possibility to cancel or modify reservations up to 24 h beforehand. Multichannel communication is also appropriately supported.

We encountered a problem related to the COVID and post-COVID periods: no information as to the type of breakfast (English, French or grab-n-go) or how the rooms are or are not cleaned (only on request to avoid housekeeper entry, only towels and hygienic products provided, classical cleaning). This information was either given during check-in or observed at the end of the day when the room was not cleaned, and we wanted to know why. Another problem was with refunding delays when we decided, for an objective (pool out of order or grab-n-go breakfast) or for a personal reason, to shorten our stay. As the hotel debited the totality of our stay on our arrival, the bank took 10 to 15 days to refund us.

3.5 Bank Role

The role of banks in mobility is mainly Credit Card management-oriented. These cards offer an implicitly limited amount for payment and withdrawal, which can be more or less easily modified. Self-modification on the customer space is an appropriate solution, but not necessarily totally generalized (as the maximum amount fixed by the bank cannot be exceeded).

Greater modification possibilities necessitate participation by bank employees, which is not always easy and is mostly dependent on their availability during vacation periods and bank opening hours (not necessarily compatible with worldwide tourist locations). The main channel used is still the phone, which is not necessarily appropriate for overseas calls. Multichannel and appropriate reactivity are the objectives.

3.6 Insurance Company Participation

In the tourist industry, mobility insurance companies offer at least assistance and rescue services, but also services for car rental, hotel and transportation reservation. In these cases, reactivity and continuity of the customer relationship are essential. The problems of rental car breakdown or accidents need appropriate reactivity and a relationship with the rental car company, which is not necessarily always the case. For various reservations (hotel, trip, etc.), it is not easy to offer the same level of services as that offered by specialized providers.

3.7 In-site Access Tools and Assistance

During our trip we observed several specific digitalized services for particular situations. We noted a timed entry reservation that was required to enter the park even if you already have a pass. Annual passes cover the payment of entrance fees, so annual or senior pass holders only need to pay the additional $2 reservation fee. This timed entry reservation is used in several US National Parks in order to manage rush hours of access to the parks during the summer season (Fig. 6). A website is devoted to the distribution of timed entry tickets in relation to the number of potential visitors [23].

We found similar services for museum access management for the Getty Villa and Getty museum in Los Angeles. Comparable services were also used to manage entry and car parking at the Gallup Inter-Tribal Indian Ceremonial and Santa Fe Opera. Generic tools such as Uber-Eat, etc., can easily be added to the above.

Fig. 6. Arches National Park access control.

3.8   Chatbot Solicitation

Textual-and speech-oriented chatbots are helpful if they work correctly. Unfortunately, this is not generally the case. In particular, in situations in which the chatbot tries to answer open-ended situations, its capacity of understanding is limited. Often it is unable to contextualize correctly the question (the problem).

An approach in which the chatbot leads the interaction is more success-oriented. In other words, the chatbot manages the interaction by proposing a list of situations (FAQ) and asking users to choose one. The hierarchical structure can then be used to manage the exchange process to discover the contextualized situation and find the appropriate answer. Of course, this approach can prove too complicated if the selection tree is too large or deep.

An interesting solution would be to combine the chatbot with human interaction, where the chatbot orients the problem formulation and the human answers the targeted problem, if the chatbot is unable to understand and answer.

3.9   Human Implication

Human implication in the digitalized process is a major contribution that must be studied and answered appropriately. Of course, for cost reasons all measures are taken at present to do away with human participation as it is both time and cost consuming. However, the latter participation is also the ultimate solution for a large number of user situations that are not correctly treated by chatbots. A speedy response to their questions is essential to ensure that users do not give up.

3.10   Generalization to Other Fields

The previous list of observations was based on our trip experience. Of course, we are able to add other experiences, not directly related to tourist mobility but nevertheless characterizing problems poorly treated during digitalization.

One of the official French websites aims at managing delivery of all official documents such as driving licenses, ID, passports, and vehicle identification. This service is totally digitalized, with no physical counter. There is just a phone number on which a robotized human gives only stereotyped answers as a robot. On the website you can explain your problem for predefined situations only (problem FAQ) and no open-ended questions can be formulated. After submitting your question, the system acknowledges it and informs you that you will receive an answer within 48 h. Unfortunately, reality is a little different, as you will normally receive a first, very short, answer after one week.

3.11 Non-observed Services and Situations

Unfortunately, during our trip we did not observe any contextual geolocalized services based on the IoT (Internet of Things). These services combine virtual services with in-the-field located objects such as opening closed roads based on special authorization transmitted from the assistive system to remote objects via a personal device such as a smartphone.

Another example is dynamic lane management making it possible to open or close a lane to a category of vehicles. The lane can be closed to private vehicles and dynamically devoted to buses or emergency vehicles [24]. The objective is to devote these lanes dynamically only when buses are present and to leave all lanes open to general traffic when buses are not present. This allows management of general traffic speed. The main technologies used are: a Location-Based Service integrating bus detection sensors; an intermediation platform collecting sensor information and determining dynamic bus section activation and deactivation; in-the-field infrastructure and/or embedded vehicle interface receiving instantaneous information on selected situations (Fig. 7). From the HCI point of view, it is important to indicate the present situation on in-the-field indicators, as well as on the screen in the vehicle.

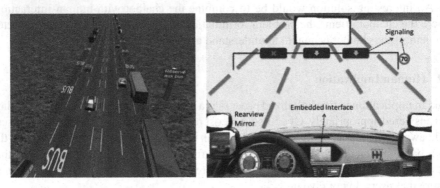

Fig. 7. Dynamic lane management [24].

We did not meet Metaverse services, which are still in the experimentation stage, and the objective of which is to mix reality and virtuality based on VR (Virtual Reality), AR (Augmented Reality), and Mixed Reality [25], using appropriate devices. Smartphone-based applications are the lowest configuration available.

4   Assistive System Structures

We can summarize all assistive services described (and used) during our trip as follows: Airlines, Rental companies, Hotels, Banks, State parks, Museums, Event structures (Santa Fe Opera, *Gallup* Intertribal *Ceremonial, restaurants, etc.*), GPS, food delivery, etc.

These are either elementary services or more sophisticated multi-services. They are made up of different aspects, namely: user interface, working data, one or more services (algorithms working on data), and remote access mechanisms using internet or other network services.

From an integration point of view, we find either elementary configurations, limited to one issue, or more or less integrated multi-issue configurations. It is clear that this is not an integrated system but only a partially integrated approach. Naturally, banks are connected / interfaced with a large majority of proposed services, mainly for payment needs. Banks also propose related activities that require integration or at least data exchange with other applications (hotel reservation, rental cars, etc.). Just like banks, airlines also propose related services such as hotel reservation, rental cars, etc. Hotel reservations are more or less homogeneous, either individual or hotel chains: homogeneous (Hilton, Best Western, etc.), heterogeneous (Choice Privilege, etc.) or totally generic (Booking, Hotels, etc.) with appropriate management of reservations, modifications, and so on. Individual services are also used for State Park access, museums, and even structures.

It appears that the notion of silo continues to be an integration model, with well-defined main operations able to exchange easily and rapidly. As we will see later, some less frequently used operations require more or less major improvements.

Inter-silo relations are in the same situation with appropriate major exchanges and the necessity to improve less frequently used operations.

Total integration seems impossible to obtain, as the global system is not closed but rather open-ended, able to receive (each day) new services and applications supporting new visions of digitalized / virtual tourist activities.

Introduction of a new digitalized service can be either independent or need to collaborate with other services already present. These services must provide an API (Application Programming Interface), a standardized approach for their inter-connection.

As we can see in Fig. 8, GPS is an independent digitalized service, while Bank is a more common digitalized service, mainly for payment. Collaboration between services means that a service can be used while another major service is being used, such as hotel reservation from an airline digitalized service. It would appear that integration based on systematic data standardization is not appropriate, as data used by different services are at least partially different, and their integration could be too complicated, necessitating complex exchanges between services. A mixed approach, with a local data store for specific data and a shared store in a chosen location for shared data would seem appropriate (payment data stored in a Bank service).

Two technics can constitute a major breakthrough by the use of available data stored in the system. Using the big data approach [26], it is possible to extract from these data more general behaviors increasing the possibility of new services. In another direction, reuse of available data in Machine Data learning and Deep Learning approaches can add important Artificial Intelligence-based behaviors to the system [27].

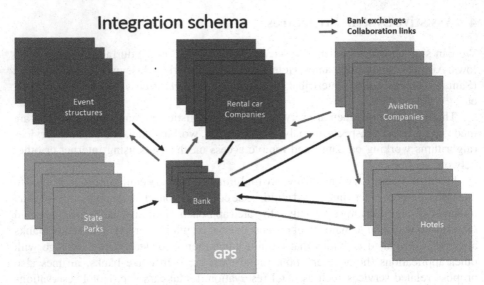

Fig. 8. Silo and inter-silo exchanges and collaborations.

5   Identified Improvements to Be Considered

After our trip and the corresponding documentation of appropriate and inappropriate behaviors of digitalized and virtual services, we can now propose a list of improvements, which we have decided to divide into 3 sections:

- Application-oriented:

  1. **Avoid inappropriate use of certain concepts**, such as "paperless" in those contexts in which it is mandatory to collaborate in real time and exchange persistent documents (contract validation by all partners, state of vehicle, etc.).
  2. Be able to update the websites to **integrate UpToDate information** as soon as possible (COVID restrictions, change of type of breakfast, pool out of order, etc.).
  3. **Avoid incompatibility between the user/consumer data model** and the in-house enterprise model (Data model compatibility – alignment), i.e. one month-limited rental which is only an in-house choice, not visible or understandable by the client.
  4. Banks, hotels, and airlines, which propose appropriate behaviors in mainstream activities, need to improve **related and peripheral activities** for which additional efforts are needed, mainly with respect to the user/customer relationship. There is a need for more efforts to provide customers with explanations and support.
  5. **Maintain coherence** of information during exchanges: price changes during the check-in process (small at the beginning and larger on actual payment).

- UI-orientated:

  6. **Allow multi-device and multi-channel interactions** to provide continuity of service in sequential and/or parallel manner.
  7. **Provide systematically answerable e-mails** easily accessible and providing answers within an appropriate time.

8. **Avoid inappropriately low reactivity**: Example - a month and more for e-mails, even for questions needing immediate answers (could I rent a car with Miles?).
9. **Avoid only pre-established / selected questions** available on digitalized forms on the Internet. Open-ended questions are also mandatory.
10. **The human relationship disappears totally** with virtualization (no window / information office available). Whenever necessary, **provide a User Interface** to maintain the relationship with the user / consumer.
11. **Provide appropriate chatbots** (textual and/or vocal) with an appropriate structure generally competent or thematic with progressive question selection. If necessary, provide real-time human intervention for more detailed discussions.
12. The **chatbot** can be used as the first step, but must be replaced by the **human** each time the former fails or is not able to provide appropriate help.
13. The **human must be ready to participate** whenever the user/consumer needs him/her.

- Evolution-oriented:

14. Integration of new services / applications requires a global study with appropriate collaboration between services to determine appropriate inter-relationships of shared data based on digitalization methodological processes such as Design thinking [28].
15. Appropriate UX for all users / customers is mandatory.
16. IoT for in-city (in the field) geographically situated services
17. Metaverse as a promising evolution that is not yet sufficiently mastered and proposed.

6 Conclusion

In this paper we studied digitalization and virtualization of operations related to tourist mobility. We based our approach on a concrete case study of a 7-week trip to the United States of America. After an initial contextualization of digitalization and virtualization in general and, more specifically, to tourist activities, we drew up a list of experiences related to our trip covering preparation, flight management, rental car experience, hotel reservation, and other specific services such as State Park access reservation. Then, we discussed more generally the level of integration of proposed services and our user experience, which led us to elaborate a list of suggestions for appropriate behaviors that tourism-oriented digitalization and virtualization could provide in order to obtain *a priori* higher level of user satisfaction.

Our conclusions are that, globally, as the proposed digitalization and virtualization for tourist mobility are appropriate for mainstream operations such as flight management, hotel and rental car reservation and management. The majority of problems occur for more specific demands and, in particular, situations such as modifications/adjustments or requests for specific services. In these situations, the lack of possibility to communicate using multiple channels is a problem, as only a mobile phone number is systematically present, but not an email. Reactivity and speed of reply are insufficient. Chatbots are not yet fully perfected, and synchronous User Interfaces are generally missing. It is our opinion that these aspects must be improved in order to provide appropriate User Experience in these peripheral, but very important, stress-generating situations.

Total integration is not a solution. Conversely, an open-ended approach could be preferred in order to integrate new services, which can be utilized on a daily basis and would communicate easily with existing services. The modularized approach is an appropriate solution.

During our trip we did not find any contextual applications connecting the user's application with in-the-field elements (opening of a door, information readable on a precise location, management of highways for specific cars, etc.) using the IoT (Internet of Things) to deploy this approach. In the future, this technology will be increasingly deployed. Another major field of tools now being extensively investigated by many companies is Metaverse [12], an approach to virtual and augmented reality allowing an interesting, rich contextualization and in sofa sitting tourism.

Generalized use of Big data and AI technics (Machine Learning and Deep Learning) [26, 27], as mentioned, can increase the capacity of these assistive systems.

Naturally, in order to confirm these recommendations, this experiment would need to be carried out with more users and at different times of the year. These tests could reveal other problems and therefore lead to new recommendations.

It would be interesting to conduct the same kind of trip with disabled users and collect their observations, difficulties, and recommendations. This could start with a comparison between a physical and a virtual travel agency for able-bodied people and for people with disabilities, before going on to address other tourist mobility services.

References

1. Bhatia A.K.: The business of Travel Agency & Tour operations Management. Sterling Publishers (2012)
2. Kautson, B.J.: Handbook of consumer behavior, tourism and the internet. J. Travel Tourism Market. **17**(2/3) (2004)
3. Chanias, S., Hess, T.: How digital are we? Maturity models for the assessment of a company's status in the digital transformation [Management Report]. Institut fur Wirtschaftsinformatik und Neue Medien, 2, 1–14 (2016). https://www.wim.bwl.uni-muenchen.de/download/epub/mreport_2016_2.pdf
4. Frolov, D.P., Strekalova, A.S.: Digital cities and interactive stakeholder policy. In: International conference: Ecosystems without borders. IOP Conf. Series: Earth and Environm. Sci. **689** (2021)
5. Camilleri, M.A.: Travel Marketing, Tourism Economics and the Airline Product, An Introduction to Theory and Practice. Springer (2018). https://doi.org/10.1007/978-3-319-498 49-2
6. Ionita, I.M.: Digitalization influence on shopping centers strategic management (2017). https://doi.org/10.1515/picbe-2017-0079
7. Massari, F.S., Del Vecchio, P.: Past for Future – museums as a digitalized "interaction platform" for value co-creation in tourism destinations. Europ. J. Innovat. Manag. (2022), ISSN: 1460–1060
8. Chen, S., Law, R., Zhang, M., Si, Y.: Mobile communications for tourism and hospitality: a review of historical evolution, present status, and future trends. Electron. 10, 1804 (2021). https://doi.org/10.3390/electronics10151804

9. Minnaert, L., Maitland, R., Miller, G.: Tourism and social policy: the value of social tourism. Annals Tourism Res. **36**(2), 316–334 (2009). http://www.sciencedirect.com/science/journal/01607383

10. Hamid, R.A., et al.: How smart is e-tourism? A systematic review of smart tourism recommendation system applying data management. Comput. Sci. Rev. **39**, art. 100337 (2021)

11. Standing, C., Tang-Taye, J.P., Boyer, M.: The impact of the Internet in travel and tourism: A research review 2001–2010. J. Travel Tourism Market. – tandfonline.com (2014)

12. Metaverse. https://en.wikipedia.org/wiki/Metaverse. (Accessed 1 May 2023)

13. El Saddik, A.: Digital Twins, The convergence of multimedia technologies. In: IEEE MultiMedia, April–June 2018. IEEE Computer Society (2018). https://doi.org/10.1109/MMUL.2018.023121167

14. Ayat Ayman Abdel-Aziz, A.A., Abdel-Salam, H.A., El-Sayad, Z.: The role of ICTs in creating the new social public place of the digital era. Alexandria Eng. J. **55**(1), 487–493 (2016)

15. Balandina, E., Sergey Balandin, S., Koucheryavy, Y., Mouromtsev, D.: IoT use cases in healthcare and tourism. In: Proceedings of 2015 IEEE 17th Conference on Business Informatics (2015)

16. Firican, G.: What is the difference between digitization, digitalization, and digital transformation?. https://www.lightsondata.com/what-is-the-difference-between-digitization-digitalization-and-digital-transformation/. (Accessed 2023/01/05)

17. Shaqrah, A.: Analyzing business intelligence systems based on 7s Model of McKinsey. Int. J. Bus. Intell. Res. 9(1) (2018)

18. Big Picture. https://www.netzoekonom.de/online-kurse/digitale-transformation-big-picture/. (Accessed 1 May 2023)

19. Donald Norman, in the 90s. http://mickopedia.org/mickify?topic=Experience_design. (Accessed 1 May 2023)

20. Video Norman's UX presentation. https://youtu.be/9BdtGjoIN4E. (Accessed 1 May 2023)

21. NNGroup. https://www.nngroup.com/. (Accessed 1 May 2023)

22. Nielsen J.: Keynote the Immutable Rules of UX, https://www.youtube.com/watch?v=OtBeg5eyEHU. (Accessed 1 May 2023)

23. Plan Your Visit (U.S National Park Service). https://www.nps.gov/planyourvisit/index.htm. (Accessed 1 May 2023)

24. Wang C., David B., Chalon R., Yin C.: Dynamic road lane management study: a smart city application. J. Elsevier Trans. Res. Part E: Logist. Transport. Rev. **89**, 272–287 (2015). https://doi.org/10.1016/j.tre.2015.06.003

25. Cipresso, P., Giglioli, I.A.C., Raya, M.A., Riva G.: The Past, present, and future of virtual and augmented reality research: a network and cluster analysis of the literature. Front. Psychol. **9**, 2086 (2018). https://doi.org/10.3389/fpsyg.2018.02086

26. Tsai, C.-W., Lai, C.-F., Chao, H.-C., Vasilakos, A.V.: Big data analytics: a survey. J. Big Data **2**(1), 1–32 (2015). https://doi.org/10.1186/s40537-015-0030-3

27. Paul, C., Jay, A., Emre, S.: Deep neural networks for youtube recommendations. Recsys., 191– 198 (2016)

28. Brown, T.: Design Thinking, HBR's 10 Must Reads on Design Thinking. Harvard Business Review Press (2020)

An Experimentation to Measure the Influence of Music on Emotions

Andrea Generosi[1]([✉])[iD], Flavio Caresana[1][iD], Nefeli Dourou[2][iD],
Valeria Bruschi[2][iD], Stefania Cecchi[2][iD], and Maura Mengoni[1][iD]

[1] Department of Industrial Engineering and Mathematical Sciences,
Università Politecnica delle Marche, Marche, Italy
a.generosi@univpm.it
[2] Department of Information Engineering, Università Politecnica delle Marche,
Marche, Italy

Abstract. Several emotion-adaptive systems frameworks have been proposed to enable listeners' emotional regulation through music reproduction. However, the majority of these frameworks has been implemented only under in-Lab or in-car conditions, in the second case focusing on improving driving performance. Therefore, to the authors' best knowledge, no research has been conducted for mobility settings, such as trains, planes, yacht, etc. Focusing on this aspect, the proposed approach reports the results obtained from the study of relationship between listener's induced emotion and music reproduction exploiting an advanced audio system and an innovative technology for face expressions' recognition. Starting from an experiment in a university lab scenario, with 15 listeners, and a yacht cabin scenario, with 11 listeners, participants' emotional variability has been deeply investigated reproducing 4 audio enhanced music tracks, to evaluate the listeners' emotional "sensitivity" to music stimuli. The experimental results indicated that, during the reproduction in the university lab, listeners' "happiness" and "anger" states were highly affected by the music stimuli and highlighted a possible relationship between music and listeners' compound emotions. Furthermore, listeners' emotional engagement was proven to be more affected by music stimuli in the yacht cabin, rather than the university lab.

Keywords: Audio Enhancement · Affective Computing · Facial Emotion Recognition · Deep Learning

1 Introduction

Many studies have focused on investigating the relationship between music and emotion, due to the power of music to express and induce emotion [10]. So

This work is supported by Marche Region in implementation of the financial programme POR MARCHE FESR 2014–2020, project *"Miracle"* (Marche Innovation and Research fAcilities for Connected and sustainable Living Environments), CUP B28I19000330007.

far, the majority of the research is related to the emotion expressed in music, i.e., perceived emotion. However, concern has raised, as well, for the emotion experienced by the listener during music reproduction, i.e., induced emotion, leading to intensifying the research around this topic [34].

Different methods have been used to measure the emotional induction from music. Several studies have considered self-reporting evaluation procedures (e.g., Likert ratings, adjective lists), in which participant is asked to assess his own emotional reaction. Moreover, peripheral (e.g., skin conductance, heart rate, muscle tension, respiration, electrocardiogram, blood pressure, etc.) and central nervous system (e.g., EEG, MEG, fMRI, PET) measures have been widely used to evaluate the emotion induced by music [10]. Regarding the emotional models used in research, both dimensional and categorical ones have been considered. For the former, the most commonly used is Russel's circumplex model, which suggests that emotions are distributed on a two-dimensional circular space, containing arousal (activation-deactivation continuum) and valence (pleasure-displeasure continuum) dimensions [10,32]. Alternatively, Thayer's multidimensional model of activation, which describes arousal in terms of two activation dimensions, namely energy arousal and tension arousal, and a single continuum of energy expenditure, has been used [38]. For the later, the Ekmans' model, which considers that all emotions can be described on the basis of a small number of universal and innate basic emotions, i.e., happiness, surprise, sadness, anger, disgust, fear and neutral is widely used [10,11]. Several studies have considered miscellaneous models consisting of the above well-established emotional models, e.g., in [24] a three-dimensional model which combines Russel's and Thayer's model was used. Moreover, music domain-specific emotions have been suggested in an attempt to better characterize emotions induced by music, e.g., Geneva Emotional Music Scale (GEMS) [10,41].

Listening to music is a very common activity in automotive and mobility environments. A smart car driving environment should exploit the emotional effects of music for regulating drivers' mood, to prevent from potentially dangerous driving behavior [13–15,40]. While several studies have considered the use of biofeedback sensors for investigating the emotional effect of music on listeners, only few of them are focused on the car driving scenario, e.g., using ECG, EEG, and electrodermal responses [13,15,40,42]. However, the improvement of music listening experience during automotive and mobility should not be exclusively dedicated to the car drivers needs, but also to the non-driving listeners and the different mobility scenarios. Therefore, systems capable of monitoring listeners' emotional state on real time and effective for mobility settings could be exploited for increasing the music listening experience in these environments.

In this context, several approaches based on face expressions' recognition are proposed in the literature. Facial Expression Recognition (FER) is a rapidly growing field of technology that aims to understand and interpret human emotions through the analysis of facial movements. The distinction between facial expressions and emotions is an important aspect of FER systems. Facial expressions are outward signals that the human body uses to communicate emotions, while emotions are what the body is attempting to communicate through these signals [3]. Theoretically, all FER systems are based on Ekman's and Friesen's FACS coding [12], which involves recognizing the Action Units (AUs) of the FACS system through computer vision. Studies in various fields have adopted deep learning methods, using convolutional neural networks (CNNs), to analyze multimedia content, videos, images, and audio tracks as in [16,25], which take images of human faces as input and provide a prediction of the relevant Ekman's main emotions. With increasingly effective training data and better-designed network architectures, the accuracy of FER systems has greatly improved [22,33,36]. Since 2013, Emotion Recognition competitions like FER2013 [20] and Emotion Recognition in the Wild (EmotiW) [7] have gathered relatively high-quality training data from real-world scenarios that are challenging, which is promoting the shift of FER from controlled laboratory settings to everyday environments where conditions such as lighting and distance from the subject are not controlled.

FER systems are considered to be the least invasive method of emotion recognition. However, the effectiveness of these systems greatly depends on the quality of the training data and the accuracy of the label data. [27] lists the different accuracy rates achieved by various models. It has been noted that models with high accuracy rates are trained using datasets generated in labs like MMI [39] and CK+ [29]. On the other hand, models trained on datasets collected from the web with real-world properties have lower accuracy rates due to inaccuracies in the labeling of these datasets [5], although they often have a large number of images available. Affectnet falls into this category [30].

In this work, face emotion recognition technologies are used to investigate the relationship between listener's induced emotion and music reproduction exploiting an advanced audio system [37]. The face expressions' recognition algorithm used in the current experiment is part of an adaptive system to manage playlists and lighting scenarios based on the user's emotions, described in [1]. In previous work, the accuracy of the system prediction was tested using audiovisual stimuli to induce emotion [2]. Contrary, in the current work we introduced audio enhancement algorithms, similarly to the work of [8], to achieve increased emotional induction using only audio stimuli. Specifically, starting from an experiment considering a university lab (with N = 15 listeners) and a yacht scenario (with N = 11 listeners), and 4 music tracks of different genres, we investigated which emotions were the most "sensitive" to music stimuli and how the experimental environment (i.e., university lab and yacht cabin) affected the overall musical emotional induction of a listener, in order to make a step forward the improvement of emotionally adaptive systems and the analysis of induced emotions in mobility settings.

The paper is organized as follows. Section 2 describes the algorithm for emotion detection based on facial expression recognition. Section 3 explains the audio enhancement algorithms applied to soundtracks. Section 4 reports the experimental procedure used in the two real scenarios. Section 5 introduces the obtained results while conclusions are drawn in Sect. 6.

2   Emotion Recognition Algorithm

In this study a state-of-the-art Convolutional Neural Network (CNN) model for facial expression recognition has been used [37], focusing on a hybrid approach for the model training so as to accurately recognize human emotions in real-world contexts. To this purpose, and to optimize facial expression recognition accuracy and performance, multiple datasets were utilized. These datasets, CK+ [29], FER+ [5], and AffectNet [30], were combined to form a large dataset of over 250k images. The FER+ dataset has an accuracy rate of approximately 90% and consists of 35k images, while the AffectNet dataset contains over 1 million web-crawled images, with over 400k images labeled by experts. These datasets were chosen because they are tagged with universal Ekman's emotions labels and have a relatively high labeling accuracy. The combined dataset was divided into training and validation portions with an 80-20 split. Preprocessing steps, such as facial alignment, face centralizing, and face rotation, were performed using Dlib facial landmarks coordinates. All images were scaled to 64 × 64 pixels to maintain consistency. The implementation of this model was built using Python, Tensorflow, and Keras frameworks. The trained network models take 64 × 64 pixel grayscale face images as input and produce the classification probability of Ekman's emotions (joy, surprise, anger, disgust, sadness, fear, and neutral) as output. Different Keras model architectures, such as Inception [35], VGG13, VGG16, and VGG19 [33], were tested, and VGG13 was found to produce the best accuracy levels during the validation phase, as indicated in Table 1.

The network hyperparameters were initialized, and variations were also tested with validation splits of 0.1 and 0.2, the number of epochs 30, 50, and 100, and a dynamic learning rate. The learning rate was set at 0.025 and updated with each epoch. The VGG13 model achieved the highest accuracy percentage in the validation phase, while overfitting occurred in other deep neural network architectures when the number of epochs was greater than 30. The FER application outputs an array of seven values, each representing the probability of the respective emotion. A Python implementation has already been tested in a real environment to assess its effectiveness in detecting emotions compared to traditional video analysis, and the results are reported in [18].

Table 1. Accuracy values achieved for each architecture.

Architectures	Accuracy (%)
VGG13	**75.48**
VGG16	74.48
VGG19	73.14
InceptionV2	75.26
InceptionV3	67.20

Ekman's emotion scores, forecasted for each video frame from the camera, are normalized to a percentage value of 100. The percentage values of the individual Ekman emotions are subsequently calculated to derive two indicators on which the analyses of the average emotional performance of the participants in the conducted experimental tests will be based, i.e. Valence and Engagement. Valence returns a value about the positivity or negativity of the emotion felt at each instant, considering positive the emotions of happiness and surprise and negative the others, using a scale from -100 to 100. A neutrality value of 100, on the other hand, corresponds to a Valence value of 0, indicating a total absence of expressiveness in these cases.

The formula used, following the approach proposed in [17] is as follows:

$$Valence = Joy\% + Surprise\% - Sadness\% - Anger\% - Disgust\% - Fear\%$$

For Engagement, on the other hand, an indicator of the expressiveness of the face and thus how much the subject is "involved" moment by moment, the formula used is as follows:

$$Engagement = Joy\% + Surprise\% + Sadness\% + Anger\% + Disgust\% \\ + Fear\%$$

Face detection is a crucial component of any deep learning system that analyzes human faces. The CNN processes lightweight images that only contain the face of the detected person, but the recognition of faces within the entire frame received from the camera is what has a significant impact on the system's response time and frame processing frequency. This can pose a challenge in scenarios where faces need to be detected from distances of up to 10 m, which often requires the use of high-resolution cameras (even 4K, such as the Logitech Brio). In the proposed FER implementation, there are two different face detection models that can be configured for use in different conditions. The first model is designed to recognize a face that is positioned frontally relative to the camera, with a maximum angle of 25° laterally and vertically. This model is based on the histogram of oriented gradient (HoG) and support vector machine (SVM), and is provided by the Dlib software toolkit [5], pre-assembled and pre-trained for frontal detection with a minimum face image size of 80×80 pixels. However, this model is not suitable for detecting faces from distances greater than three meters, as the frame size is typically less than 80×80 pixels in such cases. In these situations, the Singleshot Multibox Detector (SSD) [28] model is used, which can detect faces of various scales. While SSD is better suited for detecting faces that are not completely frontal, it can affect the accuracy of emotion recognition, which requires images in which all features necessary for facial expression analysis are visible. Additionally, tests have shown that HoG's detection speed drops significantly when the camera resolution is very high (4K or FHD), while SSD is not affected by this. To address these issues, HoG can be used as a filter after face detection by SSD within the frame. This ensures that only frontal faces will be considered for emotion recognition, and the computational performance will be maintained as HoG will only be applied to the portion of the face provided by the SSD output.

Both reported face detection models, HoG for the lab experiment and SSD for the yatch cabin, were used for the two study castes reported in the experiment.

3   Audio Enhancement Algorithms

Previous research has indicated that emotions induced by music are less strong than the ones induced by audio-visual content [31]. In this work, the emotional effect induced by music reproduction is tried to be increased by the application of three different audio enhancement algorithms, i.e.,

- room equalization for a stereo system,
- multichannel system for a 4.1 configuration,
- multichannel system with room equalization for a 4.1 configuration.

Figure 1 shows the overall audio enhancement system when all the algorithms are applied, i.e., when the multichannel system with room equalization is involved for a 4.1 configuration. The room equalization allows compensating for undesired artifacts caused by room reverberation [6]. The employed equalization procedure is based on the complex smoothing approach of [21]. Starting from the measurement of the room impulse response (RIR) that defines the acoustic path between the lth loudspeaker and the microphone, located at the listener's position, the frequency smoothing operation of [21] is applied, obtaining the smoothed frequency response $H_{\mathrm{sm}l}(z)$. Complex smoothing is a non-uniform smoothing that lowers the frequency resolution of the high-frequency part of the spectrum, in accordance with the human auditory perception. Finally, the equalization curve $H_{\mathrm{EQ}l}(z)$ is obtained by the inversion of $H_{\mathrm{sm}l}(z)$ and is used to filter the audio signal during the reproduction. This procedure is repeated for all four loudspeakers, obtaining four equalization filters $H_{\mathrm{EQ}l}(z)$, with $l = 1, 2, 3, 4,$

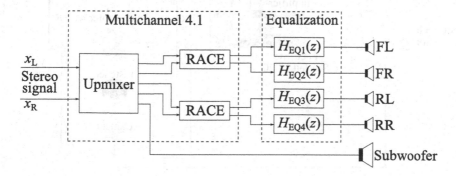

Fig. 1. Scheme of the overall system for audio enhancement.

representing the front-left (FL), front-right (FR), rear-left (RL), and rear-right (RR) loudspeaker, respectively. When the equalization is applied to a stereo system, only the two frontal loudspeakers play, so only two equalization curves are required.

The multichannel system is achieved by implementing a 4.1 configuration, i.e., four loudspeakers and a sub-woofer with a cutoff frequency of 120 Hz. The stereo signal is converted into a five-channel signal through an upmixing procedure based on the approach of [4]. Moreover, the recursive ambiophonic crosstalk elimination (RACE) algorithm of [19] is applied separately to frontal and rear loudspeakers to increase the spatialization experience [23]. The multichannel system is reproduced both without and with room equalization in order to evaluate the emotional response of the listener between non-equalized and equalized tracks. This system has been implemented on NU-Tech, a digital signal processing (DSP) platform for real-time audio signal elaboration [26].

4   Experimental Procedure

The experimental set-up is depicted in Fig. 2. The audio reproduction system consists of four line array of eight drivers each one and a Genelec 5040A active subwoofer. The loudspeaker sound levels are calibrated within 0.5 dB accuracy. The four line array are connected to a power amplifier, that is connected to a Focusrite Scarlett 18i20 sound card, while the subwoofer is directly controlled by the sound card. The sound card is managed by a computer, which uses the NU-tech software [26] for real-time reproduction. During the sound reproduction, a Logitech Brio 4K webcam records a video of the face of the listener,

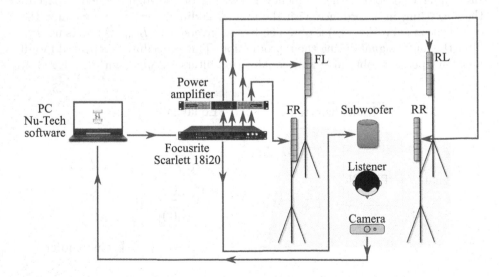

Fig. 2. Scheme of the experimental set-up used in the experiment.

Table 2. List of the music frames used in the experiment.

Genre	Track	Artist	Frame Duration
Pop	You're The One That I Want	Lo-Fang	32 s
Rock	The Word	Bettye LaVette	38 s
Jazz	I know it's over	Mario Biondi	39 s
Classical	The Nutcracker	Tchaikovsky	33 s

who is located at the center of the four loudspeakers. The experiment was conducted in both a university lab, denoted as "In-Lab", and a yacht cabin real scenario, denoted as "Yacht". The university lab was arranged to achieve a neutral environment, while the yacht cabin provided a realistic, comfortable living environment. Listeners were sitting on a chair, during the "In-Lab" case, or a sofa, during the "Yacht" case. During both reproduction conditions, a camera was recording the face of the listeners, in order to analyze the emotion recognized by the face. In both scenarios, the participant had no visual interference with the experimenters. Finally, 15 participants were involved in the "In-Lab" experiment and 11 in the "Yacht" experiment. The total duration of the experiment was almost 15 min.

The stimuli display order was the following for both environments: pop, rock, jazz, and classical track. Each track was presented in different versions, applying alternatively the three audio enhancement algorithms, while the original stereo track was used as a reference and reproduced before each algorithm, resulting in a total of six audio frames. Therefore, the final order of the audio frames for each track was the following:

1. original stereo track,
2. stereo track with room equalization,
3. original stereo track,
4. multichannel system for a 4.1 configuration,
5. original stereo track,
6. multichannel system with room equalization for a 4.1 configuration.

All music excerpts and the different elaborations have been normalized to the same loudness and modified to a fade-in (i.e., gradually increase from silence at the beginning) and fade-out (i.e., gradually reduced to silence at its end) procedure, to avoid any possible emotional reaction due to sudden onsets and offsets of the sound. Further details about the audio enhancement algorithms can be found in Sect. 3. Details about the experimental music excerpts can be found in Table 2.

5   Experimental Results

For the analysis of the results, the emotional variability in terms of standard deviation (SD) has been considered. In particular, the variability of the 6 distinct emotions (i.e., happiness, surprise, sadness, anger, disgust, and fear), valence (i.e., quality of induced emotions), and engagement (i.e., quantity of induced emotions) values, is interpreted as a measure of listeners' "sensitivity" to music stimuli. Valence and engagement are calculated by the 6 distinct emotions. Contrary, the variability of neutral is interpreted as a measure of listeners' "insensitivity" to music stimuli.

In Sect. 5.1, a visual representation of the "In-Lab" (with N = 15 listeners) results in the form of box plots is presented to investigate the variability size, for the measured emotions. Following, a within-subject correlation analysis of the "In-Lab" measurements is applied to investigate linear relationships between the emotional variability, considering emotions in a pair-wise fashion. In Sect. 5.2 a between-subjects analysis between the "In-Lab" (N = 15) and "Yacht" (N = 11) results is conducted to investigate if the environment had a significant effect on the musically induced valence and engagement variability.

5.1 "In-lab" Measurements

In Fig. 3, the emotional variability of the 15 participants during listening to the 4 experimental tracks is presented. In particular, the variability of happiness, surprise, sadness, anger, disgust, and fear, and neutral along with the valence and engagement values is depicted in the form of box plots.

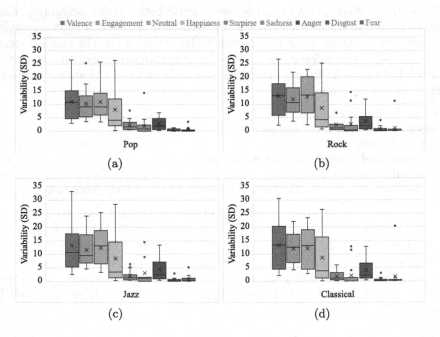

Fig. 3. Variability of valence, engagement, neutral and 6 distinct emotions (i.e., happiness, surprise, sadness, anger, disgust, and fear) in terms of standard deviation (SD) for (a) pop, (b) rock, (c) jazz, and (d) classical music, for the "In-lab" (N = 15) experiment.

Focusing on the median values of the box plots in Fig. 3, neutral emotion presented the highest variability among Ekman's basic emotions, namely happiness, surprise, sadness, anger, disgust, and fear, for all the tracks used in the experimental procedure. Following, happiness, to a greater extent, and anger, to a lesser extent, presented higher variability during the music reproduction, compared to the other emotions, for all 4 tracks. Moreover, focusing on the interquartile range (IQR), the variability values of neutral, happiness, and lesser of anger were extended to a wider range, compared to surprise, sadness, disgust, and fear. Consequently, the high and diverging variability of happiness, anger, and neutral implies that, on the total of participants, happiness and anger seemed to be the most "sensitive" emotions for measuring emotional reactions from music and "insensitivity", represented by the neutral response, was highly affected by the music stimuli, using the facial expression recognition algorithm. While the above is the general conclusion derived from the total of participants, interestingly specific participants presented, in comparison to others, remarkably high variability for some of the surprise, disgust, fear, and sadness, as denoted by the outlier values of the box plots.

Furthermore, correlations between the variability of the 6 distinct emotions (i.e., happiness, surprise, sadness, anger, disgust, and fear) were calculated in a pair-wise fashion, for the 4 tracks, employing N = 15 listeners. This analysis

Table 3. Pearson correlations between the 6 distinct emotions variability, considering the emotions in a pair-wise fashion for "In-lab" experiments, with N = 15 listeners. Only correlations r > 0.50 are reported.

Track	Emotions		Pearson Correlation
Pop	Disgust	Angry	$r = 0.73$, $p = 0.002$
	Fear	Happiness	$r = 0.53$, $p = 0.041$
		Angry	$r = 0.53$, $p = 0,043$
Rock	Disgust	Surprise	$r = 0.70$, $p = 0.003$
		Sadness	$r = 0.55$, $p = 0.033$
		Angry	$r = 0.55$, $p = 0.033$
		Fear	$r = 0.90$, $p < 0.001$
	Fear	Surprise	$r = 0.55$, $p = 0.033$
		Sadness	$r = 0.55$, $p = 0.033$
Jazz	Disgust	Surprise	$r = 0.58$, $p = 0.023$
		Sadness	$r = 0.85$, $p < 0.001$
		Fear	$r = 0.87$, $p < 0.001$
	Fear	Surprise	$r = 0.60$, $p = 0.019$
		Sadness	$r = 0.78$, $p = 0.001$
		Angry	$r = 0.56$, $p = 0.030$
Classical	Disgust	Surprise	$r = 0.58$, $p = 0.024$
		Sadness	$r = 0.91$, $p < 0.001$
		Fear	$r = 0.65$, $p = 0.009$
	Fear	Sadness	$r = 0.66$, $p = 0.007$

reported noticeable linear correlations between disgust variability and the variability of other emotions. Table 3 summarizes the best cases in which the correlation is higher than r = 0.50.

Remarkably, high linear correlations were found between disgust and fear for the Rock (r = 0.90, p < 0.001), Jazz (r = 0.87, p < 0.001), and Classical (r = 0.65, p = 0.009) track. Unlikely, the correlation between disgust and fear for the Pop track (r = 0.36, p = 0.188) was low. Likewise, linear correlations were found between disgust and sadness for Rock (r = 0.55, p = 0.033), Jazz (r = 0.85, p < 0.001), and Classical (r = 0.91, p < 0.001) track. Contrary, the correlation for the Pop track was not high (r = 0.34, p = 0.221). In addition, moderately high correlations were found between disgust and surprise for Rock (r = 0.70, p = 0.003), Jazz (r = 0.58, p = 0.023), and Classical (r = 0.58, p = 0.024). Oppositely, the corresponding correlation for the Pop track was low (r = 0.39, p = 0.149). Moreover, relatively high correlations were found between fear and sadness for Rock (r = 0.55, p = 0.033), Jazz (r = 0.78, p = 0.001) and Classical (r = 0.66, p = 0.007) music, while the correlation for the Pop track was low enough (r = 0.21, p = 0.444). The above result indicates that, for 3 out of the 4 experimental tracks, disgust variability was correlated with fear, sadness, and surprise variability while fear variability was correlated with sadness variability.

5.2 Comparison with "Yacht" Measurements

In this section, the results obtained in the yacht environment are compared with the "In-lab" ones. In Fig. 4, the valence and engagement variability during listening to the 4 experimental tracks for the "In-Lab" (N = 15) and the "Yacht" (N = 11) scenario is presented. The following between-subject analysis aims at investigating if changing the environment affected the participants' mean emotional variability, in terms of quantity of emotions (i.e., engagement) and quality of emotions (i.e., valence). An independent T-test along with the corresponding effect size analysis is applied to investigate the statistical significance and the size of the mean variability difference between the 2 scenarios, while listening to the 4 experimental tracks. The results of the statistical tests are cited in Table 4.

Focusing on Fig. 4 and Table 4, the between-subject analysis indicated that listening to music in the "Yacht" environment led to higher mean engagement variability to music, for all 4 experimental tracks. In particular, the highest increase in mean engagement variability was observed for the Rock and the Pop track. Following, Jazz presented a smaller increase in mean engagement. Finally, the Classical track presented the smallest increase in mean variability, among the 4 experimental tracks. Moreover, mean valence variability was increased for Pop, Rock, and Jazz tracks and slightly decreased for the classical track. Therefore, listeners' mean valence variability during listening to music was higher or almost equal during the "Yacht" scenario in comparison to the "In-Lab" scenario, for the same tracks. Noteworthy, the environment had a higher effect on the mean engagement variability rather than the mean valence variability for all 4 experimental tracks, as can be statistically supported by the statistical significance and effect size tests shown in Table 4.

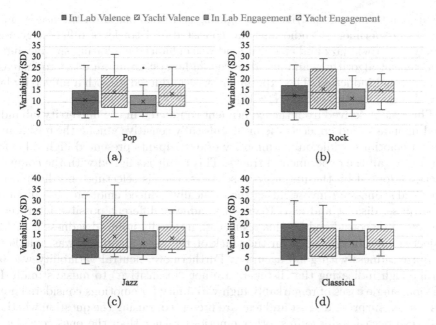

Fig. 4. Variability of valence and engagement in terms of standard deviation (SD) for (a) pop, (b) rock, (c) jazz, and (d) classical music, for the In-lab (N = 15) and yacht (N = 11) experiment.

Table 4. Independent T-Test between "In-Lab" and "Yacht" valence and engagement mean variability, for the 4 tracks. The 2-tailed probability value (p-value) and the Hedges' g effect size are reported. A positive effect size indicates that the mean variability during music reproduction in the yacht is higher than the "In-Lab" reproduction.

Track	Emotion	2-tailed p-value	Hedges' effect size g
Pop	Valence	p = 0.264	g = 0.439
	Engagement	p = 0.152	g = 0.568
Rock	Valence	p = 0.342	g = 0.373
	Engagement	p = 0.120	g = 0.621
Jazz	Valence	p = 0.682	g = 0.165
	Engagement	p = 0.350	g = 0.378
Classical	Valence	p = 0.970	g = −0.017
	Engagement	p = 0.668	g = 0.173

6   Conclusions

In this work, listeners' emotions induced by music were measured through an innovative technology for face expressions' recognition, which is part of an adaptive system to manage playlists and lighting scenarios based on the user's emotion. The experimental audio material consisted of 4 tracks of different genres, namely pop, rock, jazz and classical, and was enhanced exploiting specific audio enhancement algorithms and an advanced audio system, in an effort to increase the emotional induction. The experiment was conducted in both a university lab (N = 15) and a yacht cabin (N = 11).

The results derived from the experiment conducted in the university lab indicated happiness and anger as the most musically sensitive among the 6 Ekman's distinct emotions, while only a minority of participants presented high value for sadness, for all the experimental tracks. This result is aligned with the common practice adopted by the majority of scholars, that is selecting only happiness, anger and sadness for investigating the musically induced emotion, thus discarding surprise, disgust and fear. Regarding sadness, different possible interpretations could be considered for this result e.g., musically induced sadness is highly subjective, highly depended on the track or the used algorithm was unable to recognize sadness with good accuracy. Furthermore, neutral variability was on average high indicating that listeners are not "insensitive" to music stimuli. In addition, single cases of remarkably high variability for emotions considered non-musical i.e., surprise, disgust and fear are presented, raising the question whether music can occasionally induce other emotions rather than the ones considered musical, or this result is due to mistaken recognition of the algorithm. Moreover, the pair-wise correlation analysis between emotions variability indicated that, for 3 out of the 4 tracks, disgust variability was correlated with fear, sadness and surprise variability, while fear variability was correlated with sadness variability. The above correlations turns the interest in the theory of compound emotions (i.e., a group of distinct emotions, each one of which is constructed by combining the 6 basic emotions e.g., a fearfully disgusted emotion combines muscle movements observed in fear and disgust) [9], however these correlations may be unreal due to the small dynamic range of sadness, disgust and fear response. Furthermore, the same experiment was conducted in a yacht cabin, in addition to the university lab environment. The between-subject analysis indicated that musically induced engagement variability was noticeably higher and valence variability higher or almost equal during listening to music in the yacht environment comparing to the university lab. Finally, engagement variability was far more affected than valence variability, by changing the environment.

The above results pave the way for the improvement of adaptive systems that use the specific face emotion recognition algorithm to musically induced emotion. Towards this direction, the personalization of the system is suggested for improving the listening experience, due to the diverging dynamic range of variability responses among the participants and taking into consideration the presence of high values for "non-musical" emotions, for single cases. Moreover, the indices about the existence of compound emotions suggests that the system could consider them for achieving a more accurate emotion recognition. Finally, the higher music engagement during the yacht reproduction encourage the idea of arranging the reproduction environment to enhance the music emotional involvement.

In future works, more tracks and more participants should be involved to validate the above results. Furthermore, different real-scenario environments should be considered.

References

1. Altieri, A., et al.: An adaptive system to manage playlists and lighting scenarios based on the user's emotions. In: 2019 IEEE International Conference on Consumer Electronics (ICCE), pp. 1–2. IEEE (2019)
2. Altieri, A., Ceccacci, S., Mengoni, M.: Emotion-aware ambient intelligence: changing smart environment interaction paradigms through affective computing. In: Streitz, N., Konomi, S. (eds.) HCII 2019. LNCS, vol. 11587, pp. 258–270. Springer, Cham (2019). https://doi.org/10.1007/978-3-030-21935-2_20
3. Ambadar, Z., Cohn, J.F., Reed, L.I.: All smiles are not created equal: Morphology and timing of smiles perceived as amused, polite, and embarrassed/nervous. J. Nonverbal Behav. **33**, 17–34 (2009)
4. Bai, M.R., Shih, G.Y.: Upmixing and downmixing two-channel stereo audio for consumer electronics. IEEE Trans. Consum. Electron. **53**(3), 1011–1019 (2007)
5. Barsoum, E., Zhang, C., Ferrer, C.C., Zhang, Z.: Training deep networks for facial expression recognition with crowd-sourced label distribution. In: Proceedings of the 18th ACM International Conference on Multimodal Interaction, pp. 279–2831 (2016)
6. Cecchi, S., Carini, A., Spors, S.: Room response equalization-a review. Appl. Sci. **8**(1), 16 (2017)
7. Dhall, A., Goecke, R., Ghosh, S., Joshi, J., Hoey, J., Gedeon, T.: From individual to group-level emotion recognition: Emotiw 5.0. In: Proceedings of the 19th ACM International Conference On Multimodal Interaction, pp. 524–528 (2017)
8. Dourou, N., Bruschi, V., Spinsante, S., Cecchi, S.: The influence of listeners' mood on equalization-based listening experience. In: Acoustics, vol. 4, pp. 746–763. MDPI (2022)
9. Du, S., Tao, Y., Martinez, A.M.: Compound facial expressions of emotion. Proc. Natl. Acad. Sci. **111**(15), E1454–E1462 (2014)
10. Eerola, T., Vuoskoski, J.K.: A review of music and emotion studies: Approaches, emotion models, and stimuli. Music Perception: An Interdisc. J. **30**(3), 307–340 (2012)
11. Ekman, P.: An argument for basic emotions. Cogn. Emotion **6**(3–4), 169–200 (1992)
12. Ekman, P., Friesen, W.V.: Facial Action Coding System: A technique for the measurement of facial movement. Consulting Psychologists Press (1978)
13. FakhrHosseini, M., Jeon, M.: The effects of various music on angry drivers' subjective, behavioral, and physiological states. In: Adjunct Proceedings of the 8th International Conference on Automotive User Interfaces and Interactive Vehicular Applications, pp. 191–196 (2016)
14. Fakhrhosseini, S.M., Landry, S., Tan, Y.Y., Bhattarai, S., Jeon, M.: If you're angry, turn the music on: Music can mitigate anger effects on driving performance. In: Proceedings of the 6th International Conference on Automotive User Interfaces and Interactive Vehicular Applications, pp. 1–7 (2014)
15. Febriandirza, A., Chaozhong, W., Zhong, M., Hu, Z., Zhang, H.: The effect of natural sounds and music on driving performance and physiological. Eng. Lett. **25**(4) (2017)

16. Frescura, A., Pyoung, Jik, L.: Emotions and physiological responses elicited by neighbours sounds in wooden residential buildings. Build. Environ. **210**, 108729 (2022)

17. Generosi, A., Ceccacci, S., Faggiano, S., Giraldi, L., Mengoni, M.: A toolkit for the automatic analysis of human behavior in HCI applications in the wild. Adv. Sci. Technol. Eng. Syst. **5**(6), 185–192 (2020)

18. Generosi, A., Ceccacci, S., Mengoni, M.: A deep learning-based system to track and analyze customer behavior in retail store. In: 2018 IEEE 8th International Conference on Consumer Electronics-Berlin (ICCE-Berlin), pp. 1–6. IEEE (2018)

19. Glasgal, R.: 360° localization via 4.x RACE processing. In: Proceedings of the 123rd Audio Engineering Society Convention. New York, USA (Oct 2007)

20. Goodfellow, I.J., et al.: Challenges in representation learning: a report on three machine learning contests. In: Lee, M., Hirose, A., Hou, Z.-G., Kil, R.M. (eds.) ICONIP 2013. LNCS, vol. 8228, pp. 117–124. Springer, Heidelberg (2013). https://doi.org/10.1007/978-3-642-42051-1_16

21. Hatziantoniou, P.D., Mourjopoulos, J.N.: Generalized fractional-octave smoothing of audio and acoustic responses. J. Audio Eng. Soc. **48**, 259–280 (2000)

22. He, K., Zhang, X., Ren, S., Sun, J.: Deep residual learning for image recognition. In: Proceedings of the IEEE Conference On Computer Vision and Pattern Recognition, pp. 770–778 (2016)

23. Hohnerlein, C., Ahrens, J.: Perceptual evaluation of a multiband acoustic crosstalk canceler using a linear loudspeaker array. In: 2017 IEEE International Conference on Acoustics, Speech and Signal Processing (ICASSP), pp. 96–100. IEEE (2017)

24. Ilie, G., Thompson, W.F.: A comparison of acoustic cues in music and speech for three dimensions of affect. Music. Percept. **23**(4), 319–330 (2006)

25. Karyotis, C., Doctor, F., Iqbal, R., James, A., E.: Affect Aware Ambient Intelligence: Current and Future Directions, vol. 298 (2017)

26. Lattanzi, A., Bettarelli, F., Cecchi, S.: Nu-tech: The entry tool of the hartes toolchain for algorithms design. In: Proceedings of the 124th Audio Engineering Society Convention, pp. 1–8 (2008)

27. Li, S., Weihong, D.: Deep facial expression recognition: A survey. In: IEEE transactions on affective computing. vol. 13, pp. 1195–1215 (2020)

28. Liu, W., et al.: SSD: single shot multibox detector. In: Leibe, B., Matas, J., Sebe, N., Welling, M. (eds.) ECCV 2016. LNCS, vol. 9905, pp. 21–37. Springer, Cham (2016). https://doi.org/10.1007/978-3-319-46448-0_2

29. Lucey, P., Cohn, J.F., Kanade, T., Saragih, J., Ambadar, Z., Matthews, I.: The extended cohn-kanade dataset (ck+): A complete dataset for action unit and emotion-specified expression. In: 2010 IEEE Computer Society Conference on Computer Vision And Pattern Recognition-workshops, pp. 94–101. IEEE (2010)

30. Mollahosseini, A., Hasani, B., Mahoor, Mohammad, H.: Affectnet: A database for facial expression, valence, and arousal computing in the wild. In: IEEE Transactions on Affective Computing. vol. 10, pp. 18–31 (2017)

31. Pan, F., Zhang, L., Ou, Y., Zhang, X.: The audio-visual integration effect on music emotion: Behavioral and physiological evidence. PLoS ONE **14**(5), e0217040 (2019)

32. Russell, J.A.: A circumplex model of affect. J. Pers. Soc. Psychol. **39**(6), 1161 (1980)

33. Simonyan, K., Zisserman, A.: Very deep convolutional networks for large-scale image recognition. American Journal of Health-System Pharmacy (2015)

34. Song, Y., Dixon, S., Pearce, M.T., Halpern, A.R.: Perceived and induced emotion responses to popular music: categorical and dimensional models. Music Perception: An Interdiscip. J. **33**(4), 472–492 (2016)

35. Szegedy, C., Vanhoucke, V., Ioffe, S., Shlens, J., Wojna, Z.: Rethinking the inception architecture for computer vision. In: Proceedings of the IEEE Conference On Computer Vision And Pattern Recognition, pp. 2818–2826. IEEE (2016)
36. Szegedy, C., et al.: Going deeper with convolutions. In: Proceedings of the IEEE Conference On Computer Vision And Pattern Recognition, pp. 1–9 (2015)
37. Talipu, A., Generosi, A., Mengoni, M., Giraldi, L.: Evaluation of deep convolutional neural network architectures for emotion recognition in the wild. In: 2019 IEEE 23rd International Symposium on Consumer Technologies (ISCT), pp. 25–27. IEEE (2019)
38. Thayer, R.E.: Toward a psychological theory of multidimensional activation (arousal). Motiv. Emot. **2**, 1–34 (1978)
39. Valstar, M., Maja, P.: Induced disgust, happiness and surprise: an addition to the MMI facial expression database. In: Proceedings of 3rd Intern. Workshop on EMOTION (satellite of LREC): Corpora for Research on Emotion and Affec. p. 65 (2010)
40. Van Der Zwaag, M.D., Dijksterhuis, C., De Waard, D., Mulder, B.L., Westerink, J.H., Brookhuis, K.A.: The influence of music on mood and performance while driving. Ergonomics **55**(1), 12–22 (2012)
41. Zentner, M., Grandjean, D., Scherer, K.R.: Emotions evoked by the sound of music: characterization, classification, and measurement. Emotion **8**(4), 494 (2008)
42. Zhu, Y., Wang, Y., Li, G., Guo, X.: Recognizing and releasing drivers' negative emotions by using music: evidence from driver anger. In: Adjunct Proceedings of the 8th International Conference on Automotive User Interfaces and Interactive Vehicular Applications, pp. 173–178 (2016)

Research on Interactive Interface Design of Vehicle Warning Information Based on Context Awareness

Fusheng Jia[1], Yongkang Chen[2(✉)], and Renke He[1]

[1] School of Design, Hunan University, Changsha 410000, China
[2] College of Design and Innovation, Tongji University, Shanghai 200092, China
1195138761@qq.com

Abstract. The characteristics of situational perception of vehicles and situational cognition of users in the assisted driving context were discussed, the relationship between human, vehicle and environment in the driving context was analyzed, the types of vehicle warning information were defined, and the priority of information push was determined to improve the efficiency of drivers in obtaining information. Based on the theory of situational awareness and situational cognition, a human-car-environment relationship model composed of user situation, vehicle situation and environmental situation is constructed. By means of on-board investigation and indoor driving simulation, the warning information needs of users and vehicles in the driving process are collected, and their functions are divided and prioritized through analytic hierarchy process. The interactive design scheme and prototype design of automobile warning information are proposed, and the prototype scheme is evaluated subjectively and objectively. Based on the analysis results, the vehicle warning information interaction system was constructed, and the active corresponding strategies of warning information were summarized under various situational awareness fusion scenarios. Through the controlled experiment, it was verified that the proposed vehicle warning information interaction system has better cognitive performance and faster dangerous response speed, which provides systematic guidance for the existing design of assisted driving vehicle warning information.

Keywords: situational awareness · intelligent connected vehicle · warning information · interactive design · proactive HMI

1 Introduction

With the rapid development of intelligent connected vehicles, sensors, controllers, actuators and other devices mounted on the vehicle body can sense the surrounding environment through the network system to realize the information exchange between the vehicle and the environment, the vehicle and people, and have reached the complex environment perception, intelligent decision-making, collaborative control and other functions [1]. However, it is difficult for the driver to determine the focus of the information provided by the interactive information provided by sensors and other devices

The original version of the chapter has been revised: the second author's affiliation corrected. A correction to this chapter is available at https://doi.org/10.1007/978-3-031-35908-8_22

mounted on the body and the system. At the same time, a large number of value-added services such as infotainment services, telecommunication services and life services are integrated into the in-car system, and drivers need to deal with a large amount of information, resulting in a high cognitive load during driving. Although rich information services achieve comfortable driving experience to some extent, they will interfere with drivers' visual channel allocation and lead to operational errors during driving [2]. Based on the driver's situational cognitive characteristics and vehicle's situational perception characteristics, the human-vehicle-environment relationship in the driving situation was analyzed, the type of warning information was defined, the priority of information push was determined, and a clear interactive design of warning information was constructed to improve the driver's efficiency of obtaining information.

2 Situational Awareness and Situational Cognition

2.1 Situational Awareness Theory

Context awareness is the use of sensors and related technologies to enable computer equipment to "sense" the current situation. Contains any perceived information, which can be a role, place, or concrete thing associated with the interaction [3]. According to the development process of context, context awareness includes user context, device context, task context, environmental context, social context and spatio-temporal context [4]. Context classification can effectively define the objects involved in context awareness and its related contents [5].

Context awareness in the process of driving mainly includes the information of the car itself, the information between the car and the car, the information between the car and the environment, the information between the car and people, and the information between the car and other interactive carriers. According to the context sources, the situational perception content in the driving process can be divided into the driver's context, including physiological information such as vision, heart rate and blood pressure, ethnographic information such as gender, age and occupation, and acquired information such as driving level and cognitive ability. Vehicle situation has speed, fuel consumption, space and other state information; The task situation is mainly driving task; The environmental situation mainly includes weather conditions, traffic conditions, road laws and other external information; Spatio-temporal context contains coordinate information such as vehicle positioning and driving distance. Vehicle situational awareness can express the conditions faced by the driver in the current driving task. Different types of situational awareness information are collected by advanced sensors, processed and encoded by the vehicle system and transmitted to the information interaction interface. The transmission process is shown in Fig. 1.

2.2 Situational Cognition Theory

Situational cognition refers to the overall perception of some elements in the environment within a specific range, and the planning and prediction of the future status of the elements. In the process of situational awareness, users have to go through three stages:

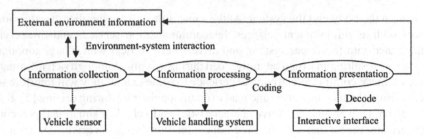

Fig. 1. Car environment perception transmission process

the reception of active and passive information, the recognition of brain information and the subsequent judgment of information [6].

In the process of driving, the situational cognition of the driver mainly comes from the traffic environment and vehicle running status, so as to maintain the steering, braking and safe distance of the vehicle. In the cognitive process, the driver obtains traffic information and running status, makes decisions and implements them. In the process of situational cognition, the driver acquires, understands, encodes and stores continuous information, which is a decision-making activity including perception, attention, memory and thinking. The driver's situational cognition process can be defined as an information processing process, including three stages: perception of receptors, decision making of nerve center and memory knowledge base, and execution of effectors, as shown in Fig. 2. In the process of information processing, the driver perceives through the sensory organs and sends the sensory information to the central system for processing. The processed signals are temporarily stored in the memory knowledge base for decision matching and then sent to the driver to complete the execution operation, which is a process of processing a huge information base. In the actual process of driving, drivers need to accept and process a large amount of inside and outside vehicle perception information, such as traffic light, road signs, moving lane and other traffic environment information, vehicle distance, lane change, overtaking and other vehicle environment information. The emergence of intelligent connected cars helps drivers shorten the process of situational cognition and greatly improves the efficiency of situational cognition

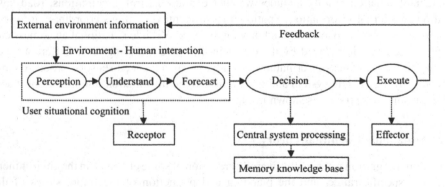

Fig. 2. Driver's situational cognition processing process

of drivers. Setting target information in an important scope is conducive to improving the level of situational cognition of drivers, which plays an important role in perceiving potential dangers of external environment [7]. The active push of automobile warning information should be synchronized with the driver's information processing during driving; otherwise, the user's situational cognition level will be reduced, resulting in the driver's cognitive load.

2.3 Human-Car-Environment Relationship in Driving Context

Based on the process of situational perception and situational cognition, driving situation influencing factors are divided into driver's situational factors, vehicle situational factors and environmental situational factors [8]. Driving level, cognitive ability, social experience and other factors belong to the driver's situational factors; Vehicle situation includes physical factors such as in-car space, in-car interaction, in-car system, etc. Environmental situation includes environmental factors such as car and car, car and person, car and other things. The three situational factors influence each other and act on the process of driving situational perception and driver situational cognition. In the actual driving process, the traffic volume sensor collects the situational information, and the results are presented after processing by the vehicle system. The encoded results conform to the driver's cognitive characteristics and are finally presented to the interactive interface to complete the delivery of the driving situational perception. The environmental situation information and the vehicle situation information then become the channel

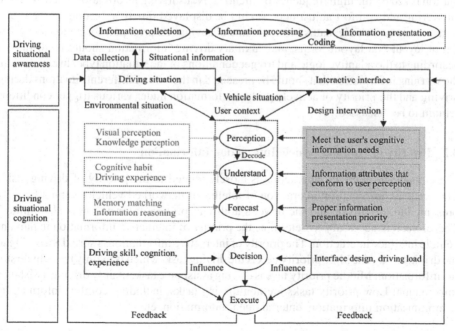

Fig. 3. Human-Vehicle-Environment relationship model based on situational perception and situational cognition

of the driver's situational cognition information and jointly act on the driver's situational cognition. After that, the driver completes the driving decision and the driving task, the information transmission cycle mechanism is established, and the driver completes the situational cognition [9]. The human-car-environment relationship model constructed by driving situational awareness and driver situational cognition is shown in Fig. 3.

3   Related Research on Vehicle Warning Information

3.1   Categories of Person-Vehicle Interaction Information

The current intelligent connected vehicle has developed from a single vehicle information interaction system to a complex human-vehicle-environment all-round interaction system, and the vehicle carries all kinds of information interacting with it [10]. In the process of driving, in addition to dealing with traffic state information and vehicle state information, the driver also needs to switch between various types of information. The interactive information module in the car can be divided into the following six categories: vehicle state information, starting information, entertainment information, driving assistance information, communication information and comfort information. The driver assistance information includes adaptive cruise, lane departure warning, lane keeping assistance, parallel assistance, automatic braking assistance, night vision assistance, driver fatigue monitoring, automatic parking assistance, etc.

The design principle of interface information display is to match high-priority information based on the high frequency of attention. Non-driving information, such as phone answering and comfort adjustment, should be displayed timely according to the user's customization. Vehicle status information and navigation prompt information show different layout and layout forms, and auxiliary driving information should be displayed according to the adaptive logic and trigger conditions of the situation type. In conclusion, the warning information that should be presented in the face of different situations during driving and the priority of actively pushed information under various trigger conditions remain to be discussed.

3.2   Priority of Human-Vehicle Interaction Information

Human-vehicle interface interaction is closely related to the priority of driving tasks. The complex information interaction of intelligent networked vehicles cannot be fully presented to the driver, and the driver's ability to process information is also limited. Therefore, it is necessary to determine the priority of interactive information in human-vehicle interface interaction. The priority of interactive information is sorted according to the driving task. The high-priority task is the driving task, which is mainly the vehicle status information. Middle priority task is driving assistance task, mainly driving assistance information; Low-priority tasks are non-driving tasks, including comfort information, communication information, entertainment information, etc.

4 Interactive Design of Automobile Warning Information

4.1 User Research and In-depth Interview on Human-Vehicle Interaction Information

This paper firstly collected the warning information needs of users and vehicles in the process of driving through on-board investigation and indoor driving simulation methods, and then conducted demand analysis of driving situations that endanger drivers' safety through in-depth interview. At present, users have little understanding of assisted driving, so it is difficult to directly obtain the warning demand. Moreover, video playback is used to compare the differences between normal driving and assisted driving situations, so as to obtain the direct demand of users.

Due to different research methods, in order to avoid the same or different demand feedback caused by differences among tested users, and in order to obtain differentiated information needs among different tested users, According to different driving experience, the tested users are divided into novice users (driving experience < 1 year), skilled users (driving experience < 10 years) and expert users (10 years) [11]. Through the analysis of user characteristics, the test users of one driving simulation are determined to be novice users, and the test users of two following car surveys are respectively skilled users and expert users. Situational cognition analysis is carried out on three types of users with different driving experiences. The feedback of novice users is comprehensive and basic, with weak perception of external environment and poor prediction of potential situations, but it is helpful to construct overall and comprehensive warning information. Skilled users have certain ability to avoid danger in driving situations, but they are powerless in sudden dangerous situations. They overlap more with novice users' information needs, so they can extract and optimize key needs. In the face of common driving situations, expert users can make timely prediction of potential dangers and have strong situational cognition ability, which can provide help for information priority allocation in emergencies.

After the observation and record of the tested users, 7 users were selected for in-depth interview, including 2 novice users, 3 skilled users and 2 expert users. According to the results of in-depth interviews and the types of daily driving roads, six types of typical driving accidents are summarized, including straight going, turning, accidents, overtaking, lane change and parking accidents. Based on the analysis of various types of traffic accidents, it is concluded that users' demands for rear-end collision accident are the most urgent, including the braking distance prompt, traffic light change prediction prompt, the distance between the car in front and the car behind, and the situation in front of the road, etc., followed by overtaking accident, including the blind area vehicle prompt, the distance between the car behind the car prompt, and the possible vehicle prediction prompt on both sides.

Based on the results of observation, analysis and interview, the information needs raised by the tested users were divided according to the situation corresponding to human-car-environment, and the information function transformation as shown in Fig. 4 below was summarized. User situational information includes driver adaptive information such as fatigue driving and driving attention. Vehicle situational information mainly involves core paths such as vehicle state, control and operation. Demand functions are transformed into vehicle distance prompts and lane departure prompts, etc. Environmental situational information includes inter-vehicle, driving regulations and environmental conditions, etc. The main transformation function includes road sign prompt, surrounding vehicle prompt, blind area pedestrian prompt and so on.

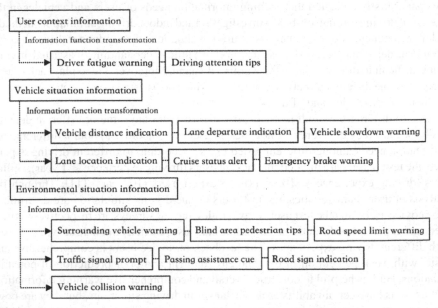

Fig. 4. User, vehicle, and environmental contextual information function transformation

4.2 Demand Analysis of Human-Vehicle Interactive Warning Information

Classification of Human-Vehicle Interaction Warning Information. Aiming at the Disordered and Complex Demand Information of Different Types of Users, Analytic Hierarchy Process is Adopted to Clarify the Division and Weight of Various Information Functions. Based on the Content of Warning Information and Supplemented by Situational Information, Five Types of Information Were Determined by Combining the Functional Attributes of Driving Situational Awareness and Human-Vehicle Interaction, Which Were User State Information, Vehicle State Information, Environmental State Information, Vehicle Braking Information and Environmental Warning Information, and a Contextual Awareness Based Hierarchical Model of Vehicle Warning Information Was Constructed, as Shown in Fig. 5.

The user status detection uses the on-board camera to detect the user's behavior and physiological state through the visual tracking, target detection and other technologies. The fatigue driving prompt and driving attention prompt belong to this kind of information, which can be prompted by text or icon. Vehicle status information refers to the driving state, operation state, etc., mainly through the form of ICONS. Vehicle braking information is to predict the state of the vehicle in advance by emergency warning. Environmental status information is presented to users through traffic signs. Environmental

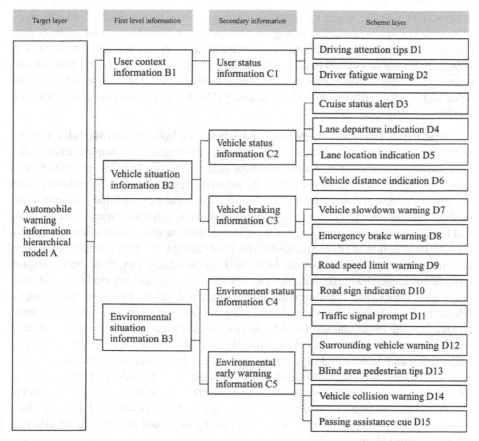

Fig. 5. Hierarchical model of automobile warning information

warning information refers to the information presented to detect surrounding vehicles and pedestrians in blind areas.

Analysis of human-vehicle interaction warning information. In the actual driving process, the user's attention is about 70% in vehicle control, road information and emergencies. Unclear information priority will cause distraction and decision-making difficulties. Determining the priority of information presentation is conducive to the establishment of information system architecture and active interaction logic, and can help users predict the key information in a specific situation in advance.

The data of 5 expert users were collected through the questionnaire. Based on the automobile warning information model, the scale was determined by pairwise comparison between various factors, and the relative weight of each indicator was calculated. The results show that the priority order of warning information type is environmental warning information (C5) > vehicle braking information (C3) > user status information (C1) > vehicle status information (C2) > environment status information (C4). Under the user status information type, driver attention prompt (D1) = driver fatigue prompt (D2); Under the vehicle status information type, vehicle distance prompt (D6) > lane departure prompt (D4) > cruise status prompt (D3) > lane position prompt (D5); Under the type of vehicle braking information, emergency braking prompt (D8) > vehicle deceleration prompt (D7); Under the environment status information type, road speed limit prompt (D9) > road sign prompt (D10) > traffic signal prompt (D11); Under the environmental warning information type, pedestrian warning in blind area (D13) > vehicle collision warning (D14) > surrounding vehicle warning (D12) > overtaking assistance warning (D15).

Construction of Human-Vehicle Interactive Warning Information Module. In order to ensure that drivers can quickly obtain effective information in complex driving situations, a large number of display information needs to be similarly integrated to form a reasonable information display module. Based on the research on attribute classification and priority ranking of warning function information, it can be found that user state information, vehicle state information and environment state information are static prompts and feedback in driving situation, while environmental warning information and vehicle braking information are dynamic scene prompts in current situation changes. Therefore, their main information functions can be divided into two categories: State change display module and scene change display module. The status change display module mainly displays user, vehicle and environment status information, including information directly related to driving operation, such as driver fatigue, lane departure and road speed limit, and tends to the important level of driving tasks, which needs to be presented according to the change of priority. Scene change display module presents environmental warning information and vehicle braking information, such as blind area pedestrian monitoring, vehicle distance and other information that may threaten the safety of drivers during the driving process, tending to the second-important level of driving tasks. To sum up, the architecture design of automobile warning information is shown in Fig. 6, in which information of the same priority is emitted side-by side, and information of different priority is emitted misaligned.

4.3 Active Responsive Interaction Design Framework

During driving, a single warning message may be triggered by multiple vehicles and environmental situational awareness, and a single situational awareness may also trigger multiple warning messages. When multiple warning messages appear in the fusion of single or multiple situational perception, the vehicle interaction system should adapt to the weight logic relationship of interactive information, display dynamic warning information on the interactive interface, and improve the situational cognition ability of the driver. Specifically, the highest priority warning information can be highlighted by changing colors, increasing brightness, flashing and other forms. In the face of lower priority warning information, the visual interference of drivers can be reduced by reducing the brightness saturation. According to the research of scholar Yang Xue, the critical value of high-priority information is 2, and that of low-priority information is 3. Higher or lower than the critical value will affect the cognitive efficiency. Based on the weight results of functional information and the information quantity control principle of analytic hierarchy Process (AHP), the active responsive interactive architecture of automobile warning information is concluded, as shown in Fig. 7.

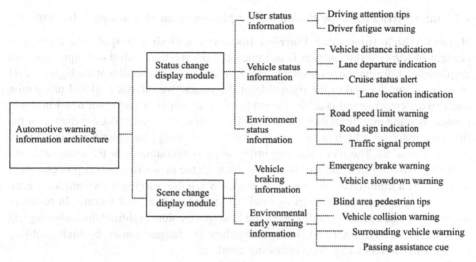

Fig. 6. Automobile warning information architecture

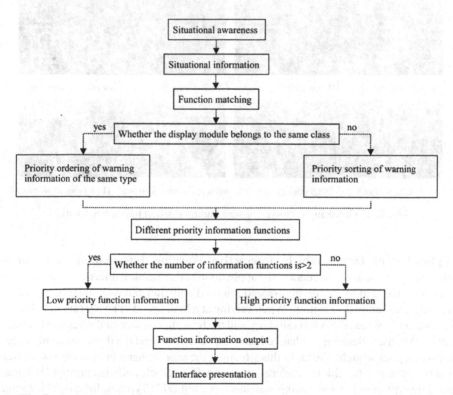

Fig. 7. Active and responsive interactive architecture for vehicle warning information

4.4 Interactive Prototype Design and Implementation of Warning Information

Human-Vehicle Interaction Warning Information Design Output The interactive content of warning information is designed in high fidelity, with flat design style and emphasis on information, as shown in Fig. 8. Warning information is at the highest level of information, located at the right side of the interactive interface, global navigation and driving environment is at the second level of information, and other non-important content is at the third level of information. The blinking red color block at the top of the interface indicates a high priority alert, while the blinking blue color block indicates a low priority alert. The more warning information is encountered at the same time, the faster the blinking frequency of the color block. In the driver interaction process, there will also be a multi-channel interaction mode. When the warning information appears, the warning sound will accompany, and the steering wheel will vibrate. In terms of interface design details, vehicle colors and lane space colors highlight the technological atmosphere through light and shade, and reduce the fatigue caused by high visibility colors when drivers enjoy good operating comfort.

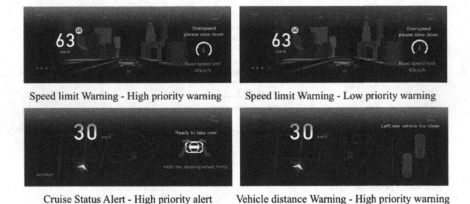

Speed limit Warning - High priority warning   Speed limit Warning - Low priority warning

Cruise Status Alert - High priority alert   Vehicle distance Warning - High priority warning

Fig. 8. High-fidelity prototype of warning information interaction (partial)

Application of Driver-Vehicle Interactive Warning Information in Complex Situations In the above scenarios, the application of the warning information interaction system is analyzed by taking high-speed rear-end collision and overtaking accidents as examples. Situation description: When driving at a high speed in heavy rain, the driver's sight becomes weak, the road is slippery and the braking distance of the vehicle becomes longer. When accelerating to change lanes, the vehicle behind the driver suddenly accelerates and goes straight ahead. In this situation, the user demand information is: vehicle distance prompt D6, vehicle deceleration prompt D7, vehicle collision prompt D14, road speed limit prompt D9, overtaking assistance prompt D15; Types of functional information: vehicle collision prompt and overtaking assistance prompt belong to environmental warning information C5, vehicle deceleration prompt belongs to vehicle braking information C3, road speed limit prompt belongs to environmental status information C4, vehicle distance prompt belongs to vehicle status information C2. Priority ordering of warning information: According to the weight of the same type of warning information, C5 > C3 > C2 > C4. According to the weight of warning information, C5 information is D14 > D15. Combined with the priority information quantity display rule, the first two priority warning information D7 and D14 are highlighted, and the remaining warning information is low priority and displayed at the second level. Vehicle system perception results: vehicle deceleration prompt and vehicle collision prompt are displayed with high priority, and the final interaction of warning information is shown in Fig. 9.

Fig. 9. Design and application of warning information interaction in a variety of situational awareness fusion scenarios

5 Design Test Evaluation

In order to verify the effectiveness of the vehicle warning information interaction system, the usability evaluation was carried out. It mainly adopts dynamic video method and fuzzy comprehensive evaluation method. Dynamic video method can objectively evaluate the cognitive performance of warning information and the ability to respond to danger. By testing the time and accuracy of users' cognition of warning information of vehicle system, it can evaluate whether the cognitive performance of the interface composed of warning information classification based on situational awareness is better than the existing interactive interface. And in a variety of situational awareness fusion scenarios, to determine the impact of differentiation of warning information levels on drivers' response ability; Fuzzy comprehensive evaluation method can help to make overall evaluation of things constrained by various factors, and can make subjective evaluation of the satisfaction of interactive interface.

The driving situation of dynamic video method comes from the simulation game "City car driving". The driving roads in the game mainly include urban roads and expressways, and the selected scene conforms to the domestic traffic conditions. The driving scene with single situational awareness scene and multiple situational awareness fusion scene was selected as the experimental materials, and the interactive interface of car warning information was added to the driving scene through video editing software to collect users' interface cognition and operational behavior responses in the driving situation. The single context perception scene is the road in the city. The warning time is 12 s, and the 15 warning messages are changed in turn. The video duration is 180s. The application case "high-speed rear-end collision and overtaking accident" mentioned above is selected in the fusion scenario of multiple situational awareness. 25 s is taken as the warning message warning time, and the user is required to pause the video when receiving the warning message and record the time when the user receives the message. There are two warning messages in this scene, and the video lasts for 50's. The experiment was divided into experimental group and control group. NIO Pilot assisted driving interaction system was selected as the control group. The fuzzy comprehensive evaluation method combines the indicators of the user experience instinct level, behavior level and reflection level, and scores various functional information in the form of a questionnaire to obtain the subjective evaluation of users.

Considering the small number of users of the assisted driving system and the low level of understanding, the tester mainly selected the users previously investigated, including 5 novice users, 4 skilled users and 2 expert users, among whom 7 males and 4 females, aged 20–45 years old.

Through the analysis of the cognitive speed and accuracy data of 11 users in the single situational awareness scenario of the interactive interface warning information, the results are shown in Table 1 below. In terms of the cognitive time of warning information, the cognitive speed of most information in the experimental group was higher than that in the control group, especially the user status information combined with text and text and the vehicle braking information, the vehicle status information combined with scene and text, the readability of warning information was significantly higher than that in the control group, and the cognitive time of environmental status information had

Table 1. Cognitive time and correct rate of warning information in single context perception scenarios

Warning message	Test group		Test control group	
	Cognitive time(s)	Cognitive accuracy(%)	Cognitive time(s)	Cognitive accuracy(%)
D1	1.32	100%	1.68	100%
D2	1.21	100%	2.34	87.65%
D3	1.45	100%	3.58	100%
D4	1.11	100%	2.21	100%
D5	1.65	100%	1.98	91.43%
D6	0.99	100%	2.51	100%
D7	1.43	95.4%	2.32	92.32%
D8	1.52	100%	3.46	100%
D9	2.06	96.28%	1.43	84.98%
D10	1.63	100%	3.51	81.43%
D11	1.69	100%	1.12	92.65%
D12	2.87	100%	4.65	100%
D13	2.49	100%	3.93	98.43%
D14	2.38	100%	4.65	79.27%
D15	2.31	94.32%	4.82	73.56%

little difference. The reason may be the single form of expression. The correct rate of warning information cognition in the experimental group was higher than that in the control group, which verified that the expression form of the interactive design of the system was consistent with the cognitive characteristics of users. In order to verify the validity of the results, the cognitive time significance test was conducted between the experimental group and the control group. The results showed that the cognitive time difference between the two groups was significant, and the difference increased. The automobile warning information interaction system based on situational awareness had better cognitive performance.

The cognitive speed and accuracy data of user interactive interface warning information under various situational awareness fusion scenarios are analyzed. In the multiple situational awareness fusion scenes shown in the two videos, the cognitive accuracy rate of the tested experimental group and the control group is the same, both being 100%. The two interactive systems have a good avoidance effect on users encountering dangerous accidents. From the perspective of users' cognitive time, the experimental group has a significantly higher time than the control group. In conclusion, the situational awareness based automobile warning information interaction system can bring users faster danger perception speed under the condition that the cognitive accuracy is 100%.

Based on the analysis of the questionnaire results, the ratio between the score of a single evaluation index and the total score of the evaluation index was calculated, and the fuzzy evaluation matrix of the warning information interaction system was constructed. Among the 7 evaluation indexes, the maximum value of fuzzy evaluation is 0.331. According to the maximum membership criterion, the final evaluation result of the warning information interaction design system is "satisfactory". In summary, through the comparison between the experimental group and the control group, users have good acceptance and satisfaction with the context-aware automobile warning information interaction system.

6 Conclusion

Based on the theory of situational awareness and situational cognition, this paper explores the characteristics of vehicle situational awareness and user situational cognition in driving situations, constructs a human-car-environment relationship model composed of user context, vehicle context and environmental context, and studies the interaction design requirements of automobile warning information. On this basis, the function division and priority ranking of the warning information are discussed, and the interactive design scheme and prototype design of the automobile warning information are proposed. The subjective and objective evaluation of the preliminary prototype scheme is carried out to verify that it has a high degree of acceptance and satisfaction of users. In the next research, it is necessary to combine multi-channel interaction, such as touch, hearing, smell, etc., to conduct in-depth research on the vehicle warning information interaction system.

References

1. Tan, Z.Y., et al.: Overview and perspectives on human-computer in. intelligent and connected vehicles. Comput. Integr. Manufact. Syst. **26**(10), 2615–2632 (2016)
2. Li, Z., Zhou, X., Zheng, Y.S.: Design and research of automobile driving assistant system based on AR-HUD. J. Wuhan Univ. Technol. (traffic science and Engineering Edition) **41**(06), 924–928 (2017)
3. Dey, A.K., Abowd, G.D., Salber, D.A.: Conceptual framework and a toolkit for supporting the rapid prototyping of context-aware applications. Human-Comput. Interact. **16**(2–4), 70 (2001)
4. Gwizdka, J.: What's in the context? A Position Paper presented at the Workshop on The What, Who, Where, When, Why and How of Context-Awareness. In: ACM SIGCHI Conference on Human Factors in Computing Systems. The Hague Netherlands, 89 (2008)
5. He, X.Q.: Research on mobile phone adaptive user interface design based on context awareness. Beijing University of Posts and Telecommunications (2012)
6. Jia, F.S., He, R.K.: Research on experience design of shopping chat robot based on cognitive mechanism. Design, **35**(04), 18–21 (2022)
7. Li, K.G.: Research on the design of HUD safety information interface based on context awareness. China University of Mining and Technology (2020)
8. Wang, J.M., et al.: Research on HMI design in ACC cut in scenario based on control strategy data. Packag. Eng. **42**(18), 9–17 (2021). https://doi.org/10.19554/j.cnki.1001-3563.2021.18.002
9. Cheng, Z.H., Tan, H.: User safe driving handover experience based on process characteristics. Packag. Eng. **41**(02), 50–56 (2020). https://doi.org/10.19554/j.cnki.1001-3563.2020.02.008
10. Wang, W.J., et al.: Research on human-machine interface design of exterior screen of driverless delivery car. J. Graph. **41**(03), 335–341 (2020)
11. Jia, F.S., He, R.K.: Research on user cognition oriented shopping chat robot interface design. Packag. Eng. **43**(16), 92–100 (2022). https://doi.org/10.19554/j.cnki.1001-3563.2022.16.009

MOSI APP - A Motion Cueing Application to Mitigate Car Sickness While Performing Non-driving Task

Daofei Li[1]([✉]) [ID], Biao Xu[1] [ID], Linhui Chen[1] [ID], Binbin Tang[1] [ID], Tingzhe Yu[1] [ID], Keyuan Zhou[2] [ID], Nan Qie[2] [ID], Yilei Shi[2] [ID], Cheng Lu[2] [ID], and Haimo Zhang[2] [ID]

[1] Institute of Power Machinery and Vehicular Engineering, Zhejiang University, Hangzhou, China
{dfli,22060070,22160126,22260076,22260079}@zju.edu.cn
[2] Oppo Mobile Communication Co., Ltd., Dongguan, Guangdong, China
{zhoukeyuan,qienan,shiyilei,lucheng1,zhanghaimo}@oppo.com

Abstract. It is common for passengers to perform non-driving tasks such as reading or gaming on smart phones when riding, which may easily lead to car sickness. The underlying cause is a possible sensory conflict between the signals received by the passenger's visual processing system and those received by the vestibular system. To reduce the conflict and mitigate car sickness, we propose the *MOSI APP*, a smart phone application to present the vehicle motion as a visual cue. Four arrows on phone screen are used for cueing, while the direction and color-filling indicate the vehicle motion direction and magnitude, respectively. When using the smart phone for non-driving tasks, passengers can observe and understand motion cues to gain situation awareness through the *MOSI APP*. A total of 30 participants of passenger were recruited for a series of driving simulator experiments, and reports of their sickness level, non-driving task performances and subjective evaluation on our application were collected and analyzed. Results show that: 1) for all 26 participants with valid data, comparing to the case without motion cues, the *MOSI APP* could contribute to motion sickness mitigation (though with no significance, $p = 0.157$), and it could delay the aggravation of moderate or severe sickness level (by 8 to 10 min in our experiment setting); 2) for the 15 participants who were more susceptible to motion sickness in the experiments, the APP could significantly ($p = 0.047$) alleviate their sickness level, while their non-driving task performances were not notably affected. This study may provide useful insights in how to mitigate car sickness, especially in the era of ubiquitous computing and connectivity on smart devices.

Keywords: Motion Sickness · Non-driving Tasks · Sickness Mitigation · Smart Phone Application · Visual Cue · Intelligent Vehicle

---

Supported by Guangdong OPPO Mobile Telecommunications Corp., Ltd.

H. Krömker (Ed.): HCII 2023, LNCS 14049, pp. 173–190, 2023.
https://doi.org/10.1007/978-3-031-35908-8_13

1   Introduction

The rapid development of smart phones and automotive infotainment systems has greatly stimulated the demand of passengers to use them for non-driving tasks (NDT) in a moving vehicle. However, reading or gaming during the journey may cause a conflict between motion states perceived by the vestibular system and that from visual receptors. Researches show that such sensory conflict is an important contribution to evoking motion sickness [16,17]. Passengers that suffer from motion sickness may exhibit several symptoms: yawning, distraction, dizziness, headache, nausea, vomiting, etc [10]. Unfortunately, compared to observing the surrounding traffic environment, reading or gaming is more likely to evoke motion sickness, which may further affect the NDT performances of passengers. Facing with these contradictory needs, research is necessary to be done on how to mitigate passengers' motion sickness when they are using phones in car, especially for those highly susceptible to motion sickness.

In addition to the general pharmacological countermeasures for motion sickness, current mitigation countermeasures mainly focus on smoothing vehicle motion or providing passengers with motion cues. Regulating the motion to be sustained by passengers, either its direction, magnitude or frequential distribution, can effectively restrict the vibrational stimuli transmitted to passengers from the vehicle chassis. For instance, passenger seats can be designed to recline [1], while the seat direction should try to avoid facing rearward [18]. The perspective area around the vehicle should be enlarged under the premise of ensuring safety to provide passengers with sufficient situation awareness [6]. For autonomous vehicles, motion sickness may be alleviated by smoothing the planned trajectory. For example, frequency-shaped filtering can be used to reduce the low-frequency accelerations that are likely to induce sickness [12]. Motion sickness model can even be directly incorporated into automated speed control [13]. Moreover, additional strategies focus on vehicle suspension control, such as adaptive adjustment of suspension stiffness based on road conditions [4] and active suspension tilting during cornering [20], etc.

On the other hand, it is also quite helpful to provide passengers with cues about what future motion is coming. A good example would be playing audio cues to the passengers with relevant motion information [9]. Passengers can then obtain a strong mental anticipation or expectation about the coming movement, thus can minimize motion sickness. An alternative way is to use haptic interface of cues for better user experience, in which a seat cushion with vibrational motor array to provide informing cues about the motion direction and intensity to the occupant [11]. Compared to these two cueing approaches, visual cues are used more extensively if considering the availability of screens in modern vehicles. For example, when passengers are reading on their smartphones, a visual cueing approach can be to set the live video stream of the road ahead as the background of reading screen, which was proven beneficial in alleviating motion sickness [15]. As an alternative, some floating bubbles on the peripheral margin of the smartphone screen can also provide cues of the vehicle motion [14].

To sum up: 1) motion-regulation based approaches are direct and straightforward in terms of car sickness mitigation, but they may be only applicable in specific vehicles with available controllability of motion; 2) interaction-based cueing approaches, especially those realizable in intelligent vehicle cabin or smart phones, can also achieve effective sickness-mitigation performances.

Considering the rapidly-developing trend of connectivity and ubiquitous computing, either on built-in-cabin or handheld screens, the cueing effectiveness and also the size of the population that could potentially benefit from this design can be hopefully further scaled up. Therefore, this study focuses on designing a smart phone application for motion cueing, which aims to mitigate users' sickness with limited interference on their NDT performances.

The rest of the paper is organized as follows. Section 2 briefs the cueing application design, and Sect. 3 gives the detailed experiment setting-up. Then Sect. 4 presents the passenger sickness level results, NDT performances and also their subjective evaluation results of the proposed smart phone application. Section 5 discusses the results and current design, and finally Sect. 6 concludes the study.

2   MOSI APP Design

According to motion sickness mechanisms, either sensory conflict theory [16] or postural instability theory [19], providing the cues of coming motion stimuli can help passengers better predict and prepare for the motion, mentally (neurally) for expectation or physically for body posture adaptation. If a smart phone has access to the precise acceleration of vehicle in a ride, the stimuli information can be conveyed to the passenger as quickly as possible through motion cues. Understanding such motion cues can not require much cognitive cost of users, and it is also necessary to minimize their negative impacts on users' non-driving tasks such as using social media on the phone.

Based on these insights, a MOSI APP is designed to mitigate passenger motion sickness, which has a simple and intuitive metaphor for vehicle motion stimuli. Inspired by the widely used traffic signs (i.e. red, green and yellow light), and also direction indicator lamp in vehicles, a four-direction "arrow" as shown in Fig. 1 is chosen to present passengers the appropriate cues of coming motion stimuli.

The design ideas are briefed as follow.

(1) Four arrows for four directions. The arrows are located in the center of the screen of the smartphone and the four directions of the arrows represent vehicle acceleration in four directions. The vertical two correspond to the longitudinal acceleration of the vehicle, and the horizontal two correspond to the lateral acceleration. For example, if the vehicle accelerates to the left, the left arrow will be filled.

(2) Red-yellow-green color coding. In order for the passenger to subconsciously understand the current movement intensity of vehicle, the green color-filling indicates the forward acceleration; the red color-filling indicates the backward deceleration; and the yellow color-filling indicates the lateral acceleration.

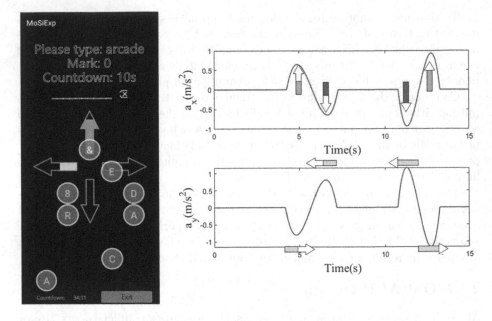

Fig. 1. Arrow-shaped motion cue and spelling task

(3) Magnitude mapping and thresholds. The mapping from actual acceleration amplitude to the length of color-filling is linear. A minimum threshold acceleration is set such that an arrow only starts to be filled when the motion stimulus is harmful enough, which can help to avoid frequent cueing refreshing in screen animation caused by low or noisy acceleration values.

(4) NDT interference minimization. To reduce the negative impact of cueing on NDT, the arrow border and the color-filling are set translucent to avoid possible occlusion.

3   Experimental Setup

To validate the effectiveness of the proposed APP, car ride experiments were carried out on a moving-base driving simulator. The experiments include two conditions, *Arrow-cue condition* and *No-cue condition*, corresponding to whether the arrow-shaped motion cue is enabled or not. The dependent variables are participants' motion sickness level and non-driving task performances.

3.1   Participants

The study invited 30 healthy participants (14 males and 16 females), but four of them were excluded due to data recording problems that led to interruption of experiment or subjective reporting. The valid data for 26 participants, aged between 19 and 28 (mean=22.73, SD=2.46), were obtained. Before the experiment, their motion sickness susceptibility evaluation was collected via the Motion Sickness Susceptibility Questionnaire(MSSQ) [5]. Their average MSSQ score was 83.08±31.04, while all scores were greater than the 50th percentile of what would normally be expected in the general population. All participants indicated they were overall in good health and had no vestibular disorders. They were instructed to refrain from alcohol in the 24 h before the experiment.

3.2   Apparatus and Stimuli Profile

A six-degree-of-freedom driving simulator in Fig. 2 was used to simulate the real ride conditions. The platform as well as the simulator windows were surrounded by black curtains, creating a closed environment free of distractions.

The IMU sensor at cabin center of gravity, computers for experiment control and simulator, and their communication are shown in Fig. 3. Passengers used an OPPO Reno 7 smartphone, with a 6.43 in. screen 90 Hz refresh rate, which could well support the software computing. The acceleration information of the simulator cabin is instantly collected by the IMU sensor, and was transmitted to the phone via TCP/IP at a frequency 100 Hz.

Two typical vehicle movements, i.e. longitudinal acceleration/deceleration and lane change, were simulated, with the maximum longitudinal and lateral accelerations as 0.2 g and 0.12 g, respectively. Figure 4 shows a portion of the simulated accelerations. To ensure the unpredictability of stimuli, the amplitude, sequence and interval of simulated motion acceleration are randomized.

Fig. 2. Moving-base driving simulator with 6 degrees of freedom

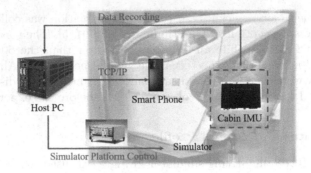

Fig. 3. Equipments used in experiment

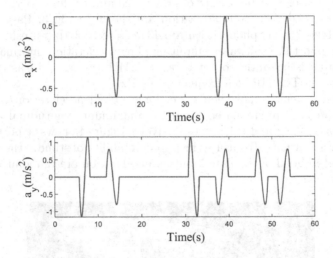

Fig. 4. A typical segment of simulator cabin acceleration stimuli

3.3 Non-driving Task

The non-driving task for participants during a simulator ride was to spell letters in the correct order according to a given letter combination. The spelling task requires cognitive processes (e.g. visual perception, memory, visual search) and finger operations. It is a typical kind of smart phone tasks, e.g. when texting messages. As shown in Fig. 1, the string of letters at the top of the current screen is "arcade", so participants need to find and click the letter buttons corresponding to "A R C A D E" at the bottom of screen. The position of letter buttons is random and there may be unnecessary letters as disturbances. For one string correctly spelled within a fixed time limit, one point of task score is recorded. This time limit is set between 8–12 s, which is consistent under the two experimental conditions and confirmed by each participant before the formal experiment.

3.4 Procedure

Each participant needs to participate in the experiments under the *No-cue* condition and *Arrow-cue* condition, each taking 35 min of simulator ride. The two experiments were spaced one week apart to eliminate the coupling effects of the post-experimental discomfort state. The order of two conditions was randomized and balanced.

A brief flowchart of experiment is shown in Fig. 5. Before the experiment, the procedure was explained to the participants and they signed an informed consent form. Then in the pre-study preparation, participants were given sufficient time to familiarize themselves with the spelling task and to determine the time limits in spelling. Besides, before the Arrow-cue conditional experiment, participants were told about what the motion cues mean.

After participants were ready, they entered into the simulator and sat quietly for three minutes. They were asked to fill out the pre-SSQ questionnaire [8]. Then the simulator began to move, and participants were required to use smart phones to complete the non-driving tasks as shown in Fig. 1 and detailed in Sect. 3.3, as they usually did in moving vehicles, as shown in Fig. 6. Also participants needed to fill out questionnaires for every 1 min of using phone, which included Misery Scale (MISC) [3] and NDT involvement questionnaires. After 35 min simulator ride, participants should finish the post-SSQ and APP evaluation questionnaires before leaving. Once a subject reported a MISC score reaching 7 (meaning moderate nausea) during the ride, the experiment would be ended earlier than 35-minute limit to avoid any negative effects on the participant health.

Fig. 5. Brief flowchart of the experiment

Fig. 6. Participant in the experiment: using a smart phone with or without MOSI APP

4   Results

4.1   Motion Sickness Level

For the total 26 participants, the MISC results in two condition groups are shown in Fig. 7. Repeated measures ANOVA on all MISC values showed no significant effect of condition ($F(1, 25) = 1.694$, $p = 0.199$), but a significant effect of time ($F(29, 725) = 68.81, p < 0.001$). For the No-cue condition, the average MISC when finishing experiment was 3.42 ($SD = 2.30$), while that for the Arrow-cue condition was 2.92 ($SD = 1.47$). Considering the final MISC, the Wilcoxon signed rank test indicated no significant difference between No-cue and Arrow-cue conditions ($z = -1.415$, $p = 0.157$).

Although all participants scored above the 50th percentile on the MSSQ questionnaire, actually a big proportion of participants did not report a high level of actual MISC scoring in the end of experiments. For health consideration, the maximum MISC score could only reach 7, while MISC scores of 3 and below represent only mild motion sickness symptoms. Therefore, the participants having an ending MISC score higher than 3 in any of the two experimental conditions can be regarded as the real sickness-susceptible group. Then a total of 15 susceptible participants were analyzed, and their MISC results are shown in Fig. 8.

Repeated measures ANOVA on all MISC values obtained on 15 participants of susceptible group showed a significant effect of condition ($F(1, 14) = 7.171$, $p = 0.012$), and of time ($F(29, 406) = 78.99$, $p < 0.001$). For the No-cue condition, the average MISC when finishing experiment was 4.87 ($SD = 1.88$), while that for the Arrow-cue condition was 3.80 ($SD = 1.01$). Considering the final MISC, the Wilcoxon signed rank test indicated a significant difference between No-cue and Arrow-cue conditions ($z = -1.98$, $p = 0.047$).

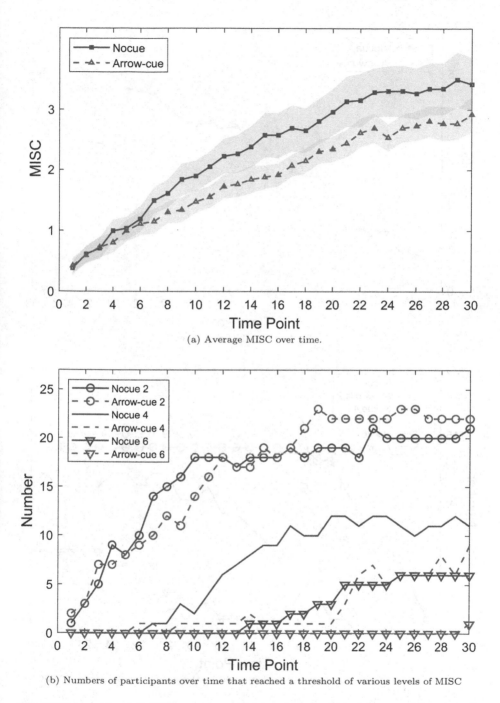

(a) Average MISC over time.

(b) Numbers of participants over time that reached a threshold of various levels of MISC

Fig. 7. Sickness level (MISC) results in No-Cue and Arrow-cue conditions for all 26 participants

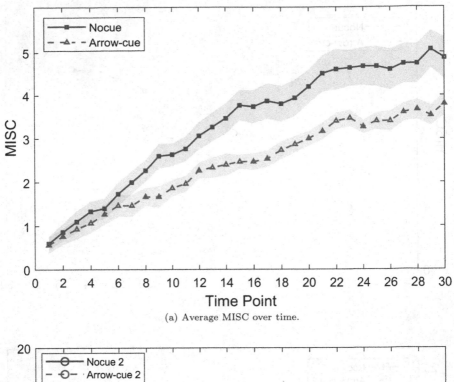

(a) Average MISC over time.

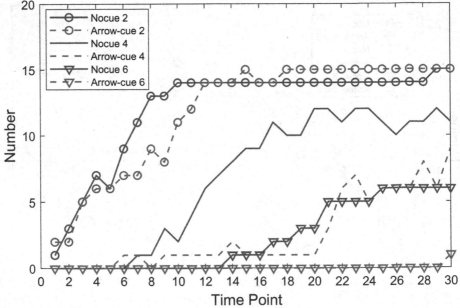

(b) Numbers of participants over time that reached a threshold of various levels of MISC

Fig. 8. Sickness level (MISC) results in No-Cue and Arrow-cue conditions for the susceptible group

In addition to MISC results, the difference between pre- and post-SSQ scores, ΔSSQ, was also used to assess the severity of the participants' motion sickness in two experimental conditions. One participant's post-SSQ record failed, so only the rest of 25 SSQ data were analyzed. Similar to the MISC analysis, for the results of 25 participants as shown in Table 1, there were no significant differences in ΔSSQ scores between two conditions for any of the four constructs. The ΔSSQ score regarding to oculomotor construct in Arrow-cue condition is the same as that in No-cue condition. As for the results of 14 susceptible participants, there was a statistically significant difference between the two conditions in total scores of ΔSSQ ($z = -2.480$, $p = 0.013$). In more details, except for Oculomotor construct, the other two SSQ's constructs (i.e. Nausea(N), Disorientation(D)) were also significantly different between the two conditions.

Table 1. Wilcoxon signed rank test (WSRT) for ΔSSQ with mean and standard deviation between the two conditions (No-cue and Arrow-cue).

ΔSSQ (#)	Condition	Mean(SD)	WSRT
Nausea(25)	No-cue	57.2(45.2)	$z = -1.733, p = 0.083$
	Arrow-cue	44.3(23.8)	
Oculomotor(25)	No-cue	44.0(34.5)	$z = -0.000, p = 1.000$
	Arrow-cue	43.7(26.7)	
Disorientation(25)	No-cue	69.0(68.6)	$z = -1.089, p = 0.276$
	Arrow-cue	59.6(47.3)	
Total Score(25)	No-cue	30.7(26.7)	$z = -1.386, p = 0.166$
	Arrow-cue	26.4(17.7)	
Nausea(14)	No-cue	84.5(42.1)	$z = -2.488, p = 0.013^*$
	Arrow-cue	56.6(23.8)	
Oculomotor(14)	No-cue	62.8(32.9)	$z = -1.344, p = 0.179$
	Arrow-cue	56.3(27.7)	
Disorientation(14)	No-cue	101.4(70.6)	$z = -1.974, p = 0.048^*$
	Arrow-cue	78.5(52.8)	
Total Score(14)	No-cue	45.1(25.8)	$z = -2.480, p = 0.013^*$
	Arrow-cue	34.5(19.1)	

\# The numbers in brackets represent the number of participants;
* Indicates significance, p<0.05.

4.2 NDT Performance

The NDT performance results for 26 participants are shown in Fig. 9. Repeated measures ANOVA on all involvement scores represented no significant effect of condition ($F(1, 25) = 0.823$, $p = 0.369$), but showed significant effect in the time dimension ($F(29, 725) = 23.401$, $p < 0.001$). The motion sickness degree was negatively correlated with the involvement scores. Similarly, repeated measures

ANOVA on all NDT task scores between time points indicated no significant effect of condition ($F(1, 25) = 0.233$, $p = 0.631$) and of time ($F(28, 700) = 0.965$, $p = 0.518$).

The NDT performances for 15 susceptible participants are shown in Fig. 10. Repeated measures ANOVA on all involvement scores represented no significant effect of condition ($F(1, 14) = 1.232$, $p = 0.276$), but showed significant effect in the time dimension($F(29, 406) = 18.918$, $p < 0.001$). The motion sickness degree was negatively correlated with the involvement scores. Similarly, repeated measures ANOVA on all main task scores between time points indicated no significant effect of condition ($F(1, 14) = 1.037$, $p = 0.317$) and of time ($F(28, 392) = 1.072$, $p = 0.366$).

4.3 Subjective Questionnaire

Upon completion of the experiment in the Arrow-cue condition, participants completed a questionnaire on a scale of 1-5 for subjective evaluations on the MOSI APP.

According to the 26 participants' feedback, they thought that the arrow cues were relatively easy to notice ($M = 3.69$, $SD = 0.93$); the arrow cues could relatively deliver accurate vehicle motion to them ($M = 3.62$, $SD = 0.85$); the arrow cues had less impact on the main task ($M = 3.54$, $SD = 0.95$) and the simulator was able to simulate real riding scenarios with sufficient fidelity. For the 15 more susceptible participants, most evaluation results are similar to that in the entire group.

Table 2. Subjective evaluation on MOSI APP: Mean (SD)

Questions	Scoring (1 ~ 5)	26 participants	15 participants
Q1. Is the cue position reasonable?	unreasonable ~ reasonable	3.85(0.88)	3.87(0.92)
Q2. Are the cues easy to notice?	not easy ~ easy	3.69(0.93)	3.53(0.92)
Q3. Is the cue timing appropriate?	too late ~ too early	3.04(0.60)	3.20(0.68)
Q4. Can the cues remind you of accurate motion stimuli?	inaccurate ~ accurate	3.62(0.85)	3.67(0.72)
Q5. Is there any side effect of the cues when spelling?	yes ~ not at all	3.54(0.95)	3.33(0.98)
Q6. Is the cognitive cost of the cues high?	high ~ low	3.23(1.24)	3.33(1.40)
Q7. Do the cues change frequently?	frequent ~ infrequently	3.12(0.71)	2.93(0.59)
Q8. Can the simulator simulate a real ride scenario?	unreal ~ real	3.58(0.76)	3.67(0.72)

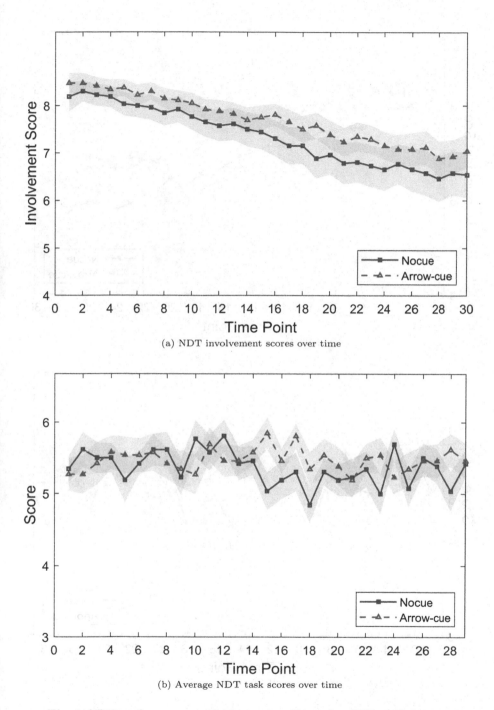

(a) NDT involvement scores over time

(b) Average NDT task scores over time

Fig. 9. NDT performance results in two conditions for all 26 participants

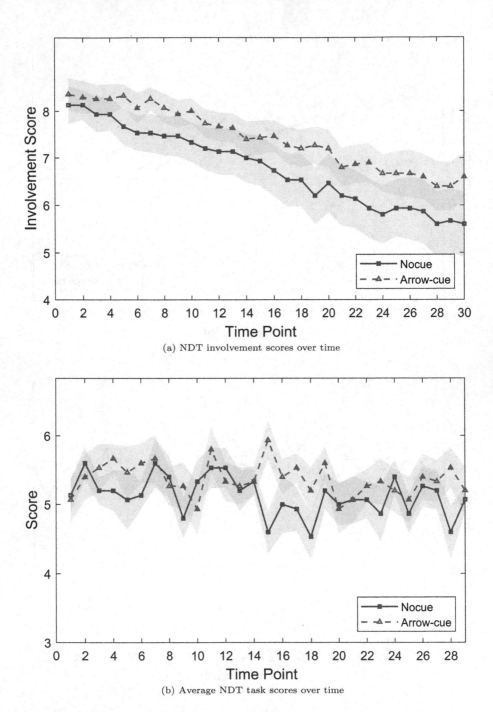

(a) NDT involvement scores over time

(b) Average NDT task scores over time

Fig. 10. NDT performance results in two conditions for the susceptible group

5   Discussion

5.1   Overall Evaluations

According to their MSSQ scores, the total 26 participants can be considered all susceptible to motion sickness, but 11 of them only reported mild motion sickness symptoms in experiments. This is probably due to the shortcomings of the MSSQ questionnaire, which evaluates sickness susceptibility largely based on participants' past experiences but may not accurately reflect the participants' current level of motion sickness susceptibility in specific ride conditions.

For the 26 participants, the APP did not significantly reduce the level of motion sickness, but it did delay the aggravation of motion sickness. For the defacto sickness-susceptible group of 15 participants, their MISC results indicate that the MOSI APP had a satisfactory effect of delaying and relieving motion sickness for more susceptible passengers.

In terms of SSQ results, the difference in ΔSSQ was significantly lower in the Arrow-cue condition for the sickness-susceptible group. But the O-class ΔSSQ scores in the Arrow-cue condition were not significantly different from the No-cue condition, although the mean values of all the participants were reduced. Oculomotor discomfort symptoms mainly include eyestrain, difficulty in focusing, blurred vision and headache. As seen in Table 2, the mean score of Q7 for all participants is 3.12, the cue refreshing may have increased eye fatigue, thus making no significant difference between the two conditions for O-class SSQ scores.

In general, participants' NDT performance decreases as their motion sickness levels increase [2]. However, our results showed that with the increase of MISC rating, the spelling performance did not significantly decrease, while their involvement scores decreased obviously. This means that the MOSI APP did not negatively affect the participants' NDT. There was a same result in the participants' NDT involvement. And it is notable that the mean value of NDT involvement in the Arrow-cue condition was slightly higher than that in the No-cue condition. In other words, the MOSI APP did not negatively affect the participants' NDT and plays a role in increasing the user's involvement because it mitigated the participants' motion sickness. For the current NDT task of spelling, the performance did not significantly reduce. However, the mean score of Q5 in Table 2 for all participants was 3.54, meaning that the APP still had some side effects on the NDT. If the NDT becomes more difficult and requires more cognitive workload, further study is needed to assess the APP's negative impacts. According to the subjective evaluation on the APP, the APP's motion cues were easy to be noticed by users even when they were doing NDT. Such cues can give them more accurate motion information, while the additional cognitive costs and work burden were low.

To sum up, the experiments have validated the positive benefits of MOSI APP: 1) it was significantly effective for sickness-susceptible participants; 2) it could successfully defer the aggravation process of sickness; 3) it had no degrading effect on the users' NDT task; 4) users reported that it could accurately provide cues for the vehicle motion.

5.2   Limitations and Possible Improvements

The experiment was conducted in a simulator rather than in a real vehicle. Although the participants thought that the simulator could simulate the real riding experience (the mean score of Q8 in Table 2 was 3.58), the magnitude and frequency of stimuli were still limited due to simulator stroke constraints. Again, it is still needed to further assess the APP's performances in relieving motion sickness in real road tests.

For rigorous experiments, here the arrow cues' showing position, size and pattern were fixed for all subjects. Such setting is unlikely to be favored by everyone, and if users do not accept such setting, it is impossible to show whether the cues can contribute to relieving their motion sickness. In theory the cue position and pattern can affect users' perception and cognition about cueing. Our method has shown effectiveness in sickness mitigation as previous research, e.g. peripherally-located bubbles [14], or a ball-and-spring Gizmo on the right [7]. Differently from them, our design adopts an intuitive metaphor of traffic light and direction lamp, which are both driving or car ride related, and thus is easier to be understood. The arrow cues are placed in the screen center and laid under non-driving task's interface, which can improve the ability to convey clear and precise information about vehicle movements, especially the acceleration magnitude in challenging scenarios. However, the personalization of cue presentation, including cue pattern, motion mapping and position, may have further potential in sickness mitigation, and thus needs more investigation before real applications.

Additionally, the provision of motion information cues is intended to generate motion anticipation for users. The cues still take time to be understood since they are provided, so the timing of cue onset can also be an interesting research question. It has been shown that audio cues [9] or vibration cues [11] that precede real motion can be effective in reducing occupant motion sickness. If such motion information can be obtained in advance, it is also worth investigating whether our design can further improve the sickness mitigation effect.

6   Conclusion

A smart phone application, MOSI APP, was designed and its effects of motion sickness mitigation were validated through driving simulator experiments. Note that it was developed based on Android operating system, so in theory it could also be easily migrated to different personal electronic devices (e.g. smartphones, tablets) or vehicle infotainment systems. In the preliminary experiments of this study, the vehicle movement information required by MOSI APP was from an acceleration sensor on the simulator cabin. With the development of vehicle intelligence and connectivity, this motion information is hopefully available from on-board cabin systems, maybe more detailed and real-time, even with advance prediction and planning of vehicle movements.

References

1. Bohrmann, D., Bengler, K.: Reclined posture for enabling autonomous driving. In: Ahram, T., Karwowski, W., Pickl, S., Taiar, R. (eds.) Human Systems Engineering and Design II, vol. 1026, chap. 27, pp. 169–175. Springer International Publishing, Cham (2020). https://doi.org/10.1007/978-3-030-27928-8_26
2. Bos, J.E.: Less sickness with more motion and/or mental distraction. J. Vestib. Res. **25**(1), 23–33 (2015). https://doi.org/10.3233/VES-150541
3. Bos, J.E., MacKinnon, S.N., Patterson, A.: Motion sickness symptoms in a ship motion simulator: effects of inside, outside, and no view. Aviat. Space Environ. Med. **76**(12), 9 (2005)
4. Ekchian, J., et al.: A high-bandwidth active suspension for motion sickness mitigation in autonomous vehicles. In: SAE 2016 World Congress and Exhibition, pp. 2016–01-1555 (2016). https://doi.org/10.4271/2016-01-1555
5. Golding, J.F.: Motion sickness susceptibility questionnaire revised and its relationship to other forms of sickness. Brain Res. Bull. **47**(5), 507–516 (1998). https://doi.org/10.1016/S0361-9230(98)00091-4
6. Griffin, M.J., Newman, M.M.: Visual field effects on motion sickness in cars. Aviat. Space Environ. Med. **75**(9), 739–748 (2004)
7. Hanau, E., Popescu, V.: Motionreader: Visual acceleration cues for alleviating passenger e-reader motion sickness. In: Proceedings of the 9th International Conference on Automotive User Interfaces and Interactive Vehicular Applications Adjunct, pp. 72–76. ACM, Oldenburg Germany (2017). https://doi.org/10.1145/3131726.3131741
8. Kennedy, R.S., Lane, N.E., Berbaum, K.S., Lilienthal, M.G.: Simulator sickness questionnaire: an enhanced method for quantifying simulator sickness. Int. J. Aviat. Psychol. **3**(3), 203–220 (1993). https://doi.org/10.1207/s15327108ijap0303_3
9. Kuiper, O.X., Bos, J.E., Diels, C., Schmidt, E.A.: Knowing what's coming: anticipatory audio cues can mitigate motion sickness. Appl. Ergon. **85**, 103068 (2020). https://doi.org/10.1016/j.apergo.2020.103068
10. Leung, A.K., Hon, K.L.: Motion sickness: an overview. Drugs Context **8**, 1–11 (2019). https://doi.org/10.7573/dic.2019-9-4
11. Li, D., Chen, L.: Mitigating motion sickness in automated vehicles with vibration cue system. Ergonomics pp. 1–13 (2022). https://doi.org/10.1080/00140139.2022.2028902
12. Li, D., Hu, J.: Mitigating motion sickness in automated vehicles with frequency-shaping approach to motion planning. IEEE Robot. Autom. Lett. **6**(4), 7714–7720 (2021). https://doi.org/10.1109/LRA.2021.3101050
13. Li, D., Xu, B., Chen, L., Hu, J.: Automated car-following algorithm considering passenger motion sickness. In: The 6th International Symposium on Future Active Safety Technology Toward zero traffic accidents (FAST-zero'21). Kanazawa, Japan (2021)
14. Meschtscherjakov, A., Strumegger, S., Trösterer, S.: Bubble margin: Motion sickness prevention while reading on smartphones in vehicles. In: Lamas, D., Loizides, F., Nacke, L., Petrie, H., Winckler, M., Zaphiris, P. (eds.) Human-Computer Interaction – INTERACT 2019, vol. 11747, chap. 39, pp. 660–677. Springer International Publishing, Cham (2019). https://doi.org/10.1007/978-3-030-29384-0_39

15. Miksch, M., Steiner, M., Miksch, M., Meschtscherjakov, A.: Motion sickness prevention system (msps): Reading between the lines. In: Adjunct Proceedings of the 8th International Conference on Automotive User Interfaces and Interactive Vehicular Applications, pp. 147–152. ACM, Ann Arbor MI USA (2016). https://doi.org/10.1145/3004323.3004340
16. Reason, J.T.: Motion sickness adaptation: a neural mismatch model. J. R. Soc. Med. **71**(11), 819–829 (1978). https://doi.org/10.1177/014107687807101109
17. Reason, J.T., Brand, J.J.: Motion Sickness. Academic Press, Oxford, England, Motion Sickness. (1975)
18. Salter, S., Diels, C., Herriotts, P., Kanarachos, S., Thake, D.: Motion sickness in automated vehicles with forward and rearward facing seating orientations. Appl. Ergon. **78**, 54–61 (2019). https://doi.org/10.1016/j.apergo.2019.02.001
19. Stoffregen, T.A., Smart, L.: Postural instability precedes motion sickness. Brain Res. Bull. **47**(5), 437–448 (1998). https://doi.org/10.1016/S0361-9230(98)00102-6
20. Zheng, Y., Shyrokau, B., Keviczky, T., Sakka, M.A., Dhaens, M.: Curve tilting with nonlinear model predictive control for enhancing motion comfort. IEEE Trans. Control Syst. Technol. **30**(4), 1538–1549 (2022). https://doi.org/10.1109/TCST.2021.3113037

BROOK Dataset: A Playground for Exploiting Data-Driven Techniques in Human-Vehicle Interactive Designs

Junyu Liu[✉], Yicun Duan, Zhuoran Bi, Xiaoxing Ming, Wangkai Jin,
Zilin Song, and Xiangjun Peng

User-Centric Computing Group, Ningbo, China
ljunyu381@gmail.com

Abstract. Emerging Autonomous Vehicles (AV) breed great potential
to exploit data-driven techniques for adaptive and personalized Human-
Vehicle Interactions. However, the lack of high-quality and rich data supports limits the opportunities to explore the design space of data-driven
techniques, and validate the effectiveness of concrete mechanisms. Our
goal is to initialize the efforts to deliver the building block for exploring
data-driven Human-Vehicle Interaction designs. To this end, we present
BROOK dataset, a multi-modal dataset with facial video records. We
first brief our rationales to build BROOK dataset. Then, we elaborate
on how to build the current version of BROOK dataset via a year-long
study, and give an overview of the dataset. Next, we present two example
studies using BROOK to justify the applicability of BROOK dataset. We
also believe BROOK dataset can foster an extensive amount of follow-
up studies, and it's available at https://github.com/Junyu-Liu-Nate/
BROOK-dataset.

Keywords: Human-Vehicle Interaction · Datasets · Data-Driven
Techniques

1 Introduction

Recent advances in data-driven techniques (e.g. Neural Networks) breed
an extensive amount of opportunities, to enable adaptive and personalized
Human-Vehicle Interaction (HVI). More interestingly, the emerging trends of
Autonomous Vehicles relax the conventional burdens of driving, and in-vehicle
drivers/passengers are capable to obtain better user experiences through more
complex HVI. Incorporating data-driven approaches can greatly improve user
experiences during driving processes. For instance, unobtrusive monitors of
multi-modal statuses (i.e. by taking alternative data sources as input), rather
than directly equipping biosensors, are more user-friendly as data sources for
personalized interaction between drivers and vehicles [3, 19, 39, 40, 44, 48]. Therefore, combining data-driven techniques with HVI becomes promising in the near
future, and relevant datasets become essential and highly demanded.

© The Author(s), under exclusive license to Springer Nature Switzerland AG 2023
H. Krömker (Ed.): HCII 2023, LNCS 14049, pp. 191–209, 2023.
https://doi.org/10.1007/978-3-031-35908-8_14

Though there are already several datasets for HVI, there are four major limitations of existing datasets. First, the design purposes of existing datasets focus on a specific type of study purposes (e.g. Stress Detection [18], Driving Workload [37] and etc.), which naturally limits their potential due to the narrowly-selected types of data streams. Second, existing datasets only collect information when drivers are in manual mode, which is not suitable for the emerging trends of Autonomous Vehicles. Third, existing datasets don't have sufficient coverage of the contextual data, especially the statistics of Vehicle Statuses; and fourth, existing datasets don't account for advanced bio-sensors (e.g. Eye-Tracking devices), which restricts the volumes of physiological data (Sect. 2).

Therefore, a comprehensive dataset can unleash the opportunities to leverage data-driven designs for HVI and validate their effectiveness in practice. Data-driven designs usually demand large-scale datasets as the building block, for training, testing, and validating their performance and efficiency. Our goal in this work is to initialize constructing such a building block to stimulate diverse types of explorations, and allow reasonable comparisons across different designs. We believe that achieving such a goal is two-folded. First, we aim to build such a dataset as comprehensively as possible; and second, to justify the applicability of such a dataset via concrete demonstrations of example studies.

To this end, we present BROOK[1] dataset, a multi-modal and facial video dataset for data-driven HVI. The aim of BROOK dataset is to provide a **publicly available**[2] and **large-scale** dataset within the domain of HVI. The outcome of building BROOK dataset is the most comprehensive dataset in HVI to date, which contains much more kinds of data streams than previous datasets. The key idea, behind BROOK, is to **map as many kinds of data flows as possible in the same timeline**, when participants are under both manual and fully-automated driving scenarios. Such guiding philosophy allows BROOK to retrieve statistics from 34 drivers in 11 dimensions (i.e. 4 physiological and 7 contextual). We provide sufficient details regarding how BROOK is organized, and how we build BROOK dataset, via a year-long study. We also provide detailed descriptions and release the sources of all relevant software, when we are building BROOK dataset[3] (Sect. 3).

To justify the applicability of BROOK dataset, we design and perform two example studies to demonstrate the potential of BROOK dataset. First, we quantitatively demystify complicated interactions between driving behaviors and driving styles, and provide an adaptive approach to handle such complexity. We achieve this through Self-Clustering algorithms over BROOK dataset, by dynamically identifying the number of clusters and merging data points accordingly. Second, we identify the tradeoffs between different privacy-preserving techniques in the context of Internet-of-Vehicles. By exploiting four state-of-the-art

---

[1] The name "BROOK" refers to multiple water flows, which stands for the key idea of our dataset.

[2] The dataset is available at https://github.com/Junyu-Liu-Nate/BROOK-dataset.

[3] Software supports for BROOK dataset is released with the works [38,45]. We did attempt to use [39], but failed to do so due to the limited computational power.

Differential Privacy mechanisms, we retrieve key observations in terms of accuracy-privacy tradeoffs among different mechanisms. We believe BROOK dataset can be applied to more contexts beyond the above two examples, by taking different types of data streams as primary sources (Sect. 4).

We highlight the most significant results/findings from our two example studies. For our first example, we justify that the Self-Clustering algorithm can quantitatively classify diverse combinations of different driving styles in an adaptive manner. And for our second example, we identify the need of co-designing the applications and privacy-preserving techniques for data utility and privacy, in the context of Internet-of-Vehicles. The above results suggest the potentials of both BROOK dataset and example studies are significant, and we conjecture that future characterizations and explorations of BROOK are promising.

We position BROOK dataset as a starting point for data support of adaptive and personalized Human-Vehicle Interactions, rather than a definitive complete answer. This is because our study has several limitations, which are incapable to be resolved by a single research group. We envision that a community-wide collaboration is essential for building a complete dataset.

We make the following three key contributions in this paper:

- We present BROOK dataset, a multi-modal and facial video dataset which consists of 34 real-world drivers and their statistics in 11 dimensions. The coverage of BROOK includes driving statuses, multi-modal statistics, and video recordings during both manual and autonomous driving procedures.
- We present two example studies, to demonstrate the potential of BROOK dataset. Our empirical results confirm the quality of BROOK dataset, and suggest novel insights for using data-driven techniques in Human-Vehicle Interaction.
- We summarize important lessons when building and using BROOK dataset, and discuss potential directions for more usage of BROOK dataset. We hope that these lessons can help with future efforts to build or leverage BROOK-like datasets for Human-Vehicle Interaction.

The rest of this paper is organized as follows. Section 2 presents the background and motivation to build BROOK dataset. Section 3 gives an overview and presents details on how we collect statistics for BROOK dataset. Section 4 presents two example studies.

2   Background and Motivation

In this section, we provide background and motivation of building BROOK dataset. We first identify the growing importance of data-driven techniques in Human-Vehicle Interaction (Sect. 2.1). Then, we briefly elaborate on existing datasets (i.e and their use scenarios) for Human-Vehicle Interaction, and identify their design purposes accordingly (Sect. 2.2). Finally, we comprehensively compare the detailed compositions of existing datasets and our proposed BROOK dataset, to justify our novelty (Sect. 2.3).

2.1   Data-Driven Techniques for Human-Vehicle Interaction

With the emerging trend of Autonomous Vehicles in the industry, research interests have grown significantly in Human-Vehicle Interaction (HVI) field, trying to tackle various problems for the coming era. A recent summary categorizes existing studies into two general types, according to manual and autonomous driving contexts respectively [26]. The underlying essence of HVI study is to understand both drivers' states and the temporal events outside the vehicle, by leveraging multiple data-driven techniques. Current data-driven techniques are designed for various purposes, such as (1) tracking driver attention [4,49]; (2) analyzing driving styles [20,29,46]; (3) analyzing effects of situations outside the vehicle [13,36,50];(4) predicting drivers' drowsiness, attention, and emotion through facial expressions or eye gazes [12,22,31,32]; (5) monitoring driver's behaviors regarding distractions and temporal performance by utilizing body-based monitoring mechanisms [1,5]; (6) detecting drivers' emotion variation (e.g. neutral, sadness and happiness) [14,25,33]. To sum up, current data-driven techniques for HVI study have introduced valuable insights towards adaptive and personalized HVI systems so far, yet it is essential to explore more advanced techniques for the field with more diversified data inputs.

2.2   Existing Datasets for Human-Vehicle Interaction

There are several publicly available datasets for advanced Human-Vehicle Interaction Research. Hereby, we summarize five key datasets and discuss their application scenarios. Table 1 presents the comparison between existing datasets and BROOK dataset (our proposal). Distracted Driver Dataset [11] aims to provide a solid foundation for studies regarding stress detection and posture classification, during the driving procedure; brain4cars [21] aims to enable design exploration in terms of Maneuver Estimation; MIT driverDB [18] provides relatively more types of data streams for detailed characterizations of stress detection; HciLab [37] aims to measure driving workloads and provides empirical studies for proof-of-concepts; and AffectiveROAD [16] targets personalized Human-Vehicle Interactions, by providing relevant monitors of affective states. All the above datasets are purpose-dependent and there are certain limitations for exploring more design spaces, since their design purposes are deterministic and centralized on particular domains. Different from the above datasets, the goal of BROOK is to provide a comprehensive dataset to exploit data-driven techniques for Adaptive and Personalized Human-Vehicle Interaction.

2.3   Detailed Comparisons Among Existing Datasets and BROOK Dataset

We first brief different types of data streams within the above datasets and BROOK dataset. Table 2 presents the detailed comparisons between existing datasets and BROOK dataset. Compared to BROOK dataset, existing datasets share two major limitations: (1) most datasets are designed for study-specific

Table 1. A comparison, in terms of design purposes, between BROOK dataset and existing datasets for Human-Vehicle Interaction.

Dataset	Design Purposes
Distracted Driver Dataset [11]	Stress Detection & Posture classification
brain4cars [21]	Maneuver Estimation
MIT driverDB [18]	Stress Detection
HciLab [37]	Driving Workload Measurement
AffectiveROAD [16]	Affective State Monitoring
BROOK (Our Dataset)	Data-driven Techniques in Human-Vehicle Interaction

Table 2. A comparison, in terms of contained data types, between BROOK dataset and existing datasets for Human-Vehicle Interaction. Those types, which can only be accessed from BROOK, are underlined. Note that EDA stands for Electrodermal Activity, ECG stands for Electrocardiogram, EMG stands for Electromyogram, and HR stands for Heart Rate.

	Number of Drivers	Driving Mode	Data Types	
			Driver Physiological Data	Contextual Data
Distracted Driver Dataset [11]	31	Manual	10 classes of driver postures	/
brain4cars [21]	10	Manual	Facial Video	GPS, Vehicle Speed, Road Video
MIT driverDB [18]	24	Manual	EDA, ECG, EMG, HR, Facial Video	Road Video
HciLab [37]	10	Manual	EDA, ECG, Body Temperature, Facial Video	In-vehicle Brightness, Vehicle Acceleration, Road Video
AffectiveROAD [16]	10	Manual	EDA, Body Temperature, HR, Hand Movement, Facial Video	Luminance, Temperature, GPS, Pressure, Humidity, Sound, Road Video
BROOK(Our Dataset)	34	Manual, **Automated**	HR, EDA, **Eye-Tracking**, Facial Video	Vehicle Speed, Vehicle Acceleration, Vehicle Coordinate, Distance of Vehicle Ahead, **Steering Wheel Coordinates**, **Throttle Status**, **Brake Status**, Road Video

purposes, which might be restrictive for exploring novel designs; and (2) most datasets focus on the monitoring of drivers' physiological states and analysis of affective states, but the important statistics of driving context are relatively overlooked. As Table 2 shows, existing publicly available datasets fall short of contextual information, which would be critical for the development of adaptive and personalized HVI systems for complicated driving scenarios.

We describe the compositions of existing datasets in detail. Most of them are designed to study drivers' affective states during driving procedures. The MIT driverDB dataset [18] collects drivers' physiological data (Electrodermal Activity (EDA), Electrocardiogram (ECG), Electromyogram (EDM) and etc.), together with in-vehicle and in-field videos from 20 mi of driving between the city and the highway (i.e. in the Greater Boston area). HciLab dataset [37] has similar driver physiological data (EDA,ECG, Body Temperature, and Facial Video) and different contextual data (In-vehicle Brightness, Vehicle Acceleration, and Road Video), which are collected from 10 drives in highways and freeways in Germany. AffectiveROAD dataset [16] obtains its data from 13 drives performed by 10 drivers, which covers EDA, Body Temperature, Heart Rate, Hand Movement, and Facial Video for driver statistics, with multiple situational data, such as Luminance, Temperature, Pressure, Humidity and etc. Both HciLab dataset and AffectiveROAD dataset include a self-scored, self-reported stress level from the driver after the experiments. A similar dataset to AffectiveROAD, brain4cars [21], contains drivers' facial video, road video, vehicle GPS information and a speed logger about 10 drivers with different kinds of driving maneuvers across two states, who drive their private cars for 1180 mi between natural freeways and cities. Distracted Driver dataset [11] mainly contains 10 different types of driver postures, such as safe driving, talking to the phone, adjusting radio and etc., which are collected from 31 participants.

Compared with existing datasets, BROOK dataset has three major differences: (1) the above datasets are mainly designed for driver stress level monitoring, while BROOK can support more diverse applications such as driver-style classifications; (2) the above datasets only contain manual mode driving data, while BROOK have both manual mode and auto mode driving statistics; and (3) BROOK collects more types of vehicle data (e.g. steering wheel positions, brake and throttle status) from more drivers (i.e. 34 drivers in total), which not only provides more contextual information in the driving process but also guarantees the diversity of driving statistics.

3   BROOK Dataset

In this section, we introduce BROOK dataset in detail. We first give an overview of BROOK dataset, regarding its high-level characteristics and compositions. Then we introduce the hardware and software for collecting BROOK dataset. Next, we highlight novel features under our consideration, when we are building BROOK dataset. Finally, we present the procedure of data collection for BROOK dataset.

3.1   An Overview of BROOK Dataset

The current version of BROOK dataset has facial videos and many multi-modal/driving status data of 34 drivers. Statistics of each participant are collected within a 20-minute in-lab study, which is separated into manual mode and automated mode. Figure 1 shows 11 dimensions of data for the current BROOK dataset, which includes: **Facial Video, Vehicle Speed, Vehicle Acceleration, Vehicle Coordinate, Distance of Vehicle Ahead, Steering Wheel Coordinates, Throttle Status, Brake Status, Heart Rate, Skin Conductance, and Eye Tracking.**

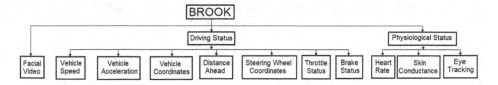

Fig. 1. The Composition of the Current BROOK dataset.

We follow the guiding principle while building BROOK dataset, so that we map all kinds of data streams to the same time flow. To summarize our collection, we use the relationships between data types and data sizes for demonstration. Table 3 shows the main characteristics of BROOK. Note that Driving Status includes Vehicle Speed, Vehicle Acceleration, Vehicle Coordinates, Distance of Vehicle Ahead, Steering Wheel Coordinates, Throttle Status, and Brake Status.

Table 3. A Summary of the Current BROOK dataset.

Data Types	Facial Video	Heart Rates	Skin Conductance	Eye-tracking	Driving Status
Size	273 GB	13.1 MB	48	27.2 GB	60.9 MB

3.2 Hardware and Software Supports

To build BROOK dataset, we combine a diverse set of hardware and software and make intensive implementations to facilitate our needs. For hardware, we apply different sensors to collect relevant data streams: (1) for Facial Video, we use an ultra-clear motion tracking camera; (2) for Hear Rates, we use a heart rate sensor; (3) for Skin Conductance, we use a leather electrical sensor; and (4) we use a Tobii eye tracker for Eye-tracking. Detailed sensor types and sample rates are shown in Table 4. As for detailed driving status, pressure sensors for throttle and brake have been integrated, and steering wheels have been monitored to obtain spatial changes from drivers during the whole period.

As for software, we use OpenDS [15,42], an open-source cognitive driving simulator, for driving scenario generations. OpenDS is widely used for in-lab simulations for both manual and automated driving. Also, we utilize ErgoLAB [43] to coordinate sensor data output for synchronization and storage. Finally, we build an in-framework tool to stream out speed and coordinates as driving status.

Table 4. Details about different sensors used when collecting BROOK dataset.

Data Types	Device Type	Frequency
Facial Video	Goldensky Webcam	30 Hz
Heart Rate	POLAR H10 Heart Rate Monitor	21 Hz
Skin Conductance	NeXus Skin Conductance Sensor	40 Hz
Eye-Tracking Data	Tobii eye tracker	60 Hz

3.3 Key Features only in BROOK

BROOK has some outstanding features, compared with existing datasets. Hereby, we summarize all three unique features of BROOK in the following paragraphs, which are Automated Mode, Region-based Customization, and Eye-Tracking Statistics.

1. Automated Mode. As shown in Table 2, BROOK has collected all types of data streams from both manual and automated driving processes. For automated driving, the reported statistics are collected during fully-automated driving processes for all participants. This is because, with the emerging trends of Autonomous Vehicles (AV), it's important to supply relevant data streams for different studies. For example, [6] suggests the need for extensive characterizations of drivers' perceived trust in AV.

2. Region-based Customization. To provide comprehensive coverage of all possible cases, we customize our scenarios [38] and scenes [45] based on the common patterns of roads in China, as shown in Table 5. These patterns are derived from our collaborations with local transportation authorities. During the procedure of data collection, all participants have experienced 4 types of driving scenarios. Note that all participants during this study have held issued driving licenses from transportation authorities, with sufficient driving experience.

Table 5. The configurations of experimental driving scenarios and scenes.

	The Number of of Lanes	The Number of Cars
1st	4	32
2nd	4	60
3rd	8	60
4th	8	92

3. Eye-Tracking Statistics. We have collected drivers' Eye-Tracking statistics via a Tobii Eye-Tracking device. The eye-Tracking analysis leverages the fixations and pupil-size variation of each driver, to study diverse types of drivers' behaviors. It's also a widely-recognized assumption that fixations and pupil-size variations are important indicators for driver attention and emotional states. In addition, Eye-Tracking statistics can be combined with road videos to locate region-of-interests/object-of-interests in the scenes, which can not only profile drivers' body movements in a new perspective but also provide important data to improve the situation awareness of Human-Vehicle Interaction systems.

3.4 Data Collection Procedure

In this section, we illustrate the whole procedure (i.e. a year-long study) while collecting BROOK database. We first prepare the necessary resources and tests to ensure the bulk of study runs smoothly. Then we perform the bulk of the study to collect necessary data streams, in various settings and modes separately.

For preparations, participants are first briefed about the study and are screened by a questionnaire to ensure that, they are at low risk of motion sickness while experiencing the driving simulation. After the introduction, participants are given five minutes to sit in the simulator by themselves and adapt to the environment. The study only begins when we ensure that participants understand the brief and are comfortable with the simulator.

The bulk of this study is as follows. The experiment includes three driving conditions, as explained in previous sections. Each participant begins with the baseline manual mode, with drivers in control of the vehicle operations. Then, they undertake the two modes of autopilot driving (L3 autonomous driving level), including both standard and personalized versions[4]. The two auto-driving versions are completed in a counterbalanced order. During the study, driving statistics are logged. After each driving session, the participants were asked to fill out a questionnaire about the cognitive aspects of their experience, including an assessment of perceived trust, comfort, and situational awareness[5]. Note that these are not collected for BROOK dataset, and are only available upon request. The study of each participant lasts approximately one hour.

4 Example Studies of BROOK Dataset

To justify the potential application fields and the feasibility of our design purposes, we present a representative example study. In the first case study, we have built an in-vehicle multi-modal predictor to estimate drivers' physiological data via facial expressions only. Our second case is the first attempt to characterize and study the feasibility and quality of Differential Privacy-enabled data protection methods in IoV. Through these two example studies, we are confident that BROOK could both support real-world applications, and open up potential research directions in HVI (and possibly beyond).

---

[4] Determined by participants' behaviors in manual mode.
[5] An early use of this dataset was for [40].

4.1   Case 1: Applying Self-clustering Methods for Adaptive Driving Style Characterizations

The first case study focuses on quantitatively classifying complicated interactions between driving styles and driving behaviours, in an adaptive way [51]. Conventional perspectives, through static partitions of driving styles, might oversimplify the classifications of driving styles. Our goal is to obtain diverse driving styles, derived from adaptive classifications of driving behaviors. To this end, we apply self-clustering algorithms to examine adaptive classification of driving styles, and how driving behaviors and styles interact in this case. We first elaborate our motivation for this study. Then we provide the detailed methodology of this study. Next, we introduce the key design choices of this study. Finally, we present key observations from the experimental results and discuss major takeaways.

Motivations. Driving styles are sets of classified driving behaviors, based on relevant activities and their measurements. According to an early summary [41], driving styles are classified into eight main categories: dissociative, anxious, risky, angry, high-velocity, distress reduction, patient, and careful. Though such classifications of different driving styles are already complicated (i.e. including the Choices of Speed, Brake Status, and etc. [10]), variations between different time intervals have been generally overlooked. Moreover, the methodologies and mechanisms to enable adaptive classifications of driving styles remain understudied. Our goal is to study such variations via a novel perspective, by equalizing clusters as driving styles. To relax the constraints of conventional approaches (i.e. predetermined number of styles), we introduce Self-Clustering Algorithms as the classifications of Driving Styles. We proof-of-concept such approaches using DBSCAN Self Clustering algorithm, and present empirical observations and takeaways.

Our Methodology: Adaptive Classifications via Self-Clustering Algorithms The driving style characterization is mainly determined by vehicle statistics in different time spots. In our experiment, we have chosen most of the vehicle data in the BROOK dataset as the source input, such as Vehicle Acceleration Rate and Steering Wheel positions. We first **pre-process Data** to reduce the noisy points for further clustering. In this stage, we apply three main methods and we elaborate them as follows: (1) we eliminate the start-up and parking stages from all data records, since our focus is to analyze the driving styles and behaviors under a stable and consecutive driving procedure; (2) we apply Min-Max Normalization [34] to regularize all statistics for a uniform format across all data streams, since the primary units and scales of the different kinds of data streams may vary greatly; and (3) we apply Principal Component Analysis (PCA) [28] to retrieve principal components from the regularized, multi-dimensional data streams.

We then perform **Self-Clustering via DBSCAN Algorithm.** To traverse all potential driving styles, we apply DBSCAN, a representative self-clustering algorithm, to enable adaptive classifications of driving styles, according to relevant driving behaviors. DBSCAN finds core samples of high density in the given dataset, and expands clusters from them. Unlike conventional approaches like K-Means, DBSCAN doesn't require a predetermined number of clusters but only takes Distance functions and Parameter Settings as input parameters.

Design Choices. Since DBSCAN can accept multiple kinds of Distance Functions, hereby we first elaborate our key design choices in terms of it. There are many kinds of widely-used distance functions, such as Euclidean Distance, Manhattan Distance and etc. Since our study aims to validate the potentials of such approaches (i.e. as proof-of-concepts), we use Euclidean Distance as our attempt. We believe it's promising to vary different Distance Functions and validate their differences, which can substantially deepen the computational understanding of relationships between driving styles and behaviors.

As for Parameter Settings, DBSCAN takes *minPoints, epsilon* as parts of Parameter Settings: (1) *minPoints* refers to the minimum numbers of data points within a cluster, and (2) *epsilon* (eps) refers to the max radius of the cluster. In this case, we follow the recommended settings and vary them slightly to observe their differences. As Table 6 shown, we attempt nine different combinations of relevant settings. We scale *eps* from 0.125 to 0.5, and scale *minPoints* from 1 to 9. Since the results suggest there are negligible differences within our clustering results, we leverage representative cases, in some scenarios, and perform the analysis.

Table 6. Groups of parameters and the corresponding clustering results.

eps	0.125	0.125	0.125	0.25	0.25	0.25	0.5	0.5	0.5
minPts	1	6	9	1	6	9	1	6	9
clusters numbers	8	4	3	6	4	3	6	4	3

Empirical Results and Observations. We introduce details about Sub-Figure 2(a): the x-axis refers to Time, the y-axis refers to Feature (i.e. the most Principal Component) and the z-axis refers to the clusters. Sub-Figure 2(b) and 2(c) presents relevant 2D planes from the presented 3D figure. We make the following three observations.

Observation 1: Feasibility for Temporal Characterizations and Adaptive Mechanisms for Driving Styles. Figure 2(a) shows the three-dimensional clustering results of all drivers when DBSCAN parameters *eps*, *minPts* are 0.5 and 9 respectively. We obtain two facts. First, the combinations of all clusters cover the whole timeline. Second, Sub-Fig. 2(c) confirms such coverage can be achieved through only the most significant feature, without including every stream of multi-modal statistics. These facts show us the needs of Temporal Characterizations and Adaptive Mechanisms for Driving Styles.

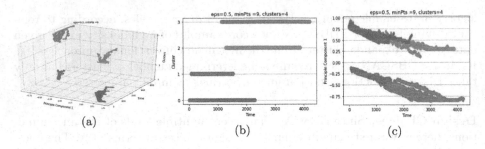

(a)   (b)   (c)

Fig. 2. Sub-Fig. 2(a) present the 3D clustering results of all drivers (i.e. eps=0.5 and minPts=9); Sub-Fig. 2(b) is the mapping in Time-Cluster plane from Sub-Fig. 2(a); Sub-Fig. 2(c) is the mapping in Time-Feature plane from Fig. 2(a).

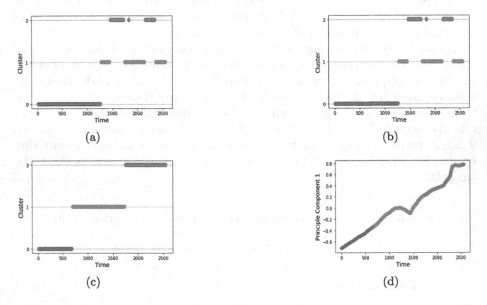

Fig. 3. Concrete Examples to Demystify the Interactions between Driving Styles and Behaviors. Sub-Figures 3(a) present one example driver set; and Sub-Figs. 3(c) present the other example driver set.

Observation 2: Classifications of Driving Styles can be Non-Deterministic. Sub-Figure 2(b) shows the results, when the plane of Time-Cluster is presented in 2D manner. We obtain the fact that, there are overlaps between different driving styles for the specific driver. This fact validates our hypothesis that the conventional partition of driving styles may not reflect these characteristics well.

Observation 3: Personalized Classifications of Driving Styles are Possible. We use concrete examples to elaborate on the possibilities of personalized classifications of driving styles. Figure 3 shows the plane of Time-Cluster from

3D figures, in a breakdown manner. We compare two example driver sets in detail. Sub-Figures 3(a) and 3(b) represent one driver set, and Sub-Figs. 3(c) and 3(d) represent the other one. We obtain the fact that these two driver sets have different driving styles, and the differences are significant. The first driver set (Sub-Figs. 3(a) and 3(b)) has more continuous and smooth driving-style transitions. But the second driver set (Sub-Figs. 3(c) and 3(d)) has more fluctuating and frequently-switched driving-style transitions.

4.2   Case 2: Differential Privacy for the Internet-of-Vehicles

Our second study aims to study the tradeoffs between different Differential Privacy (DP) approaches in the context of Internet-of-Vehicles (IoV) [8,9,27]. Data privacy is a major concern with the emergence of IoV. However, it's still unclear how to deploy efficient privacy-preserving methods for emerging IoV applications. Our goal is to characterize modern privacy-preserving techniques, namely Differential Privacy (DP), for data processing in IoV. We first elaborate our motivations of this study. Then we provide our methodology of this study for data protection using DP. Next, we introduce the design choices of this study. Finally, we summarize the key findings from the results.

Motivations. IoV is a large-scale distributed network that consists of individual vehicles as nodes. There are diverse and frequent communications within IoV, by orchestrating humans, vehicles, things, and environments, with vehicles as the interface. Such a nature of frequent communications and synchronizations, which is the same as the distributed systems, incurs the concerns of privacy issues for data processing (e.g. centralized servers). To resolve such issues, Differential Privacy has been widely-used, which is a mathematically-guaranteed mechanism for protecting private data. However, it's still unclear regarding the tradeoffs when applying such techniques in the context of IoV. To this end, we leverage BROOK dataset to characterize the impacts of DP mechanisms within the era of IoV.

Techniques: Differentially-Private Data Protection. The ϵ-Differential Privacy is achieved by adding random Laplace noises into the dataset. A mathematical expression of a randomized function N_f could be denoted as follows.

Definition 1. *(Randomized Function subject to ϵ-) We regard a randomized function K_f as satisfying ϵ-, assume that $\forall$ dataset D_1 and D_2 with at most one element difference, and $\forall S \subseteq Range(K_f)$, the K_f function obey the below in-equation:*

$$Pr[K_f(D_1) \in S] \leq exp(\epsilon) \times Pr[K_f(D_2) \in S]$$

Note that the value of ϵ is in inverse correlation with the extent of randomization (i.e. small ϵ values indicate powerful randomization).

Study Design. We focus on how to leverage DP to enable privacy-preserving data processing. To achieve this goal, we first build our DP-protected database, with the support of BROOK. We leverage the Heart Rate and Skin Conductivity data in BROOK to construct this toy database, and add random noises to the query operators via DP. In this manner, querying driving statistics in the toy database could be viewed as data processing in IoV. Second, we apply four state-of-the-art query mechanisms (i.e. Fourier [2], Wavelet [47], Datacube [7], and Hierarchical [17]) to perform a list of queries from real-world workloads. To simulate as many representative queries as possible, we consider both All-Range Query and One-Way Marginal Query. Third, we use Absolute Error and Relative Error compared with the queries on raw data as the evaluation metrics to assess DP-enabled processing on sensitive data.

DP Mechanisms for processing. First, we will introduce the four state-of-the-art query mechanisms: 1) Fourier is designed for workloads having all K-way marginals, for any given K. It transforms the cell counts, via the Fourier transformation, and obtains the values of the marginals based on the Fourier parameters; 2) Wavelet is designed to support multi-dimensional range workloads. It applies the Haar wavelet transformation to each dimension; 3) Datacube chooses a subset of input marginals to minimize the max errors of the query answer, with the aim to support marginal workloads; 4) Hierarchical is built for answering range queries. It uses a binary tree structure of queries for such a goal. The first query is the sum of all cells and the rest of the queries recursively divide the first query into subsets.

Dataset Setup and Query Types. To characterize the query errors of DP-enabled processing, we handcraft our dataset with the support of BROOK, and import it into our simulation platform[6]. In this experiment, we use the Heart Rate and Skin Conductivity as a proof-of-concept. Assume the data table is of size $m \times n$, and the constraint condition for cell (i, j) could be written as ρ_{ij}. Then, a $length - (m \times n)$ column vector $\mathbf{x} = (x_{00}, x_{01}, ..., x_{(m-1)(n-1)})^T$ where x_{ij} meets the condition ρ_{ij} could be extracted from the table. In this case, it is equal to $(12, 17, 0, 0, 2254, ..., 11)^T$, a column vector with length of 32. With the existence of this data vector $\mathbf{x}$, it is convenient to answer linear query.

Evaluation Metrics and Configurations of Workloads. We introduce two evaluation metrics to assess query errors respectively, which are Absolute Error and Relative Error. Absolute Error refers to the absolute differences between returned value by DP-enabled query approaches and real value queried on raw data. Note that when query results of DP-enabled queries r and raw query r' are all vectors, the absolute error is $\|r' - r\|_1$. Else, the absolute error is $|r' - r|$. Relative Error is the ratio of absolute error to the real value. The relative error is $|(r' - r)/r|$ when expressed in the above example.

---

[6] We emulate IoV via the remote APIs of modern distributed database systems [24,30]. This ensures our characterizations not to mistake any procedure, by overlooking the impacts of IoV.

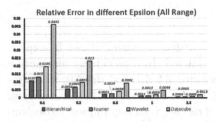
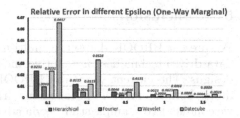

Fig. 4. Relative Error of Four Query Methods.

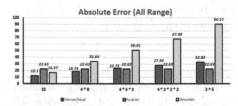
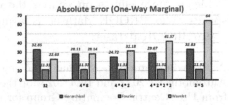

Fig. 5. Absolute Error of Four Query Methods.

Empirical Results & Observations. Figure 4 and Fig. 5 report the Relative Error and Absolute Error of the above four mechanisms for DP-enabled data sharing. We make the following two key observations.

Key Observation 1: different DP-enabled mechanisms achieve different levels of Relative Error. Figure 4 reports the query results of Relative Errors for both All-Range Query and One-Way Marginal Query. We draw two observations. First, Datacube produces more errors than the rest mechanisms in all scenarios, for both All-Range Query and One-Way Marginal Query, which suggests it may not be an ideal choice in practice. Second, Hierarchical is the optimal method in All-Range Query, but Fourier outperforms it in One-Way Marginal Query, for any ϵ settings.

Key Observation 2: different DP-enabled mechanisms achieve different levels of Absolute Error. Figure 5 shows the query results of Absolute Errors for both All-Range Query and One-Way Marginal Query. We draw four observations. First, Hierarchical and Wavelet both incur more errors when the granularity of query structures grows in both query types. Second, Fourier is the most robust mechanism when the granularity of query structures grows in both query types. It also achieves the best results in complex granularity of in All-Range Query and also the best in all the settings of One-Way Marginal Query. Third, Hierarchical obtains better results from query structure 32 and 4×8 in All-Range Query. Four, Wavelet can achieve less query errors compared with Hierarchical with a small granularity of queries in One-Way Marginal Query.

5 Conclusions

We present BROOK dataset, a multi-modal and facial video dataset to exploit data-driven techniques for adaptive and personalized Human-Vehicle Interaction designs. The current version of BROOK contains 34 drivers' driving statistics under both manual and autonomous driving modes. Through BROOK dataset, we conduct two example studies, via advanced data-driven techniques, to demonstrate its potential in practice. Our results justify the applicability and values of the BROOK dataset. We believe, with the emerging trends of Autonomous Vehicles, more efforts on similar datasets are essential to unleash the design spaces. The BROOK dataset is available at https://github.com/Junyu-Liu-Nate/BROOK-dataset.

Acknowledgements. We thank the anonymous reviewers from CHI 2021, CHI 2022, AutomotiveUI 2022 and HCI 2023 for their valuable feedback. We thank all members from User-Centric Computing Group for their valuable feedback and discussions. The position paper of this work was published at [35], and an early preprint of this paper was released at [23]. The first case study is fully exploited at [51], and the second case study is an overview of our past efforts at [8,9,27].

References

1. Abouelnaga, Y., Eraqi, H., Moustafa, M.: Real-time distracted driver posture classification (12 2018)
2. Barak, B., Chaudhuri, K., Dwork, C., Kale, S., McSherry, F., Talwar, K.: Privacy, accuracy, and consistency too: A holistic solution to contingency table release. In: Proceedings of the Twenty-Sixth ACM SIGMOD-SIGACT-SIGART Symposium on Principles of Database Systems, pp. 273–282. PODS '07, Association for Computing Machinery, New York, NY, USA (2007). https://doi.org/10.1145/1265530.1265569, https://doi.org/10.1145/1265530.1265569
3. Bi, Z., Ming, X., Liu, J., Peng, X., Jin, W.: FIGCONs: Exploiting FIne-Grained CONstructs of Facial Expressions for Efficient and Accurate Estimation of In-Vehicle Drivers' Statistics. In: International Conference on Human-Computer Interaction (2023)
4. Borghi, G.: Combining deep and depth: Deep learning and face depth maps for driver attention monitoring (2018)
5. Deo, N., Trivedi, M.M.: Looking at the driver/rider in autonomous vehicles to predict take-over readiness (2018)
6. Dikmen, M., Burns, C.M.: Autonomous driving in the real world: Experiences with tesla autopilot and summon. In: Proceedings of the 8th International Conference on Automotive User Interfaces and Interactive Vehicular Applications, pp. 225–228. Automotive'UI 16, Association for Computing Machinery, New York, NY, USA (2016). https://doi.org/10.1145/3003715.3005465
7. Ding, B., Winslett, M., Han, J., Li, Z.: Differentially private data cubes: optimizing noise sources and consistency. In: Proceedings of the 2011 ACM SIGMOD International Conference on Management of Data, pp. 217–228 (2011)
8. Duan, Y., Liu, J., Jin, W., Peng, X.: Characterizing Differentially-Private Techniques in the Era of Internet-of-Vehicles (2022)

9. Duan, Y., Liu, J., Ming, X., Jin, W., Song, Z., Peng, X.: Characterizing and Optimizing Differentially-Private Techniques for High-Utility, Privacy-Preserving Internet-of-Vehicles. In: International Conference on Human-Computer Interaction (2023)

10. Elander, J., West, R., French, D.: Behavioral correlates of individual differences in road-traffic crash risk: An examination of methods and findings. Psychol. Bull. **113**, 279–94 (04 1993). https://doi.org/10.1037/0033-2909.113.2.279

11. Eraqi, H.M., Abouelnaga, Y., Saad, M.H., Moustafa, M.N.: Detecting stress during real-world driving tasks using physiological sensors. Journal of Advanced Transportation, pp. 156–166 (2019). https://doi.org/10.1155/2019/4125865

12. Fang, J., Yan, D., Qiao, J., Xue, J.: Dada: A large-scale benchmark and model for driver attention prediction in accidental scenarios (2019)

13. Geiger, A., Lenz, P., Stiller, C., Urtasun, R.: Vision meets robotics: the kitti dataset. Int. J. Robot. Res. **32**, 1231–1237 (09 2013). https://doi.org/10.1177/0278364913491297

14. Goeleven, E., De Raedt, R., Leyman, L., Verschuere, B.: The karolinska directed emotional faces: A validation study. COGNITION AND EMOTION 22, 1094–1118 (09 2008). https://doi.org/10.1080/02699930701626582

15. Green, P.A., Jeong, H., Kang, T.: Using an opends driving simulator for car following: A first attempt. In: Boyle, L.N., Burnett, G.E., Fröhlich, P., Iqbal, S.T., Miller, E., Wu, Y. (eds.) Adjunct Proceedings of the 6th International Conference on Automotive User Interfaces and Interactive Vehicular Applications, Seattle, WA, USA, September 17–19, 2014, pp. 4:1–4:6. ACM (2014). https://doi.org/10.1145/2667239.2667295

16. Haouij, N.E., Poggi, J.M., Sevestre-Ghalila, S., Ghozi, R., Jaïdane, M.: Affectiveroad system and database to assess driver's attention. In: Proceedings of the 33rd Annual ACM Symposium on Applied Computing. p. 800–803. SAC '18, Association for Computing Machinery, New York, NY, USA (2018). https://doi.org/10.1145/3167132.3167395, https://doi.org/10.1145/3167132.3167395

17. Hay, M., Rastogi, V., Miklau, G., Suciu, D.: Boosting the accuracy of differentially private histograms through consistency. Proc. VLDB Endow. **3**(1–2), 1021–1032 (Sep 2010). https://doi.org/10.14778/1920841.1920970

18. Healey, J.A., Picard, R.W.: Detecting stress during real-world driving tasks using physiological sensors. Trans. Intell. Transport. Sys. **6**(2), 156–166 (Jun 2005). https://doi.org/10.1109/TITS.2005.848368, https://doi.org/10.1109/TITS.2005.848368

19. Huang, Z., et al.: Face2multi-modal: In-vehicle multi-modal predictors via facial expressions. In: 12th International Conference on Automotive User Interfaces and Interactive Vehicular Applications, pp. 30–33. AutomotiveUI '20, Association for Computing Machinery, New York, NY, USA (2020). https://doi.org/10.1145/3409251.3411716

20. Hooft van Huysduynen, H., Terken, J., Martens, J.b., Eggen, B.: Measuring driving styles: a validation of the multidimensional driving style inventory (09 2015). https://doi.org/10.1145/2799250.2799266

21. Jain, A., Koppula, H.S., Soh, S., Raghavan, B., Singh, A., Saxena, A.: Brain4cars: Car that knows before you do via sensory-fusion deep learning architecture (2016)

22. Jegham, I., Ben Khalifa, A., Alouani, I., Mahjoub, M.A.: A novel public dataset for multimodal multiview and multispectral driver distraction analysis: 3mdad. Signal Processing: Image Communication 88, 115960 (2020). https://doi.org/10.1016/j.image.2020.115960, http://www.sciencedirect.com/science/article/pii/S0923596520301387

23. Jin, W., Duan, Y., Liu, J., Huang, S., Xiong, Z., Peng, X.: BROOK Dataset: A Playground for Exploiting Data-Driven Techniques in Human-Vehicle Interactive Designs. Technical Report-Feb-01 at User-Centric Computing Group, University of Nottingham Ningbo China (2022)
24. Jin, W., Ming, X., Song, Z., Xiong, Z., Peng, X.: Towards Emulating Internet-of-Vehicles on a Single Machine. In: AutomotiveUI '21: 13th International Conference on Automotive User Interfaces and Interactive Vehicular Applications, Leeds, United Kingdom, September 9–14, 2021 - Adjunct Proceedings, pp. 112–114. ACM (2021). https://doi.org/10.1145/3473682.3480275
25. Kamachi, M., Lyons, M., Gyoba, J.: The japanese female facial expression (jaffe) database. Availble: http://www.kasrl.org/jaffe.html (01 1997)
26. Kun, A.L.: Human-machine interaction for vehicles: Review and outlook. Found. Trends® in Human-Comput. Interact. **11**(4), 201–293 (2018). https://doi.org/10.1561/1100000069
27. Liu, J., et al.: HUT: Enabling High-UTility, Batched Queries under Differential Privacy Protection for Internet-of-Vehicles (2022)
28. Ma, Z.: A tutorial on principal component analysis (02 2014). https://doi.org/10.13140/2.1.1593.1684
29. Md Yusof, N., Karjanto, J., Terken, J., Delbressine, F., Hassan, M., Rauterberg, M.: The exploration of autonomous vehicle driving styles: Preferred longitudinal, lateral, and vertical accelerations, pp. 245–252 (10 2016). https://doi.org/10.1145/3003715.3005455
30. Ming, X., et al.: Enabling Efficient Emulation of Internet-of-Vehicles on a Single Machine: Practices and Lessons. In: International Conference on Human-Computer Interaction (2023)
31. Ortega, J.D., et al.: Dmd: A large-scale multi-modal driver monitoring dataset for attention and alertness analysis (2020)
32. Palazzi, A., Abati, D., s. Calderara, Solera, F., Cucchiara, R.: Predicting the driver's focus of attention: The dr(eye)ve project. IEEE Trans. Pattern Anal. Mach. Intell. **41**(7), 1720–1733 (2019). https://doi.org/10.1109/TPAMI.2018.2845370
33. Pantic, M., Valstar, M., Rademaker, R., Maat, L.: Web-based database for facial expression analysis. vol. 2005, pp. 5 pp.- (08 2005). https://doi.org/10.1109/ICME.2005.1521424
34. PATRO, S.G., Sahu, K.K.: Normalization: A preprocessing stage. IARJSET (03 2015). https://doi.org/10.17148/IARJSET.2015.2305
35. Peng, X., Huang, Z., Sun, X.: Building BROOK: A Multi-modal and Facial Video Database for Human-Vehicle Interaction Research (2020)
36. Ramanishka, V., Chen, Y., Misu, T., Saenko, K.: Toward driving scene understanding: A dataset for learning driver behavior and causal reasoning. In: 2018 IEEE/CVF Conference on Computer Vision and Pattern Recognition, pp. 7699–7707 (2018). https://doi.org/10.1109/CVPR.2018.00803
37. Schneegass, S., Pfleging, B., Broy, N., Heinrich, F., Schmidt, A.: A data set of real world driving to assess driver workload. In: Proceedings of the 5th International Conference on Automotive User Interfaces and Interactive Vehicular Applications. p. 150–157. AutomotiveUI '13, Association for Computing Machinery, New York, NY, USA (2013). https://doi.org/10.1145/2516540.2516561, https://doi.org/10.1145/2516540.2516561
38. Song, Z., Duan, Y., Jin, W., Huang, S., Wang, S., Peng, X.: Omniverse-OpenDS: Enabling Agile Developments for Complex Driving Scenarios via Reconfigurable Abstractions. In: International Conference on Human-Computer Interaction (2022)

39. Song, Z., Wang, S., Kong, W., Peng, X., Sun, X.: First Attempt to Build Realistic Driving Scenes Using Video-to-Video Synthesis in OpenDS Framework. In: Adjunct Proceedings of the 11th International Conference on Automotive User Interfaces and Interactive Vehicular Applications, AutomotiveUI 2019, Utrecht, The Netherlands, September 21–25, 2019, pp. 387–391. ACM (2019). https://doi.org/10.1145/3349263.3351497, https://doi.org/10.1145/3349263.3351497
40. Sun, X., et al.: Exploring Personalised Autonomous Vehicles to Influence User Trust. Cogn, Comput (2020)
41. Taubman-Ben-Ari, O., Mikulincer, M., Gillath, O.: The multidimensional driving style inventory-scale construct and validation. Accident Anal. Prevent. **36**(3), 323–332 (2004). https://doi.org/10.1016/S0001-4575(03)00010-1, http://www.sciencedirect.com/science/article/pii/S0001457503000101
42. Team, O.D.: OpenDS - the Flexible Open Source Driving Simulation. https://opends.dfki.de/ (2017)
43. Tech, B.J.: ErgoLab: Human-Machine-Environment Sychronization Platform. Would be released if the paper is accepted
44. Wang, J., Xiong, Z., Duan, Y., Liu, J., Song, Z., Peng, X.: The Importance Distribution of Drivers' Facial Expressions Varies over Time! In: 13th International Conference on Automotive User Interfaces and Interactive Vehicular Applications, pp. 148–151 (2021)
45. Wang, S., et al.: Oneiros-OpenDS: An Interactive and Extensible Toolkit for Agile and Automated Developments of Complicated Driving Scenes. In: HCI in Mobility, Transport, and Automotive Systems: 4th International Conference, MobiTAS 2022, Held as Part of the 24th HCI International Conference, HCII 2022, Virtual Event, June 26-July 1, 2022, Proceedings, pp. 88–107. Springer (2022)
46. Wang, W., Xi, J., Zhao, D.: Driving style analysis using primitive driving patterns with bayesian nonparametric approaches. IEEE Trans. Intell. Transp. Syst. **20**(8), 2986–2998 (2019). https://doi.org/10.1109/TITS.2018.2870525
47. Xiao, X., Wang, G., Gehrke, J.: Differential privacy via wavelet transforms. IEEE Trans. Knowl. Data Eng. **23**(8), 1200–1214 (2010)
48. Xiong, Z., et al.: Face2statistics: user-friendly, low-cost and effective alternative to in-vehicle sensors/monitors for drivers. In: HCI in Mobility, Transport, and Automotive Systems: 4th International Conference, MobiTAS 2022, Held as Part of the 24th HCI International Conference, HCII 2022, Virtual Event, June 26-July 1, 2022, Proceedings, pp. 289–308. Springer (2022)
49. Yang, D., et al.: All in one network for driver attention monitoring. In: ICASSP 2020–2020 IEEE International Conference on Acoustics, Speech and Signal Processing (ICASSP). pp. 2258–2262 (2020). https://doi.org/10.1109/ICASSP40776.2020.9053659
50. Yu, F., et al.: Bdd100k: A diverse driving dataset for heterogeneous multitask learning (2020)
51. Zhang, Y., Jin, W., Xiong, Z., Li, Z., Liu, Y., Peng, X.: Demystifying interactions between driving behaviors and styles through self-clustering algorithms. In: HCI in Mobility, Transport, and Automotive Systems: Third International Conference, MobiTAS 2021, Held as Part of the 23rd HCI International Conference, HCII 2021, Virtual Event, July 24–29, 2021, Proceedings, pp. 335–350. Springer (2021)

Research on Chinese Font Size of Automobile Central Control Interface Driving

Jingfeng Shao[1,2], Zhigang Yang[1], Yanlong Li[1(✉)], Xiaolin Li[2], Yingzhi Huang[2], and Jing Deng[3]

[1] Department of Automobile, Tongji University, Shanghai, China
834542553@qq.com
[2] SAIC Innovation Research and Development Institute, Menlo Park, USA
[3] School of Art Design and Media, East China University of Science and Technology, Shanghai, China

Abstract. This paper constructs a Chinese font size evaluation index of the vehicle central control interface combined with subjective and objective data, which makes up for the lack of ignoring objective data related to driving safety in previous studies, and improves the reliability of font size recommendations. Based on the experimental design method of single factor repeated measurement, the driving data, subjective scale data and task response data are collected. Through correlation analysis and mean value comparison, the reasonable recommended value of Chinese font size is obtained. The results show that there is a significant correlation between the visual subjective score, usability, workload, task response time and speed standard deviation and the font size. The visual subjective score shows that 5.5 mm–7.5 mm is the most suitable font size, and some subjects can also accept 9.5 mm. The usability is positively correlated with the font size within a certain range, and the workload, speed standard deviation and task response time are negatively correlated with the font size, but this negative correlation trend tends to flatten after the font size is greater than 7.5 mm. Therefore, 5.5 mm–7.5 mm is a more appropriate subtitle and text size, and 9.5 mm can be used for first-class titles.

Keywords: Simulated driving · Central control interface · Font size · Usability

1 Introduction

From electronic screen reading of computer and mobile phone in static environment to rapid recognition of logos and characters on screen in dynamic driving environment, selecting appropriate text display size based on human visual characteristics has always been an important and basic subject of ergonomics. In long-term studies, scholars have put forward some standards for the application of the font size of paper media, computer and mobile phone screens based on the consideration of information transmission efficiency, article reading comfort and visual display effect [1, 2]. However, since driving is the main task of drivers, and fonts vary from country to country, in the digital age, The visual standard of automotive central control screen in different countries puts forward higher requirements for rapid recognition of content and meaning.

© The Author(s), under exclusive license to Springer Nature Switzerland AG 2023
H. Krömker (Ed.): HCII 2023, LNCS 14049, pp. 210–223, 2023.
https://doi.org/10.1007/978-3-031-35908-8_15

2 Research Status of Chinese Font Size in Central Touchscreen

The size of text on driving screens still needed to be studied. The main reason lies in the following characteristics in the usage scenario of automobile interactive interface: the distance between the human eye and the screen is relatively fixed, about 70 cm [3]; Users need to quickly browse text fragments and attach importance to visual search efficiency [4]. Most importantly, visual distraction in driving is significantly correlated with the risk of accident liability [5]. As an important source of driving distraction [6], the central touchscreen is directly related to driving safety. Therefore, the design of touchscreen interface is difficult to directly copy the traditional design specifications, but should form its own standards based on experiments. Some Chinese companies, such as Baidu and Huawei, have released their own visual white papers, which include the standards of font size, but do not disclose the evaluation methods. In existing studies, English fonts are mostly used as the experimental medium, and there are few academic studies on Chinese font size, while characters with different densities generally require different font sizes. For example, Japanese requires a larger font size to display information compared with English [7], and Chinese also has differences with Japanese and English, so it is very necessary to study simplified Chinese. In addition, in the current research, usability evaluation is mostly conducted through subjective questionnaires, and driving stability, task response time and subjective evaluation are not comprehensively analyzed.

In summary, based on the usage context of simplified Chinese, this paper explores the influence of different font sizes on the central touchscreen on the driver's rapid information acquisition and the influence of the information transfer efficiency of the central control screen on the driving stability. Through comprehensive analysis of subjective and objective data, the suitable font size range for the central control interface is determined, providing a theoretical basis for the selection of text size for the central control interface.

3 Construction of Interface Chinese Font Size Evaluation System

3.1 Selection of Evaluation Indicator

Driving Data Collection. In the evaluation research on vehicle human-machine interface, Muslim E et al. studied the ergonomic design of the instrument panel of electric vehicle, and collected the response time of the driver searching the number of viewpoints of the eye tracker and the saccades as objective indicators [8]. Wittmann M et al. studied the influence of display position of in-car visual tasks on simulated driving, and took lane departure time and average braking reaction time as important indicators to measure drivers' driving quality [9]. Ma J et al. used data mining technology to evaluate the driving distraction effect of vehicle human-machine interface display, and collected driving speed and lane departure data as well as residence time and average saccade times of eye tracker to evaluate users' driving behaviors [6]. In addition to the above indicators, Jin Xin et al. also measured the standard deviation of vehicle speed as an evaluation indicator of the touch-screen buttons of the center control. Combined with previous studies, the driving data collected in this paper mainly includes vehicle speed and horizontal lane deviation.

Central Touchscreen Click Data Selection. On the other hand, from the perspective of user experience measurement, the completion time and success rate of the central touchscreen click task are also indicators to measure whether the font size is appropriate. The response speed of the click speed of the central control can directly reflect the effect of the size on the efficiency of the click task of the central control [14]. Vivek D Bhise proposed that indicators to measure drivers' performance and behavior during vehicle running include lane departure, task completion time, steering wheel movement, etc.. Combined with previous studies, the data collection of interactive tasks in this paper mainly includes task response time and task completion degree.

Subjective Scale Selection. In this paper, the subjective questionnaire was established by reading related literature of simulated driving task and interface evaluation. Wittmann et al. [9] asked the subjects to evaluate the visibility of the red light signal in each simulated driving task, and also filled in the subjective Workload Assessment test (SWAT) scale to evaluate the time load, mental effort load and psychological stress load. Kusumawati et al. [10] used the System Usability Scale (SUS) and User Interface Satisfaction Scale (QUIS) to measure the subjective usability of the touch screen of the electronic newspaper machine. Jakus et al. [11] used the NASA-TLX multidimensional scale and user experience questionnaire (UEQ) in the study of vehicle-vehicle information system human-vehicle interaction to obtain the evaluation of subjects' subjective workload and usability of interaction modes, and also investigated the subjective preference scores of different schemes.

In this paper, referring to Jin Xin et al.'s [3, 12] evaluation system, this paper uses 7-point Likert scale to score the Visual Subjective Rating, Usability and Workload of different font sizes. Visual subjective rating refers to the driver's perception of font size, such as looking extremely small or looking just right. Referring to ISO-9241, the usability score is divided into three indicators: Effectiveness, Efficiency and Satisfaction. Workload evaluation refers to SWAT scale (used to measure psychological workload) [13] and NASA-TLX multidimensional scale (used to measure overall workload) [14], including five dimensions of heart, attention, physical load, time pressure and frustration. The higher the score, the more positive the evaluation (in the design of workload questionnaire, The lower the load evaluated by users, the higher the final score and the more positive the evaluation). The specific items and meanings are shown in Table 1.

In this paper, there are three types of data to be analyzed. The first is driving data, including driving speed and lateral lane position. Driving stability was assessed by standard deviation of vehicle speed and standard deviation of lane departure (the standard deviation of the lateral distance between the center of the vehicle and the center of the lane). Smaller values for both indicators meant less distraction from screen tasks and safer driving. Second, the response time of each time the subjects completed the task of identification and clicking, which was automatically recorded by the tablet program. The shorter the response time, the higher the efficiency of information transmission. The third is the subjective questionnaire score filled in by the subjects. The visual subjective score, usability and workload evaluation of the subjects with different sizes are recorded.

SPSS software was used to analyze the correlation between these three types of data and the size of the font. Meanwhile, the size of the font was used as the independent variable to analyze these data individually.

3.2 Experimental Equipment and Environment

Based on the consideration of safety and controllable variables, simulation driving experiments have been widely used in the study of in-car visual distraction [6, 8, 9, 15]. The driving part of the vehicle in this study is composed of Sensodrive steering wheel, acceleration and deceleration pedals and internal seats of the company. The layout is based on the real vehicle, and the participants can adjust the steering wheel and seats to a comfortable state according to their own body type. Samsung G9 49-inch belt fish screen is used for the simulation scene, and VTD (Virtual Test Drive), a complex traffic scene visual simulation tool developed by VIRES, a German company, is used for the driving scene, so as to independently set road conditions such as path, traffic vehicles and sound. CANoe (CAN open environment) development environment is used to collect driving data at 0.05 s intervals, including driving speed and lateral lane position. The speed is displayed in real time on a 10.2-inch LCD display mounted on the right side of the dashboard. A 10.2-in. iPad tablet computer is fixed 70 cm from the right front of the person with reference to the interior decoration of the car in the market to simulate the real car center control screen. Video recordings of participants and driving scenes were

Table 1. Subjective scale topic setting

scale	Question	Score range
Visual subjective rating	Do you think 12 pt looks suitable	range: from -3 to 3 -3-very small; 0-suitable; 3-very big
usability	Do you think 12 pt looks clear?	ranging from 1–7
	Do you think 12 pt would make screen touching efficient	ranging from 1–7
	Do you want the central screen show information in size 12 pt	ranging from 1–7
Work load	Mental demand needed in the task, ranging from 1–7	ranging from 1–7
	Attention needed in the task,	ranging from 1–7
	Physical demands needed during central screen touching,	ranging from 1–7
	Whether you are rushed and hurried in completing tasks?	ranging from 1–7
	Do you feel pressure and anxiety in completing tasks?	ranging from 1–7

also made using cameras during the experiment. The main man-machine dimensions of the equipment are shown in Fig. 1.

According to the ergonomics of car design, the seat height H30 of class A car is 150–400 mm. The 90th percentile of human body size is selected in the experiment, and the height of 175 cm person is measured in a comfortable sitting position. The vertical distance H30 between H Point and Accelerator Heel Point (AHP) is 360 mm. The horizontal distance between AHP and the center point of the steering wheel L11 is 300 mm, the height of the steering wheel H17 is 730 mm, the horizontal distance between H point and AHP is 550–850 mm, and the horizontal distance between shoulder point and the steering wheel is 450 mm.

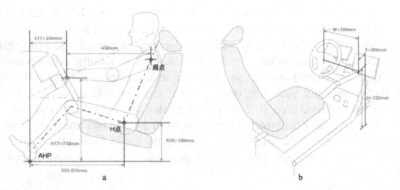

Fig. 1. Experimental equipment and man-machine size (a is side view, b is perspective view)

3.3 Screening of Subjects

Based on the study of Faulkner et al.'s sample size for usability testing (N > = 20) [20] and Wang et al.'s sample size for safety evaluation of driving simulator (N = 30) [21], 30 healthy participants were recruited to complete the driving simulator experiment with a male-female ratio of 1: 1. Be within the legal driving age requirement of China (18–70 years old), have normal eyesight, and have more than one year's driving experience. The demographic data of the subjects are shown in Table 2. Since this study is aimed at the accepted size of universal Volkswagen, there is no specific screening for specific model users.

3.4 Experimental Prototype of Center Control Screen

Experimental prototype refers to the task page displayed on the tablet, with font size as the argument. ISO 24509:2019 provides a method for estimating the minimum readable size of a single character. According to analysis, the minimum size of a sans serif font suitable for an 18-year-old person to read is 6 pt under the condition of indoor bright light and visual distance of 0.7 m. The size requirement increases with age, and a 70-year-old person needs at least 12 pt (about 4 mm) [16]. The Federal Highway

Table 2. Information of subjects

Section	class	percentage(%)
Gender	male	50
	female	50
Age	25–30 years	33.3
	30–35 years	53.3
	35–40 years	13.3
Driving year	1–2 years (include two years)	20
	2–5 years (include five years)	30
	5–10 years (include ten years)	36.7
	Over 10 years	13.3

Administration of the United States proposed the design criteria of interface factors from the perspective of human factors, and recommended that the character height should be 4.3 mm at the viewing distance of 0.7 m [17]. Crundall et al. conducted simulated driving experiments, and the results showed that 6.5 mm or smaller text should be avoided on vehicle displays [18]. In the Hicar Ecology White Paper released by Huawei, the minimum identifiable text size is defined as 5.3 mm. Based on the above literature, it can be concluded that the minimum size of the central control screen is 4 mm for the consideration of universality. Five automotive screen UI design experts were interviewed, and combined with the research on the automotive center control interface in the market, it was found that the size of common non-title information was about 4.5 mm. Therefore, this experiment presupsets five font sizes, which are respectively "3.5 mm", "5.5 mm", "7.5 mm", "9.5 mm" and "11.5 mm", covering the commonly used font sizes in the market.

The experimental prototype interface adopts white background and black text. The display content is divided into Chinese characters and numbers, and the meaningful sentences are presented in isolation and brief [4, 19]. Chinese characters use three groups of common car central control function words "navigation", "setting" and "play", while the numbers show three groups of air conditioning temperature "24 °C", "26 °C" and "28 °C". Due to the small number of phrases on the page, subjects could easily remember relative positions when the order was fixed (see Fig. 2). However, in the actual situation, the interface of the central control screen is rich in functions and the visual elements are complex, so it is difficult for the operator to operate directly by memory. In order to simulate the real driving situation, the left and right order of the three groups of characters will be randomly arranged in the experiment.

3.5 Overall Process Design

Before the experiment, the subjects first filled in the basic information form and the informed consent of the experiment. The tester introduced the experimental purpose and task to the subjects, and the subjects learned and got familiar with the simulator and the

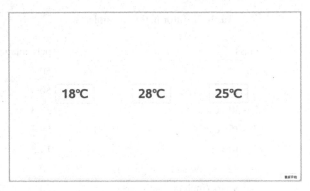

Fig. 2. Experimental prototype of central control interface

experimental task. After understanding the task requirements and adapting to the driving environment, the subjects were free to start the experiment (see Fig. 3).

Subjects were required to perform driving tasks and screen tasks. The driving task was carried out in the simulated scene of no car on the suburban highway, the road direction changed gently without sharp turns. The surrounding landscape is dominated by grass and trees to avoid other visual distractions. The participants were required to maintain the vehicle speed between 40–60 km/h and keep the right lane.

Fig. 3. Experimental scene

After the subjects entered the stable driving state, they completed the screen task on the tablet, which was automatically carried out by the preset program. In order to avoid the influence of sequence effect on the experiment, three groups of Chinese characters, navigation, setting and playing, were displayed with random size on the screen. The left and right arrangement of the characters was random. After hearing the voice command, the subjects clicked the corresponding position on the screen with their fingers, such as "please click navigation/setting/playing", the command also appeared randomly. Randomness is to ensure that subjects can read and react to the interface information immediately, so as to prevent them from completing the instructions by memory and affecting the accuracy of the experiment. In order to reduce accidental deviation, the task of Chinese character recognition and clicking was repeated for 3 times, and then the number task was performed for 3 times according to the same paradigm. After a font size is completed, the next font size is automatically tested.

The tester judges the completion degree of the task (among all the touch control tasks, one wrong touch is counted as 1/30), makes remarks to the abnormal situation, and deletes the data in the subsequent analysis process. After the task, the subjects filled in the subjective questionnaire in time.

4   Experimental Results and Data Analysis

4.1   Data Processing and Correlation

In this experiment, 150 cases of data were finally obtained. According to the task completion degree, outliers were excluded, and the mean value and standard deviation of experimental data were calculated. The final font size and subjective and objective data were shown in Table 3.

The normality test showed that the experimental data presented a non-normal distribution, so Spearman's Rank Correlation Coefficient was used to analyze the correlation between subjective and objective data and the size of the font. The correlation between the size and task response time, speed standard deviation, lane deviation standard deviation, visual subjective score, availability and workload is shown in Table 4. It can be seen that the size is significantly correlated with the standard deviation of driving speed, task response time, visual subjective score, availability and workload. However, the standard deviation of lane deviation is not significant, so it cannot be proved that its change is related to the font size. By considering the results of several related indicators, the author selected the most appropriate size of the central control font in the final box.

Table 3. Font size and subjective and objective data

Font size	Respond time of task (ms)	Standard deviation of driving speed	Lane deviation standard deviation	Visual subjective rating	Usability	Workload
3.5 mm	3225.22	4.30	0.46	−1.06	3.48	4.72
5.5 mm	2991.36	3.94	0.39	−0.20	4.82	3.54
7.5 mm	2794.54	1.68	0.37	0.62	5.13	2.99
9.5 mm	2814.52	1.65	0.32	1.10	4.86	2.70
11.5 mm	2753.26	1.54	0.37	1.79	4.80	2.37

4.2   Font Size and Subjective Score

The subjective rating shown above is significantly related to font size. Among them, the correlation coefficient between font size and visual subjective score is 0.758, indicating

Table 4. Correlation between subjective and objective data and font size

	Respond time of task (ms)	Standard deviation of driving speed	Lane deviation standard deviation	Visual subjective rating	Usability	Working load
coefficient of association	−0.349**	−0.316**	−0.117	0.758**	0.257**	−0.503**
Sig. (双尾)	0	0	0.162	0	0.002	0
N	150	150	150	150	150	150

** At level 0.01 (two-tailed), the correlation was significant

that the subjective score of user pairs has a significant positive correlation with the font size. According to the distribution of data in different sizes and the change of the average value (see Fig. 4), it can be concluded that in the range of 5.5 mm to 11.5 mm, the visual subjective score is in the appropriate range, but in the range of 7.5 mm to 11.5 mm, the visual subjective score keeps increasing, indicating that users think the size is too large on the premise of ensuring the efficiency of obtaining information. Therefore, an average score of 5.5 mm and 7.5 mm around "0" was considered most appropriate. In addition, it is worth noting that the score of 9.5 mm presents a two-stage distribution, and the scores of the subjects are concentrated around "0 - suitable" and "2 - too large", indicating that some subjects think the size of 9.5 mm is appropriate.

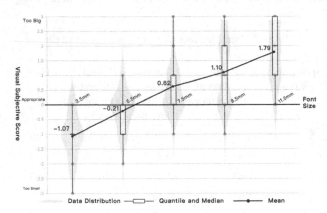

Fig. 4. Visual subjective score and font size

Secondly, the subjective index availability has a significantly weak positive correlation with font size. The usability score increases as the size increases until it starts to decrease at 9.5 mm (see Fig. 5). The 7.5 mm size has the highest usability score, followed by 9.5 mm and 5.5 mm, then 11.5 mm, and 3.5 mm has the lowest usability score because it is too small.

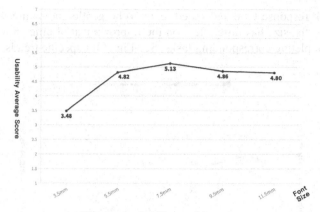

Fig. 5. Availability and font size

Finally, there is a significant moderate negative correlation between the workload and font size. The font size ranges from 3.5 mm to 7.5 mm. With the font size increasing, the workload decreases greatly. When the font size ranges from 7.5 mm to 11.5 mm, the impact of the font size change on the workload reduction becomes smaller, indicating that after the font size reaches 9.5 mm, the effect of the font size on the user workload reduction decreases (see Fig. 6).

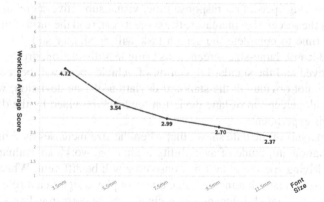

Fig. 6. Workload and font size

4.3 Font Size and Objective Data

In terms of objective data, task response time and standard deviation of speed have a significant weak positive correlation with font size. In the range of 3.5 mm to 7.5 mm, the decline of task response time and standard deviation of speed has a large change, indicating that the efficiency and stability of users in completing corresponding tasks are improved rapidly. Within the range of 7.5 mm to 11.5 mm, the standard deviation decline of task response time and speed tends to be gentle, indicating that continued enlargement of the size has little effect on the improvement of efficiency and stability of users in completing corresponding tasks. See Fig. 7 for specific trends.

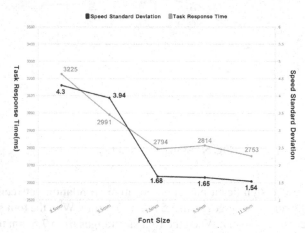

Fig. 7. Objective data and font size

Based on the above analysis, the experimental results are as follows: the Chinese font size of the car center control screen is significantly correlated with the standard deviation of driving speed, task response time, visual subjective score, availability and workload. As the screen size increases, the subjects can read the information faster, and the response time to complete the screen task will be shorter and the workload will be reduced. Because large-size screen tasks bring less distraction, driving stability will also be improved and the standard deviation of vehicle speed will be reduced, but the significance is not obvious in the standard deviation of lane deviation, which may be due to the weak perception of lane deviation feedback provided by the driving system adopted in this experiment.

Further analysis of the data shows that when the size increases by 2 mm, the corresponding variation amplitude of availability evaluation, workload evaluation, standard deviation of driving speed and task response time will be different. When the fold line changes gently, it means that the font size does not play a significant role in this range; When the change is rapid, it means that a change in the corresponding size range will result in a large increase. Combined with the above analysis, the final test results are shown in Fig. 8. The subtitle and body size of the Chinese font size recommended in this paper range from 5.5 mm to 7.5 mm, and 9.5 mm can be used in the title of high-level, which can not only maintain high user experience and safety, but also avoid space congestion in the interface design of the vehicle and engine system.

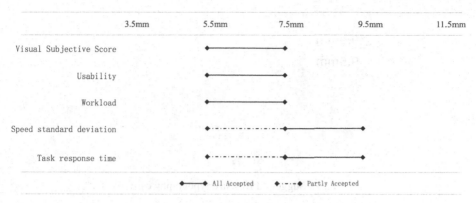

Fig. 8. Recommended font size

5 Experimental Application

In the interview with interactive experts, the author found that in the design tools and design process of the current human-computer interaction interface, the traditional font size unit "pt" has been gradually abandoned, and the pixel unit "px" has been directly used for standard formulation and prototype design. The reasons for this are: a) The definition of "pt" is closely bound to the production mode of traditional paper printing, that is, 1 pt = 1/72 inch and 1 inch = 25.4 mm. However, in the digital era, the presentation mode of information has changed from paper to screens of various resolutions, and the conversion of "pt" and "mm" has become more complicated. Correspondingly, the conversion of pixel units "px" and "mm" is more direct. b) Another reason for the large-scale application of "px" is that a large number of pictures, buttons, graphics and other elements in interactive interface design need to use "px" to measure the size, so it is more necessary to unify the font unit with the unit of other elements. In interactive interface design, Pixel Pitch is often used to describe the relationship between "px" and "mm". For example, formula (1) gives the screen pixel pitch based on the ratio of the number of pixels on the long side of the horizontal screen to the screen width.

$$P = W/w \tag{1}$$

$$Fp = Fs/P \tag{2}$$

In the formula, P is the screen pixel spacing; W is the screen width; w is the number of pixels on the screen; Fp is the number of font pixels and Fs is the physical size of the font.

According to formula (2), the recommended size of 5.5 mm to 7.5 mm can be converted to 33–45 px, and the recommended size of 9.5 mm for the big title can be converted to 57 px. The layout diagram given according to the above suggested size is shown in Fig. 9.

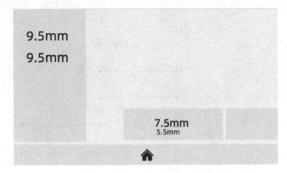

Fig. 9. Suggested font size typography case

6 Conclusion

This paper proposes a method to evaluate the Chinese font size of automotive central touch screen by combining subjective and objective data, which makes up for the previously neglecting the task response efficiency and driving behavior of the Chinese font size when the driver touches the auto center control screen, and reveals the relationship between the size of the Chinese font size and the driver's touch control behavior efficiency, subjective cognition, work load and driving behavior. The conclusion has strong practical significance. In the design practice, the Chinese font size range with high availability, reasonable workload and within the safe range is proposed, and the practical application method is suggested. The relationship between the recommended font size in the digital age and the absolute physical size is clearly elaborated, which solves the problem rarely mentioned in the past literature. It provides a scientific way to formulate visual specification of automotive central control interface design.

In this study, the size of the size was increased by 2 mm, and a preliminary size range was obtained. Further experiments can be carried out within this range to improve the accuracy of the size range. Secondly, this paper did not consider the influence of the same font size and different word weight on drivers' cognition in the experiment, so further experimental measurement is needed in the future.

References

1. Huang, S.-M.: Effects of font size and font style of traditional Chinese characters on readability on smartphones. Int. J. Ind. Ergon. **69**, 66–72 (2019)
2. Chan, A., Lee, P.: Effect of display factors on Chinese reading times, comprehension scores and preferences. Behav. Inf. Technol. **24**, 81–91 (2005)
3. Jin, X., Li, L., Yang, Y., Fu, M., Li, Y., You, F.: Touch key of in-vehicle display and control screen based on vehicle HMI evaluation. Package. Eng. **42**, 151–158 (2021)
4. Dobres, J., Wolfe, B., Chahine, N., Reimer, B.: The effects of visual crowding, text size, and positional uncertainty on text legibility at a glance. Appl. Ergon. **70**, 240–246 (2018)
5. Née, M., Contrand, B., Orriols, L., Gil-Jardiné, C., Galéra, C., Lagarde, E.: Road safety and distraction, results from a responsibility case-control study among a sample of road users interviewed at the emergency room. Accid. Anal. Prev. **122**, 19–24 (2019)

6. Ma, J., Gong, Z., Tan, J., Zhang, Q., Zuo, Y.: Assessing the driving distraction effect of vehicle HMI displays using data mining techniques. Transp. Res. F: Traffic Psychol. Behav. **69**, 235–250 (2020)
7. Fujikake, K., Hasegawa, S., Omori, M., Takada, H., Miyao, M.: Readability of character size for car navigation systems. In: Smith, M.J., Salvendy, G. (eds.) Human Interface 2007. LNCS, vol. 4558, pp. 503–509. Springer, Heidelberg (2007). https://doi.org/10.1007/978-3-540-73354-6_55
8. Muslim, E., Moch, B.N., Lestari, R.A., Shabrina, G., Ramardhiani, R.: Ergonomic design of electric vehicle instrument panel: a study case on Universitas Indonesia's national electric car. In: IOP Conference Series: Materials Science and Engineering, p. 012109. IOP Publishing (2019)
9. Wittmann, M., et al.: Effects of display position of a visual in-vehicle task on simulated driving. Appl. Ergon. **37**, 187–199 (2006)
10. Kusumawati, R.E., Muslim, E., Nugroho, D.: Usability testing on touchscreen based electronic kiosk machine in convenience store. In: AIP Conference Proceedings, p. 040027. AIP Publishing LLC (2020)
11. Jakus, G., Dicke, C., Sodnik, J.: A user study of auditory, head-up and multi-modal displays in vehicles. Appl. Ergon. **46**, 184–192 (2015)
12. You, F., Yang, Y.-F., Fu, M.-T., et al.: Design guidelines for the size and length of Chinese characters displayed in the intelligent vehicle's central console interface. Information. **12**(5), 213 (2021)
13. Reid, G.B., Nygren, T.E.: The subjective workload assessment technique: A scaling procedure for measuring mental workload. Adv. Psychol. **52**, 185–218 (1988)
14. Hart, S.G., Staveland, L.E.: Development of NASA-TLX (task load index): results of empirical and theoretical research. Adv. Psychol. **52**, 139–183 (1988)
15. Nakano, K., Zheng, R., Ishiko, H., et al.: Gaze measurement to evaluate safety in using vehicle navigation systems. In: Proceedings of the 2014 IEEE International Conference on Systems, Man, and Cybernetics (SMC) (2014)
16. ISO 24509:2019. Ergonomics—a method for estimating minimum legible font size for people at any age. Accessible design, p. 37 (2019)
17. Kelly, M.J.: United States. Federal Highway Administration (1999)
18. Crundall, E., Large, D.R., Burnett, G.: A driving simulator study to explore the effects of text size on the visual demand of in-vehicle displays. Displays **43**, 23–29 (2016)
19. Conradi, J.: Influence of letter size on word reading performance during walking. In: Proceedings of the 19th International Conference on Human-Computer Interaction with Mobile Devices and Services, F (2017)
20. Faulkner, L.: Beyond the five-user assumption: benefits of increased sample sizes in usability testing. Behav. Res. Meth. Instr. Comput. **35**(3), 379–383 (2003)
21. Wang, X., Liu, S., Cai, B., et al.: Application of driving simulator for freeway design safety evaluation: a sample size study. In: Transportation Research Board 98th Annual Meeting (2019)

Cueing Car Drivers with Ultrasound Skin Stimulation

Oleg Spakov[✉], Hanna Venesvirta, Ahmed Farooq, Jani Lylykangas,
Ismo Rakkolainen, Roope Raisamo, and Veikko Surakka

Faculty of Information Technology and Communication Sciences, Tampere University,
Tampere, Finland
{oleg.spakov, hanna.venesvirta, ahmed.farooq, jani.lylykangas,
ismo.rakkolainen, roope.raisamo, veikko.surakka}@tuni.fi

Abstract. The aim was to utilize ultrasound skin stimulation (UH) on a palm to inform the system state when interacting with mid-air hand gestures in the automotive context. Participants navigated a horizontal menu using touch, buttons, and hand gestures during simulated driving. Mid-air interaction and UH feedback design was tested in two studies. The first study proved the participants were able to perceive the menu interaction state from UH feedback, but the interaction felt inconvenient. After redesigning the mid-air gestures and the feedback for the second study, the participants rated it equally to the touch-based method, though the interaction with steering wheel buttons was still the most favorable. In conclusion UH feedback can be utilized for delivering the interactive system state to drivers, but this approach may cause additional cognitive load and should be further tested in a running car.

Keywords: multimodal interaction · mid-air gestures · ultrasound skin stimulation · ultrasonic haptic feedback · In-vehicle interaction · cueing drivers

1 Introduction

Skin stimulation with touchless ultrasound haptics (UH) [25, 42] is known technology for more than a decade. The UH feedback simulates pressure waves, causing a similar sensation to pneumatic feedback felt on the relatively small area of the skin. Earlier research focused mainly on what kind of sensation, either static or dynamic, can be projected onto a user's hand (e.g., [6, 22, 25]), and how efficiently this sensation can be interpreted (e.g. [12, 30]). Soon it was suggested that ultrasound haptics has potential as tactile feedback in interactive applications.

Since UltraLeap (formerly known as Ultrahaptics) released various versions of their ultrasound actuation devices, mid-air gesture interaction and subsequent research into non-contact tactile feedback has become more popular. For example, it was found that a static "hand-crosses-a-wall" sensation was useful to mark areas in the space where mid-air gestures should be performed [13, 52]. Another example is related to attempts in replacing or combining audible or visual feedbacks with ultrasound haptics feedback

H. Krömker (Ed.): HCII 2023, LNCS 14049, pp. 224–244, 2023.
https://doi.org/10.1007/978-3-031-35908-8_16

(e.g., [34, 47, 52]). It was found, however, that only a limited set of skin stimulations can be effective for this task (e.g. [23, 51, 56]).

The use of mid-air gestures in the automotive context is relatively new. Interaction using hand gestures can already be found in several premium car models (e.g., manufactured by BMW and VW) allowing to control a set of car systems, like changing sound level or navigating between music tracks. These gestures allow to decrease the number of functions available with traditional physical controls, like buttons on the steering wheel. It is expected that hand gestures will be offered more in the near future as the variety of systems and functions in a car grows rapidly and too many physical controls may be undesired to place on a cockpit or the steering wheel. Touchscreens may also be limited due to a relatively high demand for visual attention when drivers are busy with monitoring the road and traffic (e.g., [2]).

Several studies have been dedicated to exploring how ultrasound tactile feedback could benefit the use of mid-air gestures in a car. For example, Georgiou et al. (2017) [14] suggested using ultrasound tactile feedback to improve interaction experience with mid-air hand gestures to control in-vehicle infotainment systems. Harrington et al. [17] and Large et al. [29] showed that interaction with car systems using mid-air gesture with ultrasound tactile feedback was more robust compared to visual-only feedback, as it resulted in the shortest interaction times, highest number of correct responses, least 'overshoots' rate, and was more preferred by participants. These studies stimulated new research in designing UH feedback suitable for use in a car context (e.g., [5]).

If the number of car systems available for hand gesture interaction continues to grow, there might be cases when the same hand gesture is used to perform different actions, depending on the car system/interaction state(s). The state may be known in advance (like the sound level of the music that is played), but there might be multiple scenarios when the state is unknown and must be inspected prior to starting the interaction. Car system states are usually reported using a visualization on an instrument panel, center console screen, or a transparent head-up display.

To avoid visual shift from the road, non-visual cueing should be considered. Cueing is a rather common way to facilitate completion of various tasks, including driving tasks (e.g., [54]). We suggest that UH feedback may be used to communicate a system state in an eyes-free manner prior to starting the gesture-based interaction. To our knowledge, this operating mode has not been explored in previous studies. To investigate the potential of this approach, we conducted two studies where drivers were navigating an onscreen menu while its navigation state was communicated to the driver via ultrasound skin stimulation. We next describe the interaction design and its evolution tested in two studies. The gesture-based interaction was compared with the interaction using buttons located on the steering wheel and swipe gestures performed on a touchscreen.

2 Related Work

Gestures. Many studies showed that mid-air gestures have great potential as an alternative to interaction on touchscreens, which have gotten increasingly common in car dashboards and central consoles [56]. In some of these studies, horizontal swiping/sweeping gestures were used to start/stop engine [20], select autonomous functions [32], open or close windows [20, 32], turn seat heating on/off [32], control air conditioning [10], wipers and lights [20], but also to change volume [17, 20, 26, 32] and audio tracks [19, 20, 26, 31, 45, 48, 55] in a music player, and accept/reject calls [20]. In other studies, drivers were flexing their wrist to navigate a menu up or down [38], held their hand tilted up/down or made vertical hand swiping gestures to scroll a menu [37, 38]. Vertical swipes were also used for music volume control [31]. Contact list was traversed with similar swiping gestures and calling was executed with a finger pointing gesture in [48]. Pointing with index finger was also used to start a music player [45] and select objects on a central console [53] or on a HUD [1, 4]. A two-finger zoom / shrink gestures were used to control climate and windows [32], to scale UI element on a central console display [53], and to turn on/off a music player [31]. Rotational hand or index finger movements were most often used to control music volume [19, 32, 45, 53] as these gestures resemble rotating a physical knob or dialing on a rotary phone. Demonstrating open palm or fist were considered as suitable gestures to start or stop music player [17, 19, 45, 57], select an object [17, 26, 37], return to the previous level/state [37], or to navigate on a map [16]. Wide sliding gestures (arm movements / rotations) were studied when drivers were controlling a car running in an automated mode [8]. Other less common gestures were also proposed to be used for human-car interaction, as well as in-car targets of such an interaction, though not all of them were tested in controlled studies (e.g. [3, 7, 16, 41, 57]).

UH Feedback. Thus far mid-air haptics has not been used in cars widely. One reason is that the technology is relatively new. The other possible reason is that it might be technically challenging to place a relatively large matrix of ultrasound actuators inside a car, as the surfaces easily reachable by the driver with a hand are mostly occupied already (central stack, gear lever, ventilation, etc.). Nevertheless, 1729Harrington et al. and Large et al. used simple haptic pulses to confirm changes in the sound volume controlled by a swiping hand gesture [17, 29]. Also, they used four-point haptic feedback to create a sensation of touching a button. A 500ms long feedback on the tip of the middle and index fingers was presented to confirm a "V" gesture recognition in [47]. 2744Korres et al. and Rümelin et al. used either static or dynamic single-point feedback on the fingertip with one-finger pointing or sliding gestures [27, 44]. Two-point feedback on the lower palm's edge was used as feedback for similar mid-air gestures in [14].

A circular motion of the haptic feedback was used to confirm that the circular gesture is tracked by the system in [14, 47]. In both studies, a short pulse was presented to the palm once the hand entered the interaction space. In other studies, the interaction space marking feedback resembled a wall [51] or a circle that dynamically changed its size depending on how close the hand was to the center of the interaction space [12].

Young et al. [56] used circular feedback on a palm to confirm swipes (1 tap), hand twisting (2 taps), hand tap (circle expansion), grab-and-release gestures (circle shrinking) and pinch dragging (circle change its size depending on the controlled parameter). A so called "line scan" dynamic feedback (a line of haptic feedback moving from one palm's edge toward the other) was proposed to be used with gestures like swiping [47, 56] and hand twisting [56]. A wide variety of haptic feedbacks was collected in a study where participants were involved in feedback designing process with the aim to find best patterns fitting for the tasks typical in cars [5], though none were yet evaluated in controlled studies. An overview of the UH feedback used outside of the automotive context can be found in [42]: most of these are based on static or dynamic patterns utilizing the simplest geometrical configurations, like points, lines, and circles.

Cueing Driver. Visual and audible driver cueing is a well-established research area for driving studies (e.g., [9, 50]). Vibrotactile cues were studied in scenarios when drivers were interacting with in-vehicle interaction systems, IVIS (e.g. [40]), but also to notify a driver about events that require attention (e.g. [21]). Cueing with ultrasound skin stimulation was mostly used for creating the experience of touching a screen [6] or an interactive object [17, 29] or entering a mid-air interaction area with the hand [47, 51] and localizing virtual objects [52], both in the driving and non-driving contexts. By contrast, no research has been conducted yet on using UH feedback for providing the interaction context or state when drivers are performing a secondary task, although similar ideas were proposed in UH feedback designing guidelines published recently [56].

3   Study 1: Spatial Interaction

3.1   Interaction Design and Implementation

Navigation Menu
A custom menu software displayed a horizontal "carousel" list with 7 images. Navigation always started from the central image, thus, the maximum possible navigation length was 3 steps in either direction. Each navigation trial had to be confirmed with an additional action described later separately for each condition. The navigation actions were complemented by a "click" sound (300 ms), and the selection confirmation action was complemented by a distinct sound (500 ms).

Mid-air Gestures
Gesture-based interaction relied on hand position in relation to the center of the interaction area (as in [16, 23]) and used dwell time to trigger navigation actions. Dwell-time based interaction was chosen to avoid possible degradation of the ability to perceive dynamic UH feedback due to hand movements required in other solutions [47].

The interaction area was split into five zones as shown in Fig. 1 (left): the round central zone (r = 2 cm) was surrounded by four sectors, each 6 cm wide if measured from the inner-circle edge to the outer-circle edge. To start menu navigation, the participant had to place the hand over the central zone first. Participants were informed that the central zone is indicated by a single-point ultrasound feedback (like in, e.g., [49], Fig. 1a) and that it should be felt in the middle of their palm (Fig. 1b) for at least 500 ms to enable left and right interaction zones.

Next, a participant had to move the hand to the right or left navigation zone (Fig. 1c). Upon the hand entering these zones, a line-shaped ultrasound feedback was exposed either to the right or left palm side, correspondingly. The central area was shrunken to r = 1.5 cm to avoid back-forth border crossing effect. Participants had to wait for 2 s to trigger the left/right navigation action. Each navigation action was confirmed with a 700 ms "line-scan" ultrasound feedback as in [35, 47, 56], then the initial haptic feedback (palm edge stimulation) was restored. After navigating the required number of steps, participants had to enter the "confirmation" zone (Fig. 1d), and dwell for another 2 s. If the hand moved out of the interaction area (16 × 16 cm), participants had to bring it back to the central zone before continuing.

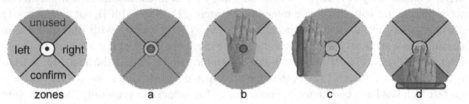

Fig. 1. Interaction zones (left) and a sequence of hand movements: a) UH feedback in the central area marks the interaction area, b) the hand enters the central zone for 0.5 s, c) the hand dwells 2 s for each navigation step over the left/right navigation zone, d) the hand dwells 2 s over the confirmation zone to finalize a trial. Grey color marks inactive zones, light purple marks available zones, and pink color marks zones currently in use. UH feedback is shown in blue.

Buttons

Top buttons on the left and right sides of the steering wheel were operated with thumbs to navigate the menu left and right accordingly, and the right-hand paddle shifter located behind the wheel was used to confirm the menu item selection by pulling it with other fingers (see Fig. 2c).

Touchscreen

In this interaction mode the menu navigation application was running on MS Surface 3 Pro tablet (see Fig. 2b). Participants navigated the menu with swipe gestures to left or right. A simple "tap" was required to confirm the selected menu item. The current menu item took most of the central zone of the viewport. The pilot tests proved this size was sufficiently large to hit it avoiding direct glances and using peripheral vision only.

3.2 Methods

Participants

Fifteen voluntary participants (3 males, 11 females, 1 non-binary; 21–49 years; $\mu = 30$ years) with a valid driving license took part in the study. Only right-handed participants were recruited due to hardware setup with hand recognition and stimulation devices located on the right. The participants were rather experienced drivers ($\mu = 11$ years of driving), driving mostly daily (33%) or weekly (53%).

Equipment

The experiment took place in a laboratory premises equipped with a fixed-base driving simulator equipped with Logitech G27 steering wheel and pedals. LCT Sim v1.2 software

[24] was shown on a display 1.2 m in front of a participant. The simulator software recorded the car position throughout the session.

Leap Motion was used as a hand-tracking device. It was assembled into a single unit together with Ultrahaptics STRATOS device with 256 ultrasonic actuators that can create a touchless haptic sensation. The unit was located close to the driver's right hand (≈30 cm), as shown on Fig. 2a. MS Surface Pro 3 device was located ≈10 cm higher on the same table (Fig. 2b), so that the hand operating space was approximately the same for both the mid-air gestures and the touchscreen swipes. We used a regular office chair with a neck rest, so that participants were seated in upright position guaranteeing they could reach the Ultrahaptics device without leaning forward. The chair and pedal locations were adjustable to accommodate participants of all heights comfortably.

We used VI myGaze eye tracker to record eye-gaze. The eye tracker was calibrated using the standard five-point calibration method provided by the eye-tracking software. The tracker was located about 15 cm behind the steering wheel on the level which enabled to track gaze of participants of all heights.

Fig. 2. Equipment and the setup to navigate the menu with a) mid-air gestures, b) touchscreen swipes, c) buttons on the steering wheel.

Procedure

After a participant signed the consent form, the supervisor introduced the study goals. Then the participant was seated in front of the steering wheel and the adjustments were made. The participant was introduced to the driving simulator and instructed to drive at the maximum speed (50 km/h) in the middle lane. The simulator was tested in a short (≈30 s) trial session.

The supervisor introduced each interaction method prior to starting the corresponding test session. Participants practiced completing a trial until they reported feeling comfortable with the given interaction method. Each trial started with an audible navigation instruction synthesized with an artificial voice. The instructions had a form of "*length direction*", where *length* was either empty (one step), *double* (2 steps) or *triple* (3 steps), and *direction* was *left* or *right*. For example, the instruction "*double left*" meant "*navigate 2 steps to the left*".

Generally, buttons and swiping gestures required very little practice, as these interaction methods are very common. Interaction with mid-air gestures, however, required more intense training. Participants were taught how ultrasound stimulation helps to find the interaction zone, how much and what direction the hand should move to enter navigation zones, and how to enter the confirmation zone.

The participants performed 24 trials with 5–7 s intervals in between during each session. The trial instructions were given only during straight driving parts of the simulator. Combinations of navigation length and direction were completely randomized within each test session, and the order of the experiment conditions was counterbalanced between the participants. After each session the participants filled in NASA Task Load Index (TLX) [18] questionnaire and a study-specific questionnaire with 13 questions on a 1–20 scale (shown in Fig. 4). Finally, we asked participants to rank the interaction methods according to their preference. A short free-form interview took place before the participants left the laboratory.

3.3 Data Analysis and Results

We further refer to the session interaction methods as *buttons* (B), *touch* (T) and *gestures* (G) conditions. Non-parametric Friedman's test and Wilcoxon signed-rank test for paired samples were applied for all subjective measurements. Objective measurements were analyzed using repeated measures analysis of variance (ANOVA) and Student t-test, unless Shapiro-Wilk test did not show the data normally distributed. In those cases, the Friedman's test with Wilcoxon test for post hoc analysis was used. Bonferroni-correction was used for all post hoc comparisons.

Prior to the data analysis, we removed the trials that were clear outliers in the participant's data for task completion time ($\pm 3\sigma$). Thus, 2.8% of the trials were removed in *touch* condition, 4.2% in *buttons* condition, and 10.6% in *gestures* condition.

Objective Measures

Attention to the Road. Eye tracking data revealed that participants' gaze was off the road for 7.5% of time when using swipes and mid-air gestures, and 6.6% of time when using buttons. All differences were not significant. Also, in the latter mode participants never directed their gaze away from the road for longer than 2 s, but did it once when using swipes, and 3 times when using gestures. Friedman's test showed that there was a significant effect of interaction method ($\chi^2 (2) = 7.0, p = .030$), but none of the post hoc pairwise comparisons were statistically significant.

Driving Outside of the Lane. If participants deviated more than 2.2 m from the center of middle lane (half of the lane width in LCT Sim), then a "lane deviation" event was recorded. Neither in *touch* nor in *buttons* conditions such events were registered. However, nine deviation events were registered in *gestures* condition. Friedman's test showed a statistically significant effect of interaction method ($\chi^2(2) = 8.0, p = .018$), but none of the post hoc pairwise comparisons were statistically significant.

Navigation Delay. This measurement refers to the time spent to start the navigation action, i.e., the time gap between the trial start and the first registered navigation action. The "navigation action" was dependent on the input method: a button press, the first display touching event, or the moment the hand settled in the central interaction zone. One-way ANOVA revealed a statistically significant effect of the interaction method ($F_{2,14} = 17.1, p < .001$). Pairwise post hoc tests showed that it was significantly faster to start using buttons ($\mu = 453$ ms) than swipes ($MD_{BT} = 270$ ms, $t(14) = 3.62, p = .003$) and mid-air gestures ($MD_{BG} = 1181$ ms, $t(14) = 4.84, p < .001$), and swipes were faster than gestures ($MD_{TG} = 911$ ms, $t(14) = 3.46, p = .004$); see Fig. 3, left.

Navigation Duration. This measurement refers to the time from the first navigation action to the selected item confirmation. One way ANOVA showed a statistically significant effect of interaction method ($F_{2,14} = 780, p < .001$). The post hoc pairwise t-tests showed no significant difference between *buttons* and *touch* conditions ($MD_{BT} = 33$ ms), but the duration was significantly longer in *gestures* condition ($MD_{BG} = 7.13$ s, $t(14) = 28.1, p < .001$ and $MD_{TG} = 7.16$ s, $t(14) = 28.8, p < .001$, respectively).

Correction Actions. This measurement was calculated as a percentage of all correctly completed trials that were recorded with a larger number of navigation steps than required. In *buttons* and *touch* condition, only 0.6% and 1.2% of the trials were corrected, correspondingly, while in *gestures* condition it was 5.4%. However, Friedman's test showed that there was no significant effect of an interaction method.

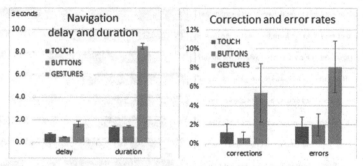

Fig. 3. Mean values ($\pm SEM$s) of navigation delay and duration (left), and mean correction and error rates (right) in Study 1.

Error Rate. Two and 1.8% of all trials were finished with an incorrect item selected in *buttons* and *touch* conditions, correspondingly, while in *gestures* condition it was 8.1% (see Fig. 3, right). Friedman's test revealed that there was a statistically significant effect of interaction method ($\chi^2(2) = 7.26, p = .027$). However, post hoc pairwise comparisons of the interaction methods were nonsignificant.

Subjective Measures

The interaction method score S was calculated based on its rank r assigned by the participants, where $r = 1$ marked the most preferable method. The following formula was used to normalize S and fit the score values into the range between 0 and 1:

$$S_{IM} = \frac{L \cdot N - \sum_i^N r_i}{(L-1)N}$$

where N is the number of participants, L is the number of interaction methods, and $r = \{1,2,3\}$ is the rank. The interaction method based on buttons was ranked as the most preferable method 14 times, and therefore received the highest score $S_B = 0.93$. Touchscreen-based interaction method was selected by 12 participants as the second best, resulting in the score $S_T = 0.40$. Mid-air gestures were usually ranked as the least preferable method (11 participants), and therefore got the lowest score: $S_G = 0.17$.

End-Session Questionnaire. Friedman's test revealed a significant effect of interaction method on the answers to all question in the end-session questionnaire, except the question #11 regarding audio feedback usefulness. According to the Wilcoxon tests, buttons-based and touch-based interaction methods were generally rated more positively (easier, faster, etc.) than gesture-based interaction. Usefulness of haptic feedback was asked only in *gestures* condition, and it was found as useful. Visual feedback was the only feedback found not much useful except when using touchscreen. Figure 4 shows mean values of the rating with the corresponding standard error of means (*SEM.*).

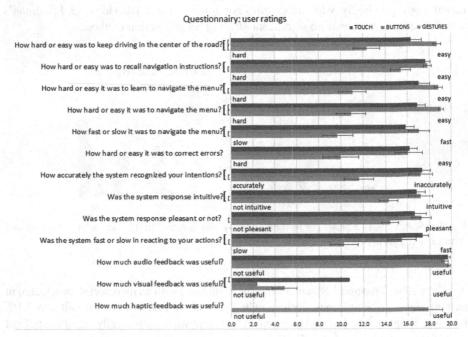

Fig. 4. Mean ratings ($\pm SEM$s) from the end-session questionnaire in Study 1. Significant differences between the interaction methods within each scale are marked using bracket-like signs.

NASA-TLX Questionnaire. Friedman's test revealed a statistically significant effect of interaction method also on evaluating the interaction with NASA-TLX questionnaire, with exception for "Temporal demand" scale. All scales had minimal values for button-based navigation, and maximum values for gesture-based navigation. The scales' values and significance of differences between them are shown in Fig. 5.

Interviews. Participants confirmed that the sensation caused by haptic feedback was new to them, and it was an interesting experience to receive this kind of feedback.

They correctly mentioned all locations on the palm this sensation was felt during the trials. However, the navigation confirmation haptic feedback ("line-scan") was mostly left unnoticed: only 3 participants (20%) mentioned it in their answers.

Observations
No major issues were observed while buttons were used for navigation with only few exceptions when the participants pressed a steering wheel button few times too quickly, resulting in one press left unrecognized by the experimental software.

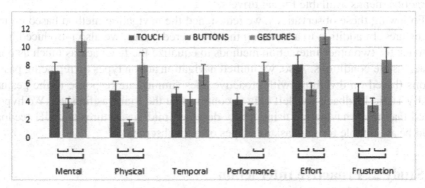

Fig. 5. Mean ratings ($\pm SEM$s) from NASA-TLX questionnaire in Study 1 (scales 1–20). Significant differences between the interaction methods within each scale are marked with brackets.

Few participants started their touchscreen session with a couple of trials navigating to the direction opposite to the one that was asked; this issue was corrected after a note from the supervisor. Another issue was observed when confirming the navigation result: too hard or too long taps could result in either the gesture being recognized as start on a sliding gesture, or as a gesture that corresponds to *mouse right-click* event.

Searching of interaction area could appear as a time-consuming action for some participants when they used mid-air gestures, if Leap Motion failed or delayed with hand recognition. Waving with the hand usually helped Leap Motion to catch up with the hand, but in rare cases participants had to remove the hand back to the steering wheel and then restart their attempt. Also, sliding with the hand from the starting location too far resulted in overshooting a navigation zone, and the participants had to restart their trial by returning to the central zone, therefore prolonging the task duration.

3.4 Discussion

Both objective measurements and subjective ratings revealed that the interaction method based on mid-air gestures was slowest and least comfortable to use. The ultrasound cueing feedback was functional nevertheless: it allowed participants to recognize that they had entered a desired navigation zone. Hand recognition issues, however, hindered the ability to evaluate UH cueing in the menu navigation context. In addition, the experiment proved that the spatial navigation design was prone to missing the interaction area.

Finally, the audio feedback played to confirm a navigation action was more dominant feedback than the line-scan haptic feedback played at the same time.

On-wheel buttons were recognized as the best method to perform the task as it did not require removing hands from the steering wheel, nor shifting driver's visual attention from the road. We suppose that any interaction method that requires physical or visual distraction from driving will be recognized as less suitable for controlling in-car systems. However, it is unlikely that all functionalities will be available in cars through the on-wheel buttons solely while complexity of the car systems continues to increase, and other input methods such as touchscreens and mid-air gestures will remain to stay among the interaction means available for car drivers.

Following these observations, we redesigned the navigation method based on mid-air gestures. In addition to addressing the discovered issues, we also introduced haptic feedback for two other interaction methods to equalize the feedback as much as possible across the conditions. Also, we unified navigation action types by replacing passive actions (based on dwelling) with proactive ones (immediate response to the gesture). Finally, audio feedback was left only to confirm that the trial was finished. We hypothesized that menu navigation using newly designed mid-air gestures and UH feedback could be as suitable for this task as swipes on a touchscreen.

4 Study 2: Proactive Interaction

4.1 Interaction Design Reconsidered

The only major change we introduced for the interaction methods based on buttons and swipes was a 300 ms vibration on the steering wheel triggered by the button press and on the touchscreen at the swipe gesture onset, accordingly. The vibration signal parameters were selected so that the vibration was quite prominent in both conditions (180 Hz, 1.2 A/14.4 W), still remaining mostly inaudible. Vibration sensation on the touchscreen, however, was very dependent on the force applied and time of contact between the device and the index finger: a very light and short touch reduced the sensation of the feedback to marginally distinguishable.

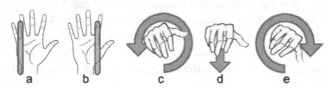

Fig. 6. UH linear stimulation reflecting the menu navigation direction (a: right; b: left) and mid-air gestures (c: change direction to "right", e: change direction to "left", d: navigate one step)

In the *gestures* condition, the menu was instantly ready to be navigated in one of two directions as soon as the user's hand entered the interaction area (enlarged to 25 × 25 × 40 cm). It reported its state (navigation direction) by forming an UH stimulation on the corresponding palm side (Fig. 6a–b). To navigate a menu toward the current direction, a user had to make a "tap" gesture (see Fig. 6d). For example, tapping twice with UH stimulation on the right palm side resulted in navigating the menu two steps to the right.

To change the navigation direction, participants had to twist their hand along the hand axis ("roll" the hand): clockwise rotation to switch it to "right" (Fig. 6c), and the opposite rotation to switch it to "left" (Fig. 6e). These rotational gestures were selected after observing several participants in Study 1 making sometimes slight rotational hand movements upon entering the navigational zone. We supposed that rotational gesture could be found to be more natural than any other gesture if it was observed in the condition when such a gesture was not required, but still made unwillingly.

To navigate, a participant had to stretch the right hand over the Ultrahaptics device and recognize the current menu navigation state. If the direction was the required one, the participant only had to "tap" with the hand as many times as needed. Otherwise, the participant first made a rotational gesture to reverse the navigation direction, and then continued with tapping. Removing the hand back to the steering wheel meant that the navigation was finished, thus no specific confirmation gesture was required.

The trial end was accompanied with the audio feedback, which was the only audio feedback across all navigation methods.

4.2 Methods

Participants
Twenty-three voluntary participants (8 males and 15 females) took part in the study (19–45 years, $\mu = 29.7$ years). The same requirements as in Study 1 were applied to all volunteers. The participants were a bit less experienced drivers than in Study 1 ($\mu = 10$ years of driving), driving mostly daily (39%) or less than weekly (35%).

Equipment
The study was moved to the lab with a fixed-base driving simulator equipped with driving controls used in Study 1 and a Volvo XC60 driver's seat. The touchscreen and the Ultrahaptics device were located approximately 5 cm more to the right from the seat than in Study 1 (see Fig. 7, left). Tectonic Elements TEAX25C10–8/HS actuators attached on top of the steering wheel column and on the rear frame of the MS Surface 3 Pro tablet were used to generate vibrotactile feedback (see Fig. 7, right). All the preparations otherwise were identical in both studies.

Procedure
The same procedure as in Study 1 was used. Training the navigation with mid-air gestures was notably shorter, though. Participants were familiarized with the way the menu reports its current navigation direction state, and then taught to reverse it and make tapping gestures. Participants performed 48 trials with a 5–7 s interval in between, twice more than in Study 1. In addition to NASA-TLX questionnaire, we asked participants to fill in Positive System Usability Scale (P-SUS) questionnaire [46], and the questionnaire used in Study 1 (see Fig. 4) without the question #10.

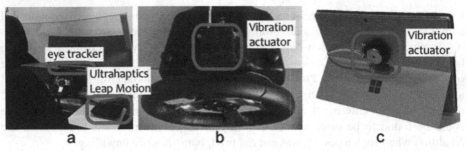

a   b   c

Fig. 7. Setup in Study 2: a) general view, b) Logitech G27 steering wheel with attached actuator, and c) MS Surface Pro 3 tablet with an actuator attached to its back lid.

4.3 Results

We did not perform data filtering due to the increased number of trials in this study. However, we excluded a few participants from some analysis due to low data quality, as mentioned below for each measurement separately. Data from one participant was excluded completely due to very poor driving performance, leaving us with the number of participants $N = 22$. We used the same statistical methods for analysis as in Study 1.

Objective Measures

Attention to the Road. Three participants were excluded from gaze data analysis due to poor tracking quality. Eye-gaze data for the rest of the participants ($n = 19$) showed that the gaze was off the road for 10.5% when interacting using touchscreen, 6.8% when using mid-air gestures, and 6.3% when using buttons. Though the touchscreen method resulted in many glances toward the screen with the menu for some participants, most of them were very short, and therefore Friedman's test did not reveal statistically significant effect of interaction method for gaze-off-the-road time, similarly as in Study 1.

The number of > 2 s-long gaze-away glances that are known to compromise driving safety [39] was ranging from 13 (*buttons*) to 36 (*gestures*) and 41 (*touch*), thus notably more often than in Study 1. For the latter two methods, many of these events ($\approx$35%) were collected from a single participant. Friedman's test did not reveal statistically significant effect of interaction method on this measurement.

Driving Outside of the Lane. Only two events of deviating more than 2.2 m from the center of the middle lane were registered when using buttons and touchscreen. The participants deviated from the center 8 times using gestures (half of them were collected from a single participant). Differently from Study 1, Friedman's test did not reveal statistically significant effect of interaction method on this measurement.

Navigation Delay. One-way ANOVA revealed a statistically significant effect of interaction method on navigation delay ($F_{2,21} = 106.7, p < .001$): it was significantly faster to start using buttons ($\mu = 387$ ms) than swipes ($MD_{BT} = 396$ ms, $t(21) = 8.17, p < .001$) and gestures ($MD_{BG} = 766$ ms, $t(21) = 13.42, p < .001$), and faster to use swipes than gestures ($MD_{TG} = 370$ ms, $t(21) = 7.2, p < .001$); see Fig. 8, left. For the

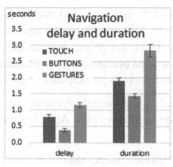
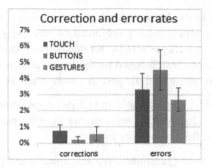

Fig. 8. Mean values (±*SEM*s) of navigation delay and duration (left) and error rate and its share due to the incorrect direction (right) in Study 2.

gesture-based method in Study 2 this delay (1.15 s) was 42% shorter than in Study 1 (1.63 s).

Navigation Duration. One-way ANOVA revealed a statistically significant effect of interaction method ($F_{2,21} = 43.9$, $p < .001$): Post hoc tests showed that button presses ($\mu = 1.44$ s) were faster than swipes ($MD_{BT} = 0.45$ s, $t(21) = 5.45$, $p = .002$) and mid-air gestures ($MD_{BG} = 1.4$ s, $t(21) = 8.13$, $p < .001$), and swipes were faster than mid-air gestures ($MD_{TG} = 0.95$ s, $t(21) = 5.2$, $p < .001$). Interaction with gestures was about three times faster than in Study 1 ($\mu = 2.84$ s vs. $\mu = 8.35$ s).

Correction Actions. Only few correction actions were recorded for each interaction method (see Fig. 8, right): 0.78% of all trials using touchscreen, 0.22% using buttons and 0.58% using gestures. Friedman's test did not show statistically significant effect of interaction method. This is in line with Study 1.

Error Rate. Four participants were recognized as outliers and their data was removed from the analysis. The average rate of uncorrected errors using buttons and touchscreen was 4.5% and 3.3%, correspondingly, thus, these were somewhat higher than in Study 1. The number of errors made when using gestures, however, decreased, and was 2.7%. Friedman's test did not show statistically significant effect of interaction method.

We calculated the share of errors caused by navigating in the wrong direction. For navigation using gestures this measurement should reveal how easy it was to confuse with the palm's stimulation side when the menu was reporting its navigation direction. Confusion with the navigation direction was the source of 32.5% errors when using touchscreen, 26.9% when using buttons, and 66% when using gestures, but Friedman's test showed that the differences were not statistically significant.

Subjective Measures

The Interaction Method Score. Buttons were selected 14 times as the most preferable method, and therefore received the highest score $S_B = 0.78$. Touchscreen-based and gesture-based interaction methods were rated very similarly and got same score: $S_T = 0$ $S_G = .35$. We divided the participants into two equal groups (N = 11), younger (19–25 years, *Y*) and older (28–45 year, *O*), and found that both groups did not differ in their ratings of the button-based interaction ($S_{B,Y} = 0.77$ and $S_{B,O} = 0.79$). However, the younger group usually preferred swipes ($S_{T,Y} = 0.59$, 4 rated it as the best) over mid-air gestures ($S_{G,Y} = 0.14$, 8 rated as the worst). The older group revealed the opposite preferences: touch-based method was rated as the least preferred by 9 participants ($S_{T,O} = 0.13$), while mid-air gestures were rated as the best by 5 participants ($S_{G,O} = 0.54$).

NASA-TLX. Friedman's and pairwise tests revealed that mental, physical and temporal demands for gestures were higher than for buttons and/or swipes. Also, buttons caused the least frustration (see Fig. 9).

End-Session Questionnaire. Friedman's and pairwise tests revealed that gestures were learned not as quickly as swipes and button presses, and they also were harder to use than buttons. Also, corrective actions with this method were harder to execute than with other interaction methods. In addition, the participants felt that it was easier to stay in the middle lane while driving when using buttons than other methods.

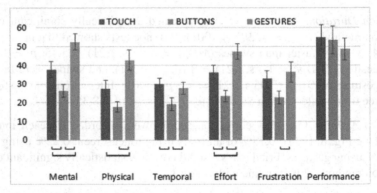

Fig. 9. Mean responses ($\pm SEM$s) for TLX rating scales (1–100) in Study 2. The significance of differences of means for each of 3 pairs within each scale is marked using bracket-like signs.

P-SUS. Friedman's and pairwise tests revealed that rating for only "Functions in this system are well integrated" and" The system is very intuitive" scales did not depend on the interaction method. Buttons were recognized as the best candidate for frequent usage. The interaction based on gestures got lower ratings than interactions based on buttons and/or swipes in the following scales: "The system is simple", "The system is easy to use", "Could use this system with a support", "Most people would learn this system quickly", "I felt confident using this system", and "No need to learn anything new".

Interviews. Some of participants noted that mid-air gestures required them to concentrate on the task notably, but with further training it got easier and easier. Few participants reported feeling the UH feedback in the middle of the palm or beyond palm edges sometimes. Others used phrases like "it felt very nice", "it was the most pleasant experience" and" it was easy to focus on driving" to describe their experience. A couple of participants noted that they were not able to verify whether they navigated in the direction as requested. One participant noted that in few trials the same UH feedback was perceived even if she changed the navigation direction with the rotational gesture.

Interaction with buttons was described as "comfortable, no need to remove hands from the steering wheel", "very easy to use" and "definitely something to expect from the car". Some participants noted that buttons were too soft and suggested adding a "click" sound. Three participants were surprised to find out that it was hard to decide which hand they should use to navigate using buttons, while there were no such issues when using other interaction methods.

Many participants were surprised to find out how much of their visual attention was occupied with looking at the screen when using swipes ("surprisingly hard", "I had to focus on the screen", "it causes distraction from the road"). Others noted that they did not look at the screen, as the menu items were large enough to hit them blindly or using peripheral vision. Some mentioned a high level of familiarity in using swipes. Also, not all participants could feel the screen vibration while swiping and requested other feedback to be provided, like sound effects.

5 Discussion

Both objective measurements and subjective ratings revealed that design issues found in Study 1 were mostly solved in Study 2: navigation delay and duration were notably shorter (up to 3 times), the haptic feedback was well perceivable (unless Leap Motion had difficulties to detect a hand), and participants ranked the touchscreen and gesture -based methods similarly. Buttons-based interaction method, as expected, remained the most preferable way to interact with the menu. Previous research shows that more complex types and designs of interaction in a car may result in totally different preferences, like touchscreen preferred over steering wheel buttons (like in [29]) or mid-air gestures preferred over touchscreen (like in [15]).

The main study focus, however, was not to find the optimal way for menu navigation using mid-air gestures, as two separate gestures for navigating left and right would result in a more efficient interaction. Instead, we were verifying the ability of ultrasound haptics feedback to deliver system state using some of the palm stimulation shapes easiest to recognize [43, 51]. The verification shows that participants were able to recognize the cueing pattern in 98.2% of the trials. Also, the error rate in *gestures* condition was similar to the one observed in *buttons* and *touch* conditions. It left unclear whether in any of the trials the participants could not correctly identify the side of the UH feedback applied to their palms. The participants, however, noted that in rare cases the feedback had an offset relative to the expected location, like closer to the palm's center, or felt only marginally at the very edge on their palm.

It is worth noting that system cueing abilities may depend not only on the ultrasound stimulation signal, but also on other technologies, like hand recognition. Participants who experienced long hand recognition delays and lower hand tracking quality complained about the system not responding to their actions as expected. These participants tended to rank the gesture-based method as the least suitable to interact with a car. Other participants who rarely faced such issues reported that cueing with ultrasound was clear and on time. These participants could prefer gestures over swipes on a touchscreen, noting that touchscreen may require some visual attention shifted from the road.

While both gestures and swipes were ranked about equally in the final questionnaire, the ratings collected using TLX questionnaire indicate that mental, physical and temporal demands were higher when using the mid-air interaction method. This is in line with previous research [36]. Similarly, ratings from P-SUS questionnaire proved that gestures required time to learn and assistance from a technical person. Rutten et al. [43] claimed that the technology novelty effect may contribute to the positive experience even though this technology results in longer and mentally more demanding interaction.

In the lab environment with little attention required to perform the driving task, no differences in deviating from the lane or prolonged gazes off the road were registered when comparing data collected using any interaction method. However, in a real car where the attention demand is higher, this may be different [51]. We suggest that the interaction with ultrasound skin stimulation used to report the system state should be further verified in a car running on a test track and equipped with a more reliable hand tracking system (refer to [55]).

An interesting effect of participants' age was discovered regarding the interaction method preferences. Younger participants familiar with touchscreens from childhood preferred this interaction method over mid-air gestures. Older participants showed the opposite attitude claiming that touchscreens require unnecessary attention shift from the road. Presumably, all our participants were rather experienced in using touch-operated devices, with smartphones being used daily. The two groups, however, had very different driving experiences, 3.5 versus 15.9 years in average. This finding shows that either participant's early interaction experience or driving experience may play a significant role when evaluating interaction methods that are still new in cars for the moment. A suggestion to separate participants by age when studying new interaction methods in automotive context was expressed also in earlier studies (e.g. [27]).

6 Conclusions

We conducted two studies to investigate the ability to use ultrasound skin stimulation to cue car drivers while interacting with IVIS using mid-air gestures. The ability to utilize UH feedback for cueing was shown in the first study. In the second study we proved that controlling IVI system that delivers its state via UH feedback may be as favorable as using swipes on a touchscreen. We showed that skin stimulation patterns projected to a palm could help drivers to stay aware of the interactive system state, and such cueing may be intuitive and require little learning efforts. Untrained drivers may require notable non-visual attention to decode the system state from the cuing signal, however. Therefore, more research is needed to ensure, that cueing with UH feedback does not deteriorate driving safety, including research in a moving car. The findings also suggest that there is room for improving the reliability of the interaction based on mid-air gestures, like increasing the robustness of the hand tracking method or ensuring the optimal UH device location. With these improvements archived, interacting with IVIS using mid-air gestures may potentially be more comfortable than using swipes on a central stack touchscreen.

Acknowledgements. This research was carried out as part of the Adaptive Multimodal In-Vehicle Interaction project (AMICI), which was funded by Business Finland (grant 1316/31/2021).

References

1. Alpern, M., Minardo, K.: Developing a car gesture interface for use as a secondary task. In: Proceedings of CHI-EA 2003, pp. 932–933. ACM (2003). https://doi.org/10.1145/765891. 766078
2. Bach, K.M., Jæger, M.G., Skov, M.B., Thomassen, N.G.: You can touch, but you can't look: interacting with in-vehicle systems. In: SIGCHI Conference on Human Factors in Computing Systems (CHI 2008), pp. 1139–1148. ACM (2008)
3. Bilius, L.B., Vatavu, R.-D.: A multistudy investigation of drivers and passengers' gesture and voice input preferences for in-vehicle interactions. J. Intell. Transport. Syst. **25**(2), 197–220 (2021). https://doi.org/10.1080/15472450.2020.1846127
4. Brand, D., Büchele, K., Meschtscherjakov, A.: Pointing at the HUD: gesture interaction using a leap motion. In: Proceedings of AutomotiveUI 2016 Adjunct, pp. 167–172. ACM (2016)
5. Brown, E., Large, D.R., Limerick, H., Burnett, G.: Ultrahapticons: "Haptifying" drivers' mental models to transform automotive mid-air haptic gesture infotainment interfaces. In: Proceedings of AutomotiveUI 2020, pp. 54–57. ACM (2020). https://doi.org/10.1145/340 9251.3411722
6. Carter, T., Seah, S.A., Long, B., Drinkwater, B., Subramanian, S.: UltraHaptics: multi-point mid-air haptic feedback for touch surfaces. In: Proceedings of UIST 2013, pp. 505–514. ACM (2013)
7. Ch, N., Tosca, D., Crump, T., Ansah, A., Kun, A., Shaer, O.: Gesture and voice commands to interact with AR windshield display in automated vehicle: a remote elicitation study. In: Proceedings of AutomotiveUI 2022, pp. 171–182. ACM (2022). https://doi.org/10.1145/354 3174.3545257
8. Detjen, H., Faltaous, S., Geisler, S., Schneegass, S.: User-defined voice and mid-air gesture commands for Maneuver-based interventions in automated vehicles. In: Proceedings of Mensch Und Computer, pp. 341–348. ACM (2019). https://doi.org/10.1145/3340764.334 0798
9. Ecker, E., Broy, V., Hertzschuch, K., Butz, A.: Visual cues supporting direct touch gesture interaction with in-vehicle information systems. In: Proceedings of AutomotiveUI 2010, pp. 80–87. ACM (2010). https://doi.org/10.1145/1969773.1969788
10. Fariman, H. J., Alyamani, H.J., Kavakli, M., Hamey, L.: Designing a user-defined gesture vocabulary for an in-vehicle climate control system. In: Proceedings of OzCHI 2016, pp. 391–395. ACM (2016). https://doi.org/10.1145/3010915.3010955
11. Farooq, A., et al.: Where's my cellphone: non-contact based hand-gestures and ultrasound haptic feedback for secondary task interaction while driving. In: 2021 IEEE Sensors, pp. 1–4. IEEE (2021)
12. Freeman, E., Vo, D.-B., Brewster, S.: HaptiGlow: helping users position their hands for better mid-air gestures and ultrasound haptic feedback. In: 2019 IEEE World Haptics Conference (WHC), pp. 289–294. IEEE (2019). https://doi.org/10.1109/WHC.2019.8816092
13. Gable, T.M., May, K.R., Walker, B.N.: Applying popular usability heuristics to gesture interaction in the vehicle. In: Adjunct Proceedings of AutomotiveUI 2014, pp. 1–7. ACM (2014)
14. Georgiou, O., et al.: Haptic in-vehicle gesture controls. In: Proceedings of AutomotiveUI 2017, pp. 233–238. ACM 2017. https://doi.org/10.1145/3131726.3132045

15. Graichen, L., Graichen, M., Krems, J.F.: Evaluation of gesture-based in-vehicle interaction: user experience and the potential to reduce driver distraction. Hum. Factors 61(5), 774–792 (2019). https://doi.org/10.1177/001872081882425

16. Graichen, L., Graichen, M., Krems, J.F.: Effects of gesture-based interaction on driving behavior: a driving simulator study using the projection-based vehicle-in-the-loop. Hum. Factors 64(2), 324–342 (2022). https://doi.org/10.1177/0018720820943284

17. Harrington, K., Large, D.R., Burnett, G., Georgiou, O.: Exploring the use of mid-air ultrasonic feedback to enhance automotive user interfaces. In: Proceedings of AutomotiveUI 2018, pp. 11–20. ACM (2018). https://doi.org/10.1145/3239060.3239089

18. Hart, S.G., Staveland, L.E.: Development of NASA-TLX (Task Load Index): results of empirical and theoretical research. In: Hancock, P. A., Meshkati, N. (eds.) Adv. in Psychology, vol. 52, pp. 139–183. North Holland Press, Amsterdam (1988)

19. Häuslschmid, R., Menrad, B. Butz, A.: Freehand vs. micro gestures in the car: driving performance and user experience. In: IEEE Symposium on 3D User Interfaces (3DUI), pp. 159–160. IEEE (2015). https://doi.org/10.1109/3DUI.2015.7131749

20. Hessam, J.F., Zancanaro, M., Kavakli, M., Billinghurst, M.: Towards optimization of mid-air gestures for in-vehicle interactions. In: Proceedings of OzCHI 2017, pp. 126–134. ACM (2017)

21. Ho, C., Tan, H.Z., Spence, C.: Using spatial vibrotactile cues to direct visual attention in driving scenes. Transport. Res. F Traffic Psychol. Behav. 8(6), 397–412 (2005)

22. Hoshi, T., Takahashi, M., Iwamoto, T., Shinoda, H.: Noncontact tactile display based on radiation pressure of airborne ultrasound. IEEE Trans. Haptics 3(3), 155–165 (2010)

23. Howard, T., Gallagher, G., Lécuyer, A., Pacchierotti, C., Marchal, M.: Investigating the recognition of local shapes using mid-air ultrasound haptics. In: Proceedings of IEEE World Haptics Conference (WHC 2019), pp. 503–508. IEEE (2019). https://doi.org/10.1109/WHC.2019.8816127

24. Huemer, A.K., Vollrath, M.: Learning the Lane change task: comparing different training regimes in semi-paced and continuous secondary tasks. Appl. Ergon. 43(5), 940–947 (2012). https://doi.org/10.1016/j.apergo.2012.01.002

25. Iwamoto, T., Tatezono, M., Shinoda, H.: Non-contact method for producing tactile sensation using airborne ultrasound. In: Ferre, M. (ed.) EuroHaptics 2008. LNCS, vol. 5024, pp. 504–513. Springer, Heidelberg (2008). https://doi.org/10.1007/978-3-540-69057-3_64

26. Kim, M., Seong, E., Jwa, Y., Lee, J., Kim, S.: A cascaded multimodal natural user interface to reduce driver distraction. IEEE Access 8, 112969–112984 (2020)

27. Korres, G., Chehabeddine, S., Eid, M.: Mid-air tactile feedback co-located with virtual touchscreen improves dual-task performance. IEEE Trans. Haptics 13(4), 825–830 (2020). https://doi.org/10.1109/TOH.2020.2972537

28. Large, D R., Burnett, G., Crundall, E., Lawson, G., Skrypchuk, L.: Twist it, touch it, push it, swipe it: evaluating secondary input devices for use with an automotive touchscreen HMI. In: Proceedings of Automotive'UI 2016, pp. 161–168. ACM (2016)

29. Large, D.R., Harrington, K., Burnett, G., Georgiou, O.: Feel the noise: Mid-air ultrasound haptics as a novel human-vehicle interaction paradigm. Appl. Ergon. 81, 102909 (2019)

30. Long, B., Seah, S.A., Carter, T., Subramanian, S.: Rendering volumetric haptic shapes in mid-air using ultrasound. ACM Trans. Graph. 33(6), 1–10 (2014)

31. Ma, J., Du, Y.: Study on the evaluation method of in-vehicle gesture control, In: IEEE International Conference on Control Science and System Engineering (ICCSSE), pp. 145–148. IEEE (2017)

32. Mahr, A., Endres, C., Müller, C., Schneeberger, T.: Determining human-centered parameters of ergonomic micro-gesture interaction for drivers using the theater approach. In: Proceedings of AutomotiveUI 2011, pp. 151–158. ACM (2011). https://doi.org/10.1145/2381416.2381441

33. Manawadu, U.E., Kamezaki, M., Ishikawa, M., Kawano, T., Sugano, S.: A hand gesture based driver-vehicle interface to control lateral and longitudinal motions of an autonomous vehicle. In: IEEE International Conference on Systems, Man, and Cybernetics (SMC), pp. 1785–1790 (2016)

34. Martinez, P.I.C., De Pirro, S., Vi, C.T., Subramanian, S.: Agency in Mid-air Interfaces. In: Proceedings of CHI 2017, pp. 2426–2439. ACM (2017). https://doi.org/10.1145/3025453.3025457

35. Maunsbach, M., Hornbæk, K., Seifi, H.: Whole-hand haptics for mid-air buttons. In: Seifi, H., et al. (eds.) Haptics: Science, Technology, Applications: 13th International Conference on Human Haptic Sensing and Touch Enabled Computer Applications, EuroHaptics 2022, Hamburg, Germany, May 22–25, 2022, Proceedings, pp. 292–300. Springer, Cham (2022). https://doi.org/10.1007/978-3-031-06249-0_33

36. May, K.R., Gable, T.M., Walker, B.N.: A multimodal air gesture interface for in vehicle menu navigation. In: Adjunct Proceedings of AutomotiveUI 2014, pp. 1–6. ACM (2014)

37. May, K.R., Gable, T.N., Walker, B.N.: Designing an in-vehicle air gesture set using elicitation methods. In: Proceedings of AutomotiveUI 2017, pp. 74–83. ACM (2017)

38. May, K.R., Gable, T.M., Wu, X., Sardesai, T.R., Walker, B.N.: Choosing the right air gesture: impacts of menu length and air gesture type on driver workload. In: Adjunct Proceedings of AutomotiveUI 2016, pp. 69–74. ACM (2016). https://doi.org/10.1145/3004323.3004330

39. NHTSA. Visual-Manual NHTSA Driver Distraction Guidelines for In-Vehicle Electronic Devices. Federal Register, vol. 77, no. 37, pp. 11200–11250 (2012)

40. Ng, A., Brewster, S.: An evaluation of touch and pressure-based scrolling and haptic feedback for in-car touchscreens. In: Proceedings of AutomotiveUI 2017, pp. 11–20. ACM (2017)

41. Riener, A., et al.: Standardization of the in-car gesture interaction space. In: Proceedings of AutomotiveUI 2013, pp. 14–21. ACM (2013)

42. Rakkolainen, I., Freeman, E., Sand, A., Raisamo, R., Brewster, S.: A survey of mid-air ultrasound haptics and its applications. IEEE Trans. Haptics **14**(1), 2–19 (2021)

43. Rutten, I., Geerts, D.: Better because it's new: the impact of perceived novelty on the added value of mid-air haptic feedback. In: Proceedings of CHI 2020, pp. 1–13. ACM (2020)

44. Rümelin, S., Gabler, T., Bellenbaum, J.: Clicks are in the air: how to support the interaction with floating objects through ultrasonic feedback. In: Proceedings of AutomotiveUI 2017, pp. 103–108. ACM (2017). https://doi.org/10.1145/3122986.3123010

45. Sauras-Perez, P., Taiber, J., Smith, J.: Variability analysis of in-car gesture interaction. In: International Conference on Connected Vehicles and Expo (ICCVE), pp. 777–780. IEEE (2014)

46. Sauro, J., Lewis, J.R.: When designing usability questionnaires, does it hurt to be positive? In: Proceedings of CHI 2011, pp. 2215–2224. ACM (2011). https://doi.org/10.1145/1978942.1979266

47. Shakeri, G., Williamson, J. H., Brewster, S.: May the force be with you: ultrasound haptic feedback for mid-air gesture interaction in cars. In: Proceedings of AutomotiveUI 2018, pp. 1–10. ACM (2018). https://doi.org/10.1145/3239060.3239081

48. Stiegemeier, D., Bringeland, S., Kraus, J., Baumann, M.: User experience of in-vehicle gesture interaction: exploring the effect of autonomy and competence in a mock-up experiment. In: Proc. of AutomotiveUI 2022, pp. 285–296. ACM (2022)

49. Sun, C., Nai, W., Sun, X.: Tactile sensitivity in ultrasonic haptics: do different parts of hand and different rendering methods have an impact on perceptual threshold? Virt. Real. Intell. Hardw. **1**(3), 265–275 (2019). https://doi.org/10.3724/SP.J.2096-5796.2019.0009

50. Swette, R., May, K.R., Gable, T.M., Walker, B.N.: Comparing three novel multimodal touch interfaces for infotainment menus. In: Proceedings of AutomotiveUI 2013, pp. 100–107. ACM (2013). https://doi.org/10.1145/2516540.2516559

51. Špakov, O., Farooq, A., Venesvirta, H., Hippula, A., Surakka, V., Raisamo, R.: Ultrasound feedback for mid-air gesture interaction in vibrating environment. In: Ahram, T., Taiar, R. (eds) IHIET-AI: Artificial Intelligence & Future Applications International Conference, vol. 23. AHFE Open Access (2022)

52. Vo, D-B., Brewster, S.: Touching the invisible: Localizing ultrasonic haptics cues. In: Proceedings of IEEE World Haptics Conference (WHC 2015), pp 368–373. IEEE (2015)

53. Weidner, F., Broll, W.: Interact with your car: a user-elicited gesture set to inform future in-car user interfaces. In: Proceedings of the 18th International Conference on Mobile and Ubiquitous Multimedia (MUM 2019), Article 11, pp. 1–12. ACM (2019). https://doi.org/10.1145/3365610.3365625

54. Wolfe, B., Kosovicheva, A., Stent, S., Rosenholtz, R.: Effects of temporal and spatiotemporal cues on detection of dynamic road hazards. Cogn. Res. Princip. Implicat. 6(1), 1–15 (2021). https://doi.org/10.1186/s41235-021-00348-4

55. Wu, S., Gable, T., May, K., Choi, Y., Walker, B.: Comparison of surface gestures and air gestures for in-vehicle menu navigation. Arch. Des. Res. 29(4), 65 (2016)

56. Young, G., Milne, H., Griffiths, D., Padfield, E., Blenkinsopp, R., Georgiou, O.: Designing mid-air haptic gesture-controlled user interfaces for cars. In: Proceedings of the ACM on Human-Computer Interaction, vol. 4(EICS), pp 1–23. ACM (2020). https://doi.org/10.1145/3397869

57. Zhang, N., Wang, W.-X., Huang, S.-Y., Luo, R.-M.: Mid-air gestures for in-vehicle media player: elicitation, segmentation, recognition, and eye-tracking testing. SN Appl. Sci. 4(4), 1–18 (2022). https://doi.org/10.1007/s42452-022-04992-3

Direct or Indirect: A Video Experiment for In-vehicle Alert Systems

Fengyusheng Wang$^{(\boxtimes)}$, Chia-Ming Chang, and Takeo Igarashi

The University of Tokyo, Tokyo, Japan
wangfys@outlook.com, info@chiamingchang.com, takeo@acm.org

Abstract. Advancements in Vehicle-to-Infrastructure (V2I) and Vehicle-to-Everything (V2X) have improved in-vehicle information systems' ability to provide drivers with global information. Although most global information does not worth a high priority, such as how many empty parking lots are available, the prediction of a potential traffic accident in the next few seconds is of great significance. Our study compares user preferences and performance between two types of accident warning systems which delivers urgent global information in different ways (direct vs indirect). Direct information points out where the potential danger is coming from, while indirect information only implies its presence without further details. Our online video experiment aimed to determine which approach leads to better decision-making by drivers. Results showed that participants preferred direct information, but indirect information led to better decision-making and these findings warrant further discussion.

Keywords: Vehicle-to-Everything · Anthropomorphism · Traffic accident

1 Introduction

Traffic accidents often occur between vehicles and pedestrians/cyclists, with various factors contributing to the cause. With advancements in technology, trajectory prediction is becoming more accurate, enabling vehicles to take proactive measures, rather than relying solely on Automatic Emergency Braking (AEB) as a last resort. A common cause of these accidents is the blockage of view, both for human drivers and for the cameras, ultrasonic radars, and lidars equipped on the vehicle, leading to a limited understanding of the situation due to the lack of global information. Vehicle-to-Infrastructure (V2I) and Vehicle-to-Everything (V2X) technologies can help address this issue by providing vehicles with prediction information based on sensors elsewhere, such as traffic cameras. However, trajectory prediction data from blind zones is not easily understandable by human drivers. It is crucial to develop an effective method to present **direct** information on trajectory data or provide clear warnings and suggestions as **indirect** information. The main difference between direct and indirect information is whether the vehicle helps the driver directly observe a potential danger source hidden behind obstacles.

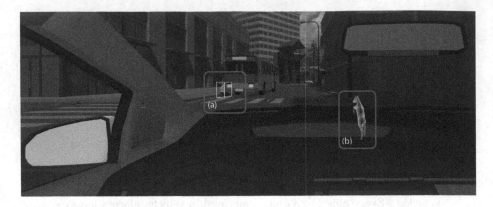

Fig. 1. Direct vs. Indirect. (a) Direct information as warning signals. (b) Indirect information as warning signals.

Enhancing human vision is a **direct** way to deliver trajectory prediction results. The potential danger can be presented to the driver in a clear and specific manner, including its location and type. This approach has been applied in other fields, such as in medicine where X-rays and ultrasonic waves assist doctors in detecting disease signals in the human body, and in video games where cheaters can see through walls. Similarly, thermal imaging systems can detect humans behind walls in real-world scenarios. However, processing this additional information while driving can cause a heavier workload and divert the driver's attention away from the visible road conditions.

In contrast, **indirect** information, which doesn't specify the source and location of danger, can still provide a different perspective for global understanding. In the past, prediction methods were passed down from generation to generation based on observations of natural events, particularly animal behavior. For instance, in some regions, folklore says that if low-flying dragonflies are seen, it will soon rain. Although not always accurate, this demonstrates the correlation between animal behavior and future events. Similarly, animals often exhibit unusual behavior before earthquakes, as evidenced by numerous video recordings. However, when incorporating this type of information into in-vehicle systems, people may not trust it as they cannot observe it directly - "seeing is believing".

In this study, we evaluated the efficacy of direct and indirect information warning systems through an online video experiment. Participants made decisions based on the information from each warning system and gave feedback using Likert Scales. As shown in Fig. 1, the direct warning utilized a Head-Up Display (HUD) or AR Display to show bounding boxes around potential hazards and the indirect warning used an irregular horse behavior to indicate danger. The results indicated that, when a suspected danger source that was predicted to cause a traffic accident within the next few seconds was completely invisible, the indirect information warning system improved participants' braking decision accuracy. There was also a discrepancy between participants' preferred warning system and their actual performance with each system. Although 10 of the 16 participants preferred the direct information warning system and the direct information warning system was deemed more helpful, intuitive, and reliable, the actual performance results were that the indirect system resulted in higher accuracy in decision-making. The indirect warning system was also deemed more understandable.

2 Background

2.1 Vehicle-to-Everything (V2X)

Advances in technology have allowed vehicles to connect with each other (V2V), infrastructure (V2I), and other devices (V2X). This expanded access to information can bring new benefits but also presents new safety risks. Lee et al. [11] studied the distractions posed by connected vehicles and proposed a model for evaluating display.

Other studies have explored advanced road safety goals in a V2X context. Baumgardner et al. [1] evaluated driver use cases for an open platform called the "V2X Hub" during rail crossing tasks. Their work demonstrated the utility of using a variety of human-factors-driven analytical approaches to identify human-machine interface (HMI) enhancements. Fank et al. [7] proposed a user-centered conception and design of an HMI for cooperative truck overtaking maneuvers on freeways. After developing with early involvement of the user, the results show that truck drivers are willing to use such a system and the HMI can support cooperative overtaking maneuvers. Weidl et al. [19] described a novel approach to situation analysis at intersections using object-oriented Bayesian networks. The network can infer the collision probability for all vehicles approaching the intersection and the environment perception is fused from communicated data, vehicles' local perception, and self-localization.

2.2 Anthropomorphic Interface

Another interesting topic is designing an anthropomorphic interface for vehicles. Loehmann et al. [12] proposed a multimodal electric vehicle information system called Heartbeat to communicate the state of the electric drive including energy flow and the energy level of batteries in a natural and experienceable way. Chang et al. [3,5,10] designed a novel interface that adds eyes to self-driving cars so that pedestrians can understand the intention of the approaching autonomous vehicles, as well as a "smile" emoticon [4] on an autonomous vehicle to communicate with pedestrians. Row et al. [15] proposed a vehicle-applicable pet-morphic design strategy and a concept validation prototype to support an emotional user experience of a personal EV. Okamoto et al. [14] introduced an AI agent that supports both internal and external HMI.

Flemisch et al. [9] provided a general ergonomic framework of cooperative guidance and control for vehicles. Cooperation is mainly done on guidance and

Fig. 2. SeeThrough

control with a haptically active interface with H-Mode, a haptic-multimodal interaction with highly automated vehicles based on the H(orse)-Metaphor. Wang et al. [18] proposed an interface that directly uses the image of horses to show warnings to drivers.

2.3   Traffic Accident Alerts

There are many kinds of traffic accidents and lots of them can be predicted and alerted. Merenda et al. [13] investigated how increasing urgency affects driver behavior while using varying levels of the pedestrian alert system(PAS). They got the conclusion that PAS can improve the drivers' braking performance where pedestrians may pose a threat but both audio- and visually-based PAS do not produce gains in the localization of pedestrians.

Brown et al. [2] examined the differences between auditory and haptic Lane Departure Warning systems used in two commercially available cars. The results showed that a haptic system leads to a faster response on average. Du et al. [6] compared different types of display information and modality under different event criticality situations. The results showed that presenting *why* only information was not sufficient compared to *what* only information and *why+what* information.

Von Sawitzky et al. [16] used a Mixed Reality environment and outlined a user study to investigate how to prevent cyclists crashed into a suddenly-opened door of a parked vehicle with the help of the "Smart Traffic" assumption (which is a term similar to V2X).

3   Warning Systems, Research Question, and Hypotheses

3.1   See Through Obstacles

The most direct information is undoubtedly providing the driver with visibility of the source of danger, even if there are obstacles. In virtual worlds like video games or VR, this is achieved by removing [8] or fading [20] obstacles based on their

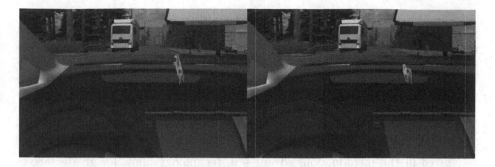

Fig. 3. VirtualHorse

z-depth. In the real world, we have to rely on synthesized content overlaid on the obstacles. To ensure safety, we must consider the dangers of synthesizing content. A visible obstacle is more crucial to notice as it is closer, so we cannot completely overlay it with synthesized content. Instead, we employ a straightforward concept proposed by Von Sawitzky et al. [17]: showing the bounding box of the potential danger. The vehicle collects data through V2X, allowing it to easily determine the size and position of hidden road users. This information is then displayed as a wireframe, as illustrated in Fig. 2.

We call this method "SeeThrough" as it enhances human vision and allows drivers to see through obstacles. Drivers must assess whether each target represents a potential threat or not and decide whether to take an action.

3.2   Virtual Horse

There are several ways to provide indirect information to the driver, including visual cues such as warning signs, ambient light, or text, auditory cues such as beeps, sirens, or speech, and haptic cues on the steering wheel or seat. It is important to consider the amount of time required for the driver to understand the information and the level of distraction from the road conditions. A balance must be struck. The method we have chosen is called VirtualHorse [18], which is an anthropomorphic interface displayed on the dashboard platform through a HUD or AR display. It shows a mini horse that represents the vehicle's state - standing still when stopped and running when in motion. In the event of potential danger, the horse jumps.

The advantage of this warning method is the historical and cultural connection between horses and traffic. Although horses are less common in modern society, their jumping behavior is widely recognized as a symbol of spook and danger in movies, stories, and video games. People can understand this warning even without prior experience with horses. Unlike commonly used signals like traffic lights and sirens, this method reduces the risk of confusion and misinterpretation of the type of danger. For example, a siren may be mistaken as a police car or ambulance approaching.

The "VirtualHorse" method uses data collected through V2X to predict the future trajectory of targets based on their current speed and position. When the target is also connected to V2X, the trajectory prediction can be more accurate if the driving intention is shared. If a target presents a high risk of intersecting with the driver's future trajectory, the mini horse will jump to imply the driver needs to take some actions.

3.3   Research Question and Hypotheses

As stated in the introduction, advances in technology allow for greater access to global information. While much of this information may not be immediately relevant to drivers, there are some critical pieces of information that must be conveyed. So the research question is: **Which type of information is more useful, direct information or indirect information?**

So we have following hypotheses:

- H1: Using indirect information helps users make braking decisions more correctly when facing potential dangers.
- H2: Users think direct information is more understandable when making decisions.
- H3: Users think direct information is more helpful when making decisions.
- H4: Users think direct information is more intuitive when making decisions.
- H5: Users think direct information is more reliable when making decisions.

H1 is our primary hypothesis. We think indirect information can prompt people to react before fully understanding the reason. The virtual horse's reaction in the presence of potential danger is intended to trigger an unconscious understanding, without requiring thorough reasoning. This can be highly beneficial in a traffic accident scenario, as it saves time and effort in the deduction.

Regarding hypotheses H2, H3, H4, and H5, it is very natural that indirect information is considered not direct enough and easy to understand. We anticipate that participants may prioritize direct information when completing the questionnaire after thoughtful consideration. However, this does not necessarily mean that direct information performs better in a traffic accident scenario.

4   User Study

We conducted a purely online video experiment to compare the efficacy of direct and indirect information warnings. Participants completed an online questionnaire that included several short videos, each followed by two corresponding questions.

16 participants (8 males, 8 females) were recruited from social media platforms, mainly university students with an average age of 25.06. While all participants had a driving license, not all drove daily. Participants were informed of the data privacy policy before they agreed to participate and were compensated with 500 JPY for the 20-minute online survey.

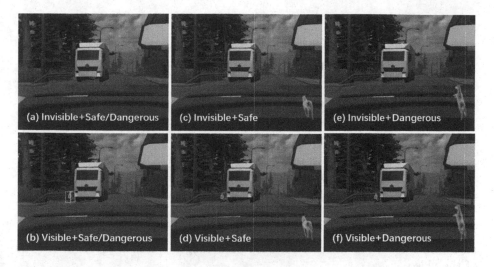

Fig. 4. Different Settings

4.1 Tasks and Conditions

The within-subject experiment has three independent variables:

- The warning system being used.
- The actual risk of the potential source of danger.
- The visibility of the potential source of danger. It refers to whether it can be seen with the naked eye, not through any enhancement of vision.

Each of the variables has two values, resulting in 8 total settings. Figure 4 displays 6 frames from the same scenario of the corresponding videos. Note that there are 8 videos and thus 8 frames, but the SeeThrough warning system provides no information about the actual risk. It's up to the driver to assess the position and velocity and make a decision. Since two pairs of videos are actually identical, they share the same frame.

The 8 settings show a car driving to a crossroads and attempting to turn left, with a truck parked as an obstacle on the opposite side. A scooter is in the scene but only becomes visible in half of the videos.

After each video, the participants are asked to answer two questions: 1) Do you think it is safe to turn left? and 2) Are you confident about your choice?

4.2 Experimental Setup

We build the experiment environment in Unity3D and captured video sequences to create the questionnaire. The participants are shown 8 videos in a round and these videos are repeated 4 times so there are 4 rounds in total. The order of videos in each round is illustrated as Table 1. "Safe" and "Dangerous" mean whether the suspected danger source is actually safe or dangerous. "Visible"

Table 1. Order of videos with varying settings in each round.

ID	Warning System	Actual Risk	Visibility
1	SeeThrough	Safe	Invisible
2	VirtualHorse	Safe	Invisible
3	SeeThrough	Dangerous	Invisible
4	VirtualHorse	Dangerous	Invisible
5	SeeThrough	Safe	Visible
6	VirtualHorse	Safe	Visible
7	SeeThrough	Dangerous	Visible
8	VirtualHorse	Dangerous	Visible

and "Invisible" mean whether the suspected danger source is visible at the end of the video.

Participants were given instructions on the questionnaire before viewing the videos. They were informed of the basic information of two warning systems, SeeThrough (a system that helps them see through obstacles) and VirtualHorse (a system that shows a mini horse that represents the vehicle). However, for SeeThrough they were shown two images as in Fig. 2, but for VirtualHorse they were only shown the right part of Fig. 3, without knowledge of the direct correlation between the horse's behavior and the correct answer. The participants were instructed to watch each video only once before answering the two corresponding questions. They were also informed that the parked truck would not pose a danger in the videos, to avoid any confusion or nervousness.

Participants were asked to fill out four Likert Scales evaluating the five properties of the two warning systems after watching all the videos. Four of the properties (understandable, helpful, intuitive, and reliable) correspond to our hypotheses, while the fifth was the visibility of the warning signal, allowing us to gauge participants' ability to detect warning information while driving. The four Likert Scales were included to capture participants' differing perspectives on visible and non-visible dangers.

5 Results

The accuracy of decision-making was recorded, while feedback was collected using Likert Scales. The Likert Scale results were analyzed using the Wilcoxon test and yielded interesting findings.

5.1 User Preference

The results indicate that 10 out of 16 participants preferred using SeeThrough as a warning system, meaning our primary hypothesis H1 is supported. As illustrated in Fig. 5, the indirect information provided by VirtualHorse is perceived

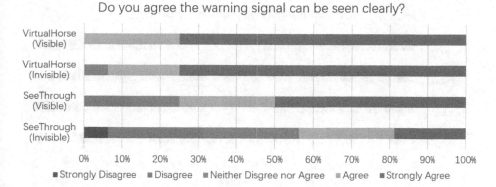

Fig. 5. Likert scale results of perceived clarity of warning information for two warning systems in relation to the visibility of the potential danger source.

as significantly clearer (p-value < 0.001) as it is displayed within the vehicle. The direct information presented as a bounding box is not as easily visible, particularly when the potential danger source is far away, resulting in a small bounding box size in the field of view.

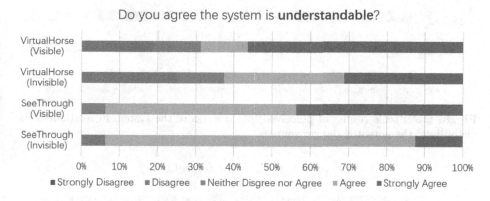

Fig. 6. Likert scale results of perceived understandability of the system in relation to the visibility of the potential danger source.

As shown in Fig. 6, the VirtualHorse system is perceived as significantly more understandable (p-value = 0.037). This supports the rejection of hypothesis H2. The remaining hypotheses (H3, H4, H5) are supported by the results shown in Fig. 7, Fig. 8, and Fig. 9, although without significance. The insights provided by participants through their detailed feedback offer a deeper understanding of the results.

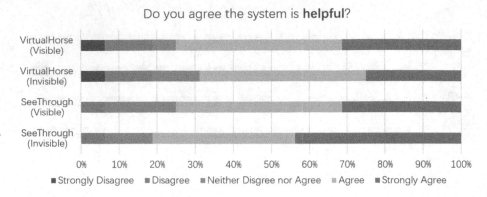

Fig. 7. Likert scale results of perceived helpfulness of the system in relation to the visibility of the potential danger source.

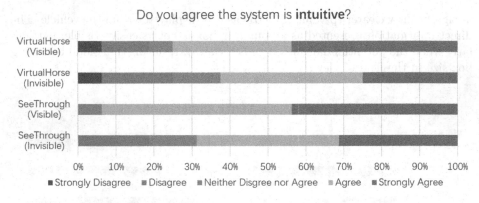

Fig. 8. Likert scale results of perceived intuitiveness of the system in relation to the visibility of the potential danger source.

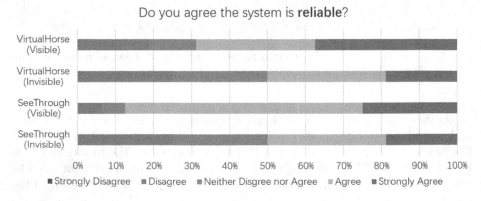

Fig. 9. Likert scale results of perceived reliability of the system in relation to the visibility of the potential danger source.

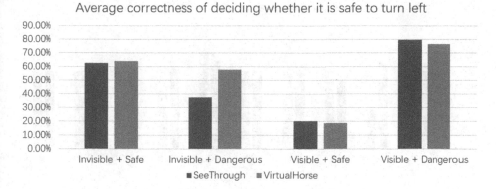

Fig. 10. The accuracy of decision-making

P6, who preferred VirtualHorse, said, "First, VirtualHorse is in motion. It is easier to detect when the state changes. Second, the movement state of the VirtualHorse is synchronized with the car, which is very vivid, and it is easy for the driver to have sympathy so that it can brake in time when a dangerous situation happens."

On the other hand, P7, who preferred SeeThrough, stated, "Although the VitualHorse system tells that it's not safe to turn left, it doesn't give the direct signal that it's safe to turn since there is no image change of the Horse. I'm not sure if the VitualHorse is still working or freezes when facing a crosswalk."

It is clear that participants, no matter which warning system they prefer, were able to understand the connection between the mini horse's behavior and the actual risk, even without receiving detailed information beyond the instruction that "the mini horse represents the car."

5.2 Actual Performance

The results in Fig. 10 contrast with participant preferences. It indicates the accuracy of their decision-making when determining if it's safe to turn left.

For visible scenarios, SeeThrough and VirtualHorse perform similarly, with most participants recognizing a visible potential danger as an actual danger. An unexpected scooter is perceived as highly risky, leading to cautionary actions such as slowing down. Choosing "not safe" when seeing a sudden scooter is not a major mistake. Being cautious while driving is always beneficial.

For invisible potential dangers, VirtualHorse helps participants make safer decisions. Despite some participants favoring SeeThrough for its ability to extend their vision, it leads to more errors. Some participants exhibit indecision when the enhanced vision detects something hidden behind a truck. It is important to note that "Invisible + Safe + SeeThrough" and "Invisible + Dangerous + SeeThrough" have exactly the same video content, so participants should make the same decisions in both cases. However, some of the participants make different choices.

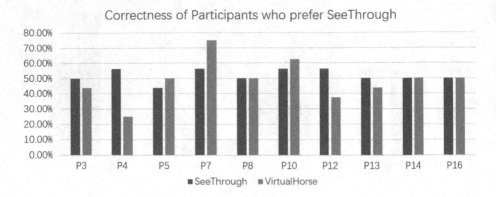

Fig. 11. The accuracy of participants preferred "SeeThrough".

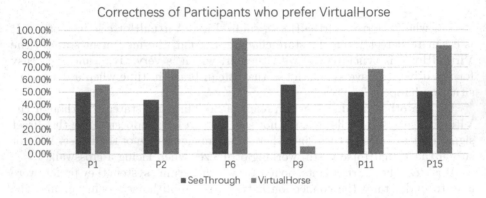

Fig. 12. The accuracy of participants preferred "VirtualHorse".

As shown in Fig. 11, even among participants who favored SeeThrough, VirtualHorse still produced comparable accuracy in decision-making. Figure 12 shows that, for participants who preferred VirtualHorse, it appeared to obviously enhance their decision-making accuracy. The accuracy of P9 appears to be very strange but maybe it is just a strange outlier.

6   Limitation and Future Work

The user study had some limitations that leave room for future improvements. Firstly, the study was conducted during the COVID-19 pandemic, making it challenging to recruit onsite participants for a VR study. We resorted to an online video questionnaire, but it was not interactive enough to gather data such as time spent on decision-making.

Secondly, the assumption of 100% accuracy in predicting traffic accidents with future technologies is unrealistic. To accurately evaluate the trustworthiness of the VirtualHorse warning system, we need to take a more realistic prediction accuracy into consideration and present the consequences of actions taken.

Lastly, the binary warning signals provided by VirtualHorse may impact the results, even though we did not explain the detailed mechanism to participants. Further research is needed to determine the importance of binary signals versus non-binary signals.

Additionally, there is a crucial phenomenon to consider: During the user study, participants may be overly cautious to avoid a traffic accident, which is not representative of a typical driving situation. Simulating the average accident rate and frequency of dangerous situations encountered in daily life is not practical. Thus, future study on how to alleviate this effect is necessary.

7 Conclusion

In this study, we evaluated the direct information warning system "SeeThrough" and the indirect information warning system "VirtualHorse" using an online video experiment. The main difference between these two warning systems is that SeeThrough helps drivers to "see" hidden road users behind obstacles while VirtualHorse implies the existence of potential danger. Despite the fact that a majority of participants preferred SeeThrough, they rated VirtualHorse as being more understandable. Our research found a mismatch between the warning system preferred by participants and the system that was more accurate in making decisions. VirtualHorse showed higher accuracy, especially in the case of a predicted traffic accident where the danger source is not visible. These findings offer a vital understanding of the trade-off between direct and indirect warning information and can guide the creation of future warning systems. The conflicting performance of these two types of warnings emphasizes the need for further study to enhance the efficiency of information warnings for drivers and other users.

Acknowledgments. This work was supported by JST CREST Grant Number JP-MJCR17A1, Japan.

References

1. Baumgardner, G., Hoekstra-Atwood, L.M., Prendez, D.M.: Concepts of connected vehicle applications: Interface lessons learned from a rail crossing implementation, pp. 280–290. AutomotiveUI '20, Association for Computing Machinery, New York, NY, USA (2020). https://doi.org/10.1145/3409120.3410640
2. Brown, D.E., Reimer, B., Mehler, B., Dobres, J.: An on-road study involving two vehicles: Observed differences between an auditory and haptic lane departure warning system. In: Adjunct Proceedings of the 7th International Conference on Automotive User Interfaces and Interactive Vehicular Applications, pp. 140–145. AutomotiveUI '15, Association for Computing Machinery, New York, NY, USA (2015). https://doi.org/10.1145/2809730.2809747

3. Chang, C.M., Toda, K., Gui, X., Seo, S.H., Igarashi, T.: Can eyes on a car reduce traffic accidents? In: Proceedings of the 14th International Conference on Automotive User Interfaces and Interactive Vehicular Applications. pp. 349–359. AutomotiveUI '22, Association for Computing Machinery, New York, NY, USA (2022). https://doi.org/10.1145/3543174.3546841

4. Chang, C.M., Toda, K., Igarashi, T., Miyata, M., Kobayashi, Y.: A video-based study comparing communication modalities between an autonomous car and a pedestrian. In: Adjunct Proceedings of the 10th International Conference on Automotive User Interfaces and Interactive Vehicular Applications, pp. 104–109. AutomotiveUI '18, Association for Computing Machinery, New York, NY, USA (2018). https://doi.org/10.1145/3239092.3265950

5. Chang, C.M., Toda, K., Sakamoto, D., Igarashi, T.: Eyes on a car: An interface design for communication between an autonomous car and a pedestrian. In: Proceedings of the 9th International Conference on Automotive User Interfaces and Interactive Vehicular Applications, pp. 65–73. AutomotiveUI '17, Association for Computing Machinery, New York, NY, USA (2017). https://doi.org/10.1145/3122986.3122989

6. Du, N., Zhou, F., Tilbury, D., Robert, L.P., Yang, X.J.: Designing alert systems in takeover transitions: The effects of display information and modality. In: 13th International Conference on Automotive User Interfaces and Interactive Vehicular Applications, pp. 173–180. AutomotiveUI '21, Association for Computing Machinery, New York, NY, USA (2021). https://doi.org/10.1145/3409118.3475155

7. Fank, J., Knies, C., Diermeyer, F.: After You! Design and Evaluation of a Human Machine Interface for Cooperative Truck Overtaking Maneuvers on Freeways, pp. 90–98. Association for Computing Machinery, New York, NY, USA (2021). https://doi.org/10.1145/3409118.3475139

8. Feiner, S.K., Seligmann, D.D.: Cutaways and ghosting: satisfying visibility constraints in dynamic 3d illustrations. Vis. Comput. **8**, 292–302 (1992). https://doi.org/10.1007/BF01897116

9. Flemisch, F.O., Bengler, K., Bubb, H., Winner, H., Bruder, R.: Towards cooperative guidance and control of highly automated vehicles: H-mode and conduct-by-wire. Ergonomics **57**(3), 343–360 (2014). https://doi.org/10.1080/00140139.2013.869355

10. Gui, X., Toda, K., Seo, S.H., Chang, C.M., Igarashi, T.: "i am going this way": Gazing eyes on self-driving car show multiple driving directions. In: Proceedings of the 14th International Conference on Automotive User Interfaces and Interactive Vehicular Applications, pp. 319–329. AutomotiveUI '22, Association for Computing Machinery, New York, NY, USA (2022). https://doi.org/10.1145/3543174.3545251

11. Lee, J., Lee, J.D., Salvucci, D.D.: Evaluating the distraction potential of connected vehicles, pp. 33–40. AutomotiveUI '12, Association for Computing Machinery, New York, NY, USA (2012). https://doi.org/10.1145/2390256.2390261

12. Loehmann, S., Landau, M., Koerber, M., Butz, A.: Heartbeat: Experience the pulse of an electric vehicle, pp. 1–10. AutomotiveUI '14, Association for Computing Machinery, New York, NY, USA (2014). https://doi.org/10.1145/2667317.2667331

13. Merenda, C., Kim, H., Gabbard, J.L., Leong, S., Large, D.R., Burnett, G.: Did you see me? assessing perceptual vs. real driving gains across multi-modal pedestrian alert systems. In: Proceedings of the 9th International Conference on Automotive User Interfaces and Interactive Vehicular Applications, pp. 40–49. AutomotiveUI

'17, Association for Computing Machinery, New York, NY, USA (2017). https://doi.org/10.1145/3122986.3123013

14. Okamoto, S., Sano, S.: Anthropomorphic ai agent mediated multimodal interactions in vehicles. In: Proceedings of the 9th International Conference on Automotive User Interfaces and Interactive Vehicular Applications Adjunct, pp. 110–114. AutomotiveUI '17, Association for Computing Machinery, New York, NY, USA (2017). https://doi.org/10.1145/3131726.3131736

15. Row, Y.K., Kim, C.M., Nam, T.J.: Dooboo: Pet-like interactive dashboard towards emotional electric vehicle. In: Proceedings of the 2016 CHI Conference Extended Abstracts on Human Factors in Computing Systems, pp 2673–2680. CHI EA '16, Association for Computing Machinery, New York, NY, USA (2016). https://doi.org/10.1145/2851581.2892460

16. von Sawitzky, T., Grauschopf, T., Riener, A.: No need to slow down! a head-up display based warning system for cyclists for safe passage of parked vehicles. In: 12th International Conference on Automotive User Interfaces and Interactive Vehicular Applications, pp. 1–3. AutomotiveUI '20, Association for Computing Machinery, New York, NY, USA (2020). https://doi.org/10.1145/3409251.3411708

17. von Sawitzky, T., Wintersberger, P., Löcken, A., Frison, A.K., Riener, A.: Augmentation concepts with huds for cyclists to improve road safety in shared spaces. In: Extended Abstracts of the 2020 CHI Conference on Human Factors in Computing Systems, pp. 1–9. CHI EA '20, Association for Computing Machinery, New York, NY, USA (2020). https://doi.org/10.1145/3334480.3383022

18. Wang, F., Chang, C.M., Igarashi, T.: Virtual Horse: An Anthropomorphic Notification Interface for Traffic Accident Reduction. pp. 16–20. Association for Computing Machinery, New York, NY, USA (2021). https://doi.org/10.1145/3473682.3480255

19. Weidl, G., Breuel, G., Singhal, V.: Collision risk prediction and warning at road intersections using an object oriented bayesian network. In: Proceedings of the 5th International Conference on Automotive User Interfaces and Interactive Vehicular Applications, pp. 270–277. AutomotiveUI '13, Association for Computing Machinery, New York, NY, USA (2013). https://doi.org/10.1145/2516540.2516577

20. Zhai, S., Buxton, W., Milgram, P.: The partial-occlusion effect: Utilizing semitransparency in 3d human-computer interaction. ACM Trans. Comput.-Hum. Interact. **3**(3), 254–284 (sep 1996). https://doi.org/10.1145/234526.234532

Accessibility and Inclusive Mobility

Understanding Driving Behaviour in Individuals with Mild Cognitive Impairments: A Naturalistic Study

Vitaveska Lanfranchi[1](✉) (ID), Muhammad Fadlian[1] (ID), Sheeba G. A. Koilpillai[1],
Lise Sproson[2] (ID), Mark Burke[3], Sam Chapman[3] (ID), and Daniel Blackburn[1]

[1] The University of Sheffield, Sheffield S10 2TN, UK
v.lanfranchi@sheffield.ac.uk
[2] NIHR Devices for Dignity MedTech Co-operative (D4D), Sheffield S10 2JF, UK
[3] The Floow Ltd., Sheffield S3 8HQ, UK

Abstract. Driving is a crucial function to maintain independence for many adults, however, changes in health may affect driving skills. Researchers have sought to understand how the risk from driving changes as people age or experience cognitive decline. Previous studies, whilst recognising a number of cognitive factors that correlate with driving, have not found statistical evidence that could support using any specific cognitive test to support fitness to drive assessments. In-car monitoring technology (telematics) can provide a low-cost way to monitor driving risk by understanding measurable aspects of driving which can be correlated to risk behaviours, for instance: smoothness of driving, locationally excessive speeds and aggressive acceleration behaviours.

In this paper we present preliminary results from a naturalistic study that uses telematics to collect driving behaviour data and investigate the relationship between measurable driving factors and neurological conditions under normal driving conditions.

Keywords: Mild Cognitive Impairment · Driver Behaviour · Telematics

1 First Section

Driving is a crucial function to maintain independence for many adults. This is especially true in rural areas where access to public transport is limited and for example in the UK 23.5% of drivers in this environment are elderly compared to 16.3% in an urban setting. This has resulted in an 30% overall increase in drivers over 70 years in the last 14 years [1]. Inability to drive contributes to loneliness, as it becomes difficult to maintain hobbies/activities (e.g. religious meetings, group activities, walking groups etc.) and the burden on carers to provide transport is increased. Driving cessation is associated with social isolation, health problems, institutionalisation and increased depression and death rates [2]. However, changes in health may affect driving skills.

A major and growing contributing factor of road safety incidents is age-related health conditions. Age is the greatest risk factor for Parkinson's disease, Alzheimer's

H. Krömker (Ed.): HCII 2023, LNCS 14049, pp. 263–274, 2023.
https://doi.org/10.1007/978-3-031-35908-8_18

Disease (AD) and preclinical or prodromal AD. Parkinson's disease (PD) and Mild Cognitive Impairments (MCI) are common diseases of the elderly. In the UK, approximately 145,000 people are affected by PD, 850,000 people by dementia and between 5% and 20% of people aged over 65 have MCI. With life expectancy constantly increasing, the number of elderly drivers will increase proportionally (people aged over 70 represent 6% of all UK driving licences) therefore these conditions are a growing area of concern for road safety. Since 1990 in the UK there has been a 5% rise in personal injury collisions involving at least one older driver (aged over 70 years), whereas in the under 70s such collisions have decreased by 48% in the same period [3].

Several studies have analysed the relationship between elderly drivers and road collisions, listing contributory factors such as "driver failed to look properly", "driver failed to judge other person's path or speed", "loss of control" etc. [4] studied the significance of health impairments for the total number of crashes and concluded that dementia statistically significantly increases crash likelihood.

Age is the greatest risk factor for Alzheimer's Disease (AD) and Mild Cognitive Impairments (MCI). When an individual is diagnosed with AD or MCI, an assessment is made by the clinician, in cooperation with family members, caregivers and the individual, about their fitness to drive. Between 2008 and 2014 there has been a huge increase (682%) in people referred for assessment and in 2019 a National Audit found that on average 17% of people attending each clinic are diagnosed as having mild cognitive difficulties, for example difficulty processing information and/or making decisions quickly and accurately.

Driving is a highly cognitive loaded task and several variables can increase the cognitive load further, such as distractions, speed of traffic etc. This in return can increase the risk of accidents. Researchers have sought to understand how the risk from driving changes as people age or experience cognitive decline. Previous studies, whilst recognising a number of cognitive factors that correlate with driving, have not found statistical evidence that could support using any specific cognitive test to support fitness to drive assessments [5, 6]. However [7] claim that driving features can be used to provide preclinical diagnosis of AD. Our study aims to augment these results by looking at a wider number of variables and providing a control for variables such as age.

In-car monitoring technology (telematics) can provide a low-cost way to monitor driving risk by understanding measurable aspects of driving which can be correlated to risk behaviours, for instance: smoothness of driving, locationally excessive speeds and aggressive acceleration behaviours.

In this paper we present preliminary results from a naturalistic study that uses telematics to collect driving behaviour data and investigate the relationship between measurable driving factors and neurological conditions under normal driving conditions.

2   Literature Review

2.1   Driving Behaviour and Cognitive Impairments

The relationship between driving behaviour and cognitive impairments has been a subject of interest in recent research, with studies carried out to assess the risk of motor vehicle crashes in patients with MCI or dementia and to understand what are the most relevant risk factors. Whilst some studies have found positive correlations between cognitive impairments and risk of crashes [4], others have published opposite results [8]. This could be explained by the fact that most of these studies did not collect detailed driving behaviour or relied on self-rating.

A retrospective study has found 'patients with dementia have a lower risk of crash compared to those without dementia' [8]. This result can be explained by the fact that they didn't account for driving exposure during the study (i.e. people simply stopped or reduced driving). Another study, focused on drivers aged 65 and older without dementia, analysed the correlation between cognitive function and crash risk, by asking participating drivers to undertake a cognitive function test using as a measure the Cognitive Abilities Screening Instrument. The results were linked with instances in a crash database and concluded there was a link between cognitive function scores and propensity to crash [9].

Other studies focused on the fact that cognitive impairments may increase driving risk as drivers have a reduced capacity to self-monitor and self-regulate [10]. [11] found however, that the behaviour of those with MCI was not significantly different to the control group when it came to avoiding left turns, night driving, or driving in bad weather.

A different approach is to utilise GPS data to examine naturalistic driving behaviour [12–14]. Studies adopting this approach have shown how people with MCI and AD drove less frequently, especially at night, and more frequently at times of low congestion. Furthermore, they had fewer aggressive behaviours such as hard braking, speeding, and sudden acceleration [15, 16].

2.2 Telematics to Assess Driving Behaviour

The usage of telematics to assess driving behaviour for risk management has increased in recent years, and several studies have analysed the advantages of such an approach [17], mostly with a focus on young drivers or fleet drivers. While there is widespread acceptance that researchers should adopt a holistic approach in trying to understand driving habits and behaviours [18], there is clear evidence that the use of telematics can help to bring a reduction in safety-related events such as harsh braking, swerving and speeding in the fleet sector [19, 20] and in younger drivers [21].

3   Methodology

3.1   Ethics

Ethical approval was obtained from London Surrey Borders Research Ethics Committee (IRAS Number: 292929, REC reference: 21/PR/0583). The project is a multi-center, NIHR portfolio-registered study (NIHR number: 49393), sponsored by The University Of Sheffield. Research Governance approval, approvals from the Research Ethics Committee and HRA/HCRW were in place before commencement of the research. The project has been conducted in accordance with Good Clinical Practice standards, the declaration of Helsinki[1], and The European Code of Conduct for Research Integrity.

---

[1] World Medical Association Declaration of Helsinki: ethical principles for medical research involving human subjects. J Am Coll Dent. 2014.

Informed consent has been obtained by a trained researcher for all participants. To provide enhanced privacy protection for participants all driving data is held separately from Personal Information of participants. The subsequent analysis has been carried out using only pseudo-anonymised information to best protect privacy and minimising special category data handling.

3.2 Participants

Fifteen participants were recruited for this phase of the study, as volunteers, from memory clinics, care groups and community settings across the UK. These participants undertook a variety of clinical cognitive tests alongside monitoring of long-term driver behaviours.

The participants were then divided in two groups: 1) Mild Cognitive Impairments Participants with associated cognitive testing scores, and 2) Healthy elderly. We recruited drivers aged over 65 years. More details on the characteristics of the participants are presented in Sect. 4.

After the participants consented to join the trial, we carried out a cognitive test using the Cambridge Cognition software[2], run at their convenience (90 min). The chosen series of test was the CANTAB battery – a series of assessments run from an iPad. We have used eight of the tests to measure memory, attention and executive function as follows:

- Motor Screening Task (MST)
- Reaction Time (RT)
- Rapid Visual Information Processing (RVIP)
- Paired Associates Learning (PAL)
- Delayed Matching to Sample (DMS)
- Pattern Recognition Memory (PRM)
- Verbal Recognition Memory (VRM)
- Spatial Working Memory (SWM)

3.3 Equipment

The study utilised 'telematic devices', telemetry gathering devices, which are used in vehicle fleet monitoring or mass market insurance programs for remote risk and logistic monitoring. There is a large variety of device types and configurations for telematics monitoring. To select an optimal device type a key requirement was a need to facilitate easy installation without mechanical and automotive knowledge to best support installation within the study. This device type review ultimately selected 'On Board Diagnostic II (OBDII) device" types for this purpose. These can be installed in a plug-and-play manner into mandated vehicle diagnostic ports (see Fig. 1).

As the project requirements highlighted the need for gather and transmit continuous driver behaviour measurements using motion (accelerometers) and positional (GPS) but no need for vehicle CANBus data (e.g. engine data), we used specific device configurations to capture additional data compared to what is standard. The data was sent to The Floow's (partners in the project) fault tolerant and secure servers for later cleansing and analysis.

---

[2] https://www.cambridgecognition.com/cantab/cognitive-tests.

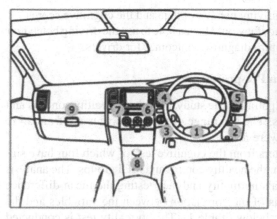

Fig. 1. This shows fitting instructions for the utilised device showing locations in the vehicle where ports can be found and the simple plug-in installation into the OBDII diagnostic port.

The selected devices are CE marked (E11, n. 10R-05 10429), approved by UK regulators and EU automotive certification and type approvals for both electronic and automotive usage.

3.4 Procedure

Invitations to patients and public to participate were shared via a number of methods: memory clinics, dementia cafes, care groups and other community settings across the UK. Potential participants contacted the study team to express interest in receiving further information. This was then shared either by phone or email as preferred by individuals, in the form of a verbal or written introduction to the project, an Information Sheet and a Consent Form to read in their own time. To support inclusivity and optimise engagement, lightly simplified, large font participant information sheets were sent to the Healthy Older Adults and an easy-read large font version made for people with a diagnosis of mild cognitive impairment or dementia. Once the participants consented to the research, they were offered the choice between being visited by the Researcher at a time and on a date that was convenient or to come and visit the Researcher in the University's premises, to carry out the cognitive test and to have the monitoring device installed.

The driving data was then pre-processed to extract driving variables and analysed alongside the pseudo anonymised cognitive data.

4 Results

Once the data was collected we carried out three types of analysis, to obtain a more comprehensive understanding of the data and their relationships:

- descriptive analysis of the cognitive test scores and the participant demographics, and separately of the driving data, summarising the basic features of the data, such as the mean, median, and standard deviation.

- correlation analysis between the driving behaviour data and the cognitive scores.
- predictive modelling, to understand if it would be feasible to use the driving behaviour data to make predictions about future diagnoses outcomes for drivers.

4.1 Descriptive statistics and Data Pre-processing

We had in total 16 individuals participating in the study, of which 7 healthy controls and 9 individuals with an MCI diagnosis. The age range of the participants are from 55 to 90 in which 6 were females and 10 were males.

In total, we collected 20 parameters from the cognitive test, of which four have significant differences in means between the healthy control and MCI groups. The analysis was conducted by assessing the data's normality and then testing the mean difference between the classes. Further analysis of the correlation between the variables and the classes is also presented in the table below (Table 1). The normality test is conducted using the Saphiro-Wilk test for normality due to the small sample size.

Table 1. Cognitive tests: descriptive statistics. In this table we show only the variables that had significant differences in means between the healthy control and MCI groups

Variable	Distribution	Mean Difference	Correlation Coefficient
VRMDRTC	Normally Distributed	Significant Difference	−0.65
VRMDRIRD	Not Normally Distributed	Significant Difference	0.73

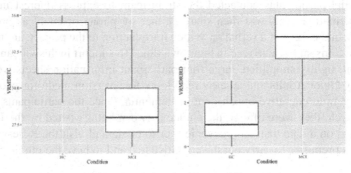

Fig. 2. Analysis of the parameters that have significant difference means between the healthy control and MCI groups

Figure 2 presents an analysis of the parameters that have significant difference means between the healthy control and MCI groups. VRMDRTC is the number of total correct answers from the delayed recognition test of the verbal recognition memory. Healthy controls tend to have higher scores compared to MCI. VRMDRIRD is a parameter that describes the total incorrect distractor from the verbal recognition memory test. Healthy controls tend to have lower incorrect numbers compared to MCI for this test.

Raw driving behaviour data was pre-processed to remove noise, such as journeys where the ignition was activated but the vehicle did not move[3]. This left a total of 8553 valid journeys collectively covering over 62,000 km of travel each with second-by-second detail of driver behaviours. Across the journeys, 25% of journeys were less than 1.7 km with a median distance of 3.7 km and a mean distance of 7.3 km with the longest journey being 260.1 km in length.

For each journey a range of analysis calculated fields of interest to older driver behaviour analysis. These included calculations of 1) risk exposures, by mileage, times and road types driven on, 2) driving factors including turning behaviours, acceleration and braking patterns, extreme events, roundabout approach dynamics and a large array of other measurement factors. In total over 660 factors were extracted to characterise behaviour of each journey. These were then used to analyse behavioural correlations to cognitive test scores.

4.2 Correlation Analysis

This analysis was aimed at determining the strength and direction of the relationship between cognitive test performance and driving behaviour in our participants, to understand the best candidate variables to further focus on in future studies.

Table 2 presents the result of correlation analysis between the driving behaviour data and medical data. We have found 5 pairs of variables (pairs of driving behaviour variable and cognitive test variable) that have a significant correlation with a strong (positive or negative) correlation coefficient.

Table 2. Correlation analysis results

Driving Variable	Medical Variable	Correlation	Correlation Coefficient
RA_Hard_Acc	RVPTM	Significant Correlation	−0.57
JA_Hard_Ret	RVPML	Significant Correlation	−0.54
RE_Hard_Ret	PALTE28	Significant Correlation	0.52
JA_Hard_Acc	SWMTE12	Significant Correlation	−0.51
LR_ratio	RTIFMMT	Significant Correlation	0.50

Overall, there is a significant correlation between the behaviour of performing hard celeration or jerk movement (>3 m/s/s) and several cognitive scores. The behaviour of performing hard acceleration on roundabout approach (RA_Hard_Acc) inversely correlated with RVPTM (Rapid Visual Information Processing Time), which suggest that

---

[3] This utilised a 100 m threshold in total distance to ensure only valid journeys where used for the study.

person that tend to perform hard acceleration on junction approach have higher (slower) time to process rapid visual information. In addition, hard retardation on roundabout exit behaviour (RE_Hard_Ret) also correlated with the number of errors in Paired Association Learning Test (PALTE28). The behaviour of performing hard retardation and acceleration in junction approach have inverse correlation with RVPML and SWMTE12 (rapid visual processing mean latency time and total errors between revisiting box). Lastly, there is a significant positive correlation between the Left-Right hand turn ratio and RTIFMMT (Mean Movement Time in Reaction Time test) which suggest that the person with higher mean movement time (slower movement and reaction) tend to perform left-hand turn compared to the right-hand turn.

Moreover we clustered journeys into daytime trips and night time driving, using a variable drive_night_perc (percentage of driving trips performed after dusk, calculated comparing time of the day and sunset time at location). Healthy controls tend to have a lower percentage of night driving compared to the MCI, as can be seen in Fig. 3.

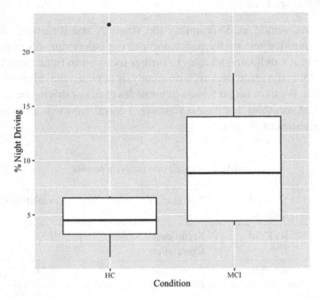

Fig. 3. Percentage of night driving between healthy controls and MCI drivers.

4.3 Predictive Modelling

One of the aims of our research is to understand if driving behaviour can be predictive of MCI or of a deterioration in cognitive capabilities. Whilst the dataset is too small to fully capture the true underlying patterns and relationships in the data, we carried out a simple classification task as a feasibility study. The classification tasks aimed at using driving behaviour data to predict the likelihood of an individual driver to be belonging to the MCI or healthy control group.

The classification task is conducted using 8553 trips from 16 users of which 9 are MCI patients and 7 are healthy controls. We are using 35 driving parameters that include the driving behaviour (acceleration, speed, duration, distance) and driving context (type of road, time of drive). In total, there are 4098 MCI trips and 4455 HC trips. The classification is conducted using XGBoost algorithm (eXtreme Gradient Boosting) with 8:2 ratio between the train and test test. XGBoost algorithm is a machine learning algorithm (optimised version of the gradient boosting decision tree algorithm) for classification and regression problems. The performance of the model was evaluated using metrics such as precision, recall, and F1 score.

The result of the classification is 84% accuracy without any fine-tuning on the algorithm's side, indicating that the algorithm is able to correctly predict the categorical labels in 84% of the cases. In addition Table 3 and Table 4 shows the detailed performance and result of the classification. The model has managed to classify correctly 755 of 891 Healthy Controls, with 136 classified incorrectly as MCI (Precision = 0.85, Recall = 0.85, F1 Score = 0.85). 689 MCI are classified correctly out of 820 total MCI samples with 131 classified incorrectly as Healthy Controls (Precision = 0.84, Recall = 0.84, F1 Score = 0.84).

Table 3. Classification Results

		Precision	Recall	F1 Score
Condition	HC	0.85	0.85	0.85
	MCI	0.84	0.84	0.84
	Overall Accuracy			0.84

Table 4. Confusion Matrix

True Table	HC	755	136
	MCI	131	689
		HC	MCI
		Predicted Table	

XGBoost provides a measure (feature importance score) for each input variable, indicating which variables are the most important predictors for the target variable (MCI or Healthy control). A variable with a high feature importance score based on the mean decrease in impurity (GINI) is an indicator that the variable is a strong predictor of the target variable. Figure 4 displays the 10 variables with the highest XGBoost feature importance score based on the mean decrease in impurity.

It is can be inferred that the highest variable is the drive_night which associated with the night time driving (after dusk hours), followed by RA_Total_Celeration (total celeration in roundabout approach), highspeedRoad & lowspeedRoad (distance travelled in either high speed road or low speed road), and RA_hard_acc_perc (hard acceleration performed during roundabout approach). RA_hard_acc_perc is particularly interesting as a candidate variable as it also showed to be significantly correlated with cognitive scores in the correlation analysis.

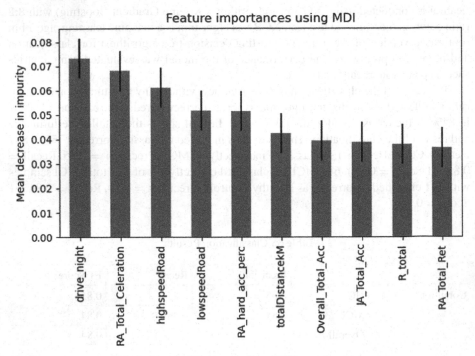

Fig. 4. Feature Importance from XGBoost algorithm.

5   Conclusions and Future Work

In this study, we investigated the relationship between driving behaviour and cognitive scores by carrying out a naturalistic study with 16 participants, recording their driving behaviour for 6 months. The results of this study provide new insights into how some aspects of driving behaviour could become digital biomarkers for the detection and monitoring of cognitive impairments, as they seem to be significantly correlated with changes in the cognitive scores. However it is important to acknowledge the limitations of our study when interpreting the results and to reflect on potential future work. Some potential limitations to consider include:

- Sample size: due to difficulties in recruiting participants, our dataset is quite small in terms of participants, therefore it may not represent the diverse range of values or scenarios that might be encountered in the real world. This has implications if we want to generalise the results.

- Whilst we were able to identify statistically significant correlations between driving behaviour variables and cognitive scores, we were not able to identify any statistically significant correlations between driving behaviour variables and MCI/healthy controls, due to the small size of the dataset. Further investigation is needed on this point.
- Our classification model produced good results, however, given the small sample size there is a risk of overfitting, for which the model might perform well on the training data but poorly on new, unseen data.

We are planning to carry out follow-up studies focusing on:

- further recruitment to increase participants numbers and therefore number/types of journeys. This will enable us to achieve statistical significance and validate against gold standard measures for monitoring cognitive decline and biomarkers of MCI.
- on the machine learning side, we will include improving the accuracy of the model by fine-tuning the hyperparameters of the XGBoost algorithm. However, it is important to keep in mind that excessive fine tuning on a small sample of data can cause overfitting. Fine-tuning should therefore be done with caution and using a rigorous cross-validation process.
- understanding how feedback about driving behaviour can be communicated to drivers and clinical/road safety professionals to aid in their decision-making.
- understanding how our approach can fit and be combined in a new ecosystem of tools for remote monitoring and diagnosis of MCI using digital biomarkers.

Overall we believe our study highlights the need for further research in this area, to better understand the relationship between driving behaviour and cognitive impairments, and to develop more effective interventions to support drivers with cognitive impairments.

Acknowledgements. This project would like to acknowledge funding from UK's Road Safety Trust (https://www.roadsafetytrust.org.uk/funded-projects/the-floow-ltd) and UKRI Ageing Catalyst grant whose support has made this collaboration and research possible.

References

1. UK Gov. Rural Population and Migration (2015). https://www.gov.uk/government/publicati ons/rural-population-and-migration/rural-population-201415. Accessed 1 Feb 2023
2. Chihuri, S., et al.: Driving cessation and health outcomes in older adults. J. Am. Geriatr. Soc. **64**, 332–341 (2016)
3. UK National Statistics (Stats19) (2019). https://data.gov.uk/dataset/cb7ae6f0-4be6-4935-9277-47e5ce24a11f/road-safety-data. Accessed 1 Feb 2023
4. Elvik, R., Vaa, T., Hoye, A., Sorensen, M.: The Handbook of Road Safety Measures, 2nd edn. Emerald Group Publishing (2014)
5. Staplin, L., Lococo, K.H., Mastromatto, T., Gish, K., Golembiewski, G., Sifrit, K.: Mild cognitive impairment and driving performance. TransAnalytics. NHTSA (2019)
6. Feng, Y.R., et al.: Driving exposure, patterns and safety critical events for older drivers with and without mild cognitive impairment: findings from a naturalistic driving study. Accid. Anal. Prev. **151**, 105965 (2021)

7. Bayat, S., et al.: GPS driving: a digital biomarker for preclinical Alzheimer disease. Alzheimer's Res. Therapy **13** (2021).
8. Fraade-Blanar, L., et al.: Diagnosed dementia and the risk of motor vehicle crash among older drivers. Accid. Anal. Prev. **113**, 47–53 (2018)
9. Fraade-Blanar, L., et al.: Cognitive decline and older driver crash risk. J. Am. Geriatrics Soc. **66** (2018)
10. Pomidor, A. (ed.): Clinician's Guide to Assessing and Counseling Older Driver, 3rd edn. American Geriatrics Society, Washington, DC (2016)
11. Horswill, M., Marrington, S., McCullough, C., et al.: The hazard perception ability of older drivers. J. Gerontol. B Psychol. Sci. Soc. Sci. (2008)
12. Eby, D., Molnar, L.: Cognitive impairment and driving safety. Accid. Anal. Prev. **49**, 261–262 (2012)
13. Molnar, L., Eby, D.: The relationship between self-regulation and driving-related abilities in older drivers: an exploratory study. Traffic Inj. Prev. **9**, 314–319 (2008)
14. Molnar, L., et al.: Factors affecting self-regulatory driving practices among older adults. Traffic Inj. Prev. **15**, 262–272 (2014)
15. Eby, D., Silverstein, N., Molnar, L., Leblanc, D., Adler, G.: Driving behaviors in early stage Dementia: a study using in-vehicle technology. Accid. Anal. Prev. **49**, 330-7 (2012)
16. Babulal, G.M., et al.: Perspectives on ethnic and racial disparities in Alzheimer's disease and related dementias: update and areas of immediate need. Alzheimer's Dement. **15**(2), 292–312 (2019)
17. Tong, S., et al.: Provision of telematics research (2015). https://assets.publishing.service.gov.uk/government/uploads/system/uploads/attachment_data/file/479202/provision-telematics-research-report.pdf. Accessed 1 Feb 2023
18. Sagberg, F., Selpi, Piccinini, G.F.B., Engström, J.: A review of research on driving styles and road safety. Human Factors **57**(7), 1248–1275 (2015)
19. Sullman, M.J.M.: Reducing risk amongst those driving during work. Prev. Accidents Work 89–94 (2017)
20. Quayle, D.J., Forder, L.E.: Reducing risk in workplace vehicles. In: Proceedings of the Australasian Road Safety Research, Policing and Education Conference. Department for Transport, Energy and Infrastructure, Adelaide (2008)
21. RoSPA. Young drivers at work (Scotland) Black Box Pilot, May 2014

Independent Travel and People with Intellectual Disabilities: Viewpoints of Support Staff About Travel Patterns, Skills and Use of Technological Solutions

Hursula Mengue-Topio[1]([✉]) [iD], Laurie Letalle[1] [iD], Yannick Courbois[1], and Philippe Pudlo[2]

[1] Univ. Lille, ULR 4072 - PSITEC - Psychologie: Interactions Temps Émotions Cognition, 59000 Lille, France
hursula.mengue-topio@univ-lille.fr
[2] LAMIH UMR CNRS 8201, Université Polytechnique Hauts-de-France, 59313 Valenciennes Cédex 9, France

Abstract. Public transportation is a vector of social inclusion. However, people with intellectual and developmental disabilities face a variety of barriers to independent travel in their communities. The results of this survey of support staff show that assistive technology systems can be a solution to improve the teaching of navigation skills in this population provided that it is truly adapted to the profile of the individuals: cognitive characteristics, needs, and initial skill level. Finally, effective use in daily life also requires the involvement of professionals in the design of such tools.

Keywords: Intellectual Disability · Independent travel · public transportation · technological solution · mobility

1 Introduction

Independent travel is a fundamental requirement to gain access to resources in the local community. As a result, teaching travel skills to persons with intellectual disabilities (ID) is a major goal of education in order to promote their independence and social inclusion [1, 2].

ID refers to significant limitations in both **intellectual functioning** (IQ < 70) and **adaptive behavior**, which covers many everyday social and practical skills. This disability originates **before the age of 22** according to the Diagnostic and Statistical Manual of Mental Disorders (DSM)-5th edition [3] and the American association on intellectual and developmental disabilities (AAIDD) [4]. According to AAIDD, intellectual functioning refers to abilities such as reasoning, planning, abstract thinking, understanding complex ideas, and problem-solving. Adaptive behavior refers to the set of conceptual skills (language, reading, writing, concepts of numbers, etc.), social skills (interpersonal skills, social responsibility, gullibility, etc.), and practical skills (personal care, use of money, occupational skills, health care, travel/transportation, etc.). All of these conceptual, social, and practical skills allow the individual to function on a daily basis in his or her usual life context. Individuals with ID face difficulties in the ability to meet expectations of autonomy and responsibility given their age and the cultural context in which they lead their lives. The causes of ID are multi-factorial and this population presents a very high heterogeneity [5]. Indeed, the levels of severity of ID are based on the description of the person's daily functioning by considering the conceptual, social, and practical domain and the level of support they need. There are then 4 levels of ID severity: mild, moderate, severe, or profound [3].

Field surveys conducted with this population show a strong restriction of independent travel, which is mainly performed in the vicinity of the home and the institutions attended [6–8]. Difficulties in planning the trips are mentioned by the individuals themselves and human assistance is often needed (families, professionals, etc.). Another complex problem concerns the management of unforeseen events or disruptions in transportation such as route changes, delays, or even the presence of crowds during rush hour transport and human errors (wrong directions, bus or metro line; failing to get off at the right stop, etc.). In these situations, people with ID do not always know how to react, may feel anxiety, and have difficulty asking for help [6, 7, 9].

Community mobility includes different dimensions such as the representation of the space in which one moves, the use of transportation, appropriate interactions with other users of public space, and the management of unforeseen events that may occur during travel [10]. These different facets of mobility make the process of helping adolescents and young adults with ID to become autonomous difficult and time-consuming [9]. A solution could then come from technologies designed for navigation. This involves teaching the affected individuals specific skills such as crossing the road safely, planning a few routes within local environments, and following the appropriate procedures for using public transport: defining the destination, determining the departure time, knowing where to get off the bus or the metro once the destination has been reached, and following the safety rules in the event of an accident or emergency. In general, studies related to the design, use and assess the effectiveness of navigation assistant shows that people with ID can learn all the behaviors required to move around safely and easily following individual training, detailing each behavior to be adopted, at the person's pace. Such training takes place over several months or even years depending on the skill level and initial experience of the person being supported [2, 11–15].

However, this approach is not without its limitations: it allows people to be taught a few fixed routes, and it is difficult to teach them how to react to unexpected events. Moreover, work on the design and use of assistive technology solutions for travel assistance for people with ID remains scarce and often involves participants with different cognitive disorders (dementia, epilepsy, stroke, etc.). This heterogeneity makes it difficult to generalize the results obtained. Finally, the acceptability and acceptance of such tools are very little evaluated by the people concerned (people with ID, professionals, and families). Through the scientific literature, many factors are now better identified as having an impact on the acceptability and acceptance of these devices by these populations. These include the perceived "need" for assistance or help, recognition of the quality of the product or usability, availability, and the cost of the technological product [16].

In addition, individual characteristics such as users' health status/level of impairment severity, and users' familiarity with assistive technology products are also other criteria to be considered in the prospective judgment toward these tools [17]. Finally, it is essential to consider the implications for individuals around the person with ID, especially support staff [18–23]. Indeed, the proper functioning of the equipment, ensuring that the person knows how to use it and that he or she carries it with him or her when traveling outside of his or her living space are additional points of vigilance for the professional entourage. In addition, the training of professionals in the use of the technological product (to access the location of the person being supported) and in alternative solutions in case of malfunctioning are necessary so that the proposed tool is not perceived as an additional burden that would prevent its effective use [20, 24]; or even contribute to its abandonment after purchase [17]. Similarly, it is essential that the person being cared for and the support staff or family entourage easily integrate the technological assistant into their daily routines and are satisfied with it. For these reasons, [17] suggests taking into account the point of view of the person being supported and his or her professional caregivers and/or family environment jointly, when making decisions about the use of technological aids. Regarding the use of navigation aid technologies (GPS) for vulnerable populations (elderly people with or without dementia, people with sensory or cognitive impairments following a stroke), very few studies specifically concern people with intellectual disabilities.

Although old, the question of independent travel for people with ID is still topical because of the importance of this dimension as a necessary condition for the social participation of the people concerned. In addition, the characteristics of the constantly changing environments hinder these people in their autonomous daily movements. Several reasons explain these difficulties: the increasing extension of territories, the complexity of transport systems, the digitalization of services related to mobility which in many cases replaces human assistance, and the flexibility needed to adapt to these changes end up excluding a large number of persons with ID [25]. As shown in various field surveys, these individuals often rely on assistance from family and friends and severely restrict their travel compared to others without disabilities [8].

Faced with these difficulties, in addition to the help provided by the family, mobility assistance systems can help to promote the daily autonomy and social inclusion of people with ID. In this descriptive study, we explore different dimensions of mobility (elaboration of spatial representations, use of transport, problem-solving during travel, use of assistive technology systems during travel) and question carers about the contribution of technological aid in the teaching of travel skills. Moreover, it is essential to rely on the professionals who accompany these people, not to replace the people concerned, some field surveys give voice to them [6, 7, 26], but to examine our object of study from another angle and obtain a diversity of viewpoints. Thus, this study relies on the perception of professionals in the medico-social sector, who teach these people to move around on a daily basis, about the resources and difficulties encountered by people with ID during their travels, the skills deemed necessary to learn to move around and the modus operandi used to structure the learning. It also seems beneficial to us to interview a large number of professionals in order to overcome the difficulties encountered in generalizing the results of work involving a small number of participants in the literature.

Thus, we interviewed support staff in order to answer the following questions: What is the level of complexity of the spatial representations (landmarks, routes, cognitive map of the environment) developed by ID individuals and what use do they make of them to get around on a daily basis? What is the degree of efficiency in the use of transport by these people? What resources and strategies do they mobilize in case of unforeseen events? What are the tools most used by these people during their travels? How is the learning of autonomous travel structured by professionals? What important principles should a technological aid dedicated to navigation and adapted to people with ID respect (presentation, content, proposed aids, functionalities, etc.)?

2 Method

2.1 Participants

115 French professionals (84 women and 31 men) working in medico-social establishments located in the Nord and Pas-de-Calais departments (Hauts de France region) voluntarily responded to the survey (occasional sample). The sample included 47.83% of specialized educators or monitors and 26.96% of paramedical professionals (psychometricians, occupational therapists, etc.). The length of professional practice in specialized institutions ranged from 1 to 37 years, with an average length of 11 years (M = 10.9; SD = 8.25). In addition, these professionals had an average of 10 years of experience (M = 9.58; SD = 7.99) in supporting the autonomy of people with ID. Most of them worked in urban areas (72.17%). 32 professionals (27.82%) worked in an institution located in a rural or semi-rural area or in an industrial zone. They accompanied people with ID aged between 10 and 60 years in medico-social establishments and services. 53 professionals (46.09%) stated that they worked in the adult sector (20–60 years), 50 professionals (43.48%) worked in the child/adolescent sector (10–20 years); 10 professionals (8.69%) worked in both the adult and child/adolescent sectors (mixed sector) and 2 participants did not answer this question. As for the level of severity of the disorders, 68 professionals mentioned that they work with a population with mild to profound ID, 17 professionals indicated that they work with a population with mild ID, and 30 professionals did not specify the severity of the ID of the public they work with. As regards associated disorders, we note mainly the presence of psychological disorders, behavioral disorders, communication and language disorders, and autism spectrum disorders (70% of disorders cited). On the other hand, sensory deficiencies, and motor disorders, concerned 5% of the disorders cited. Finally, 25% of associated disorders were not specified.

2.2 Material

The questionnaire, constructed by us, included 35 questions, 21 of which were closed questions (multiple choice, dichotomous, enumeration) and 14 open questions. The tool was based on the major dimensions of people's navigation studied in recent scientific literature and those used in orientation and mobility programs for people with ID [11, 27]. The questionnaire was proposed in two versions: a paper version completed by 77 professionals and an online digital version completed by 38 and carried out with the Lime Survey software. The paper version was submitted and distributed by e-mail to associations supporting people with ID.

3 Results

The open-ended questions were manually analyzed for content, and then the responses obtained were recoded into objective, mutually exclusive, comprehensive categories. In this paper, we present the results in relation to travel patterns, obstacles and the use of a technological aid by people with ID.

3.1 General Information About Travel Patterns

According to professionals, the majority of trips start at home (82.72% of trips identified by professionals). The main destinations are: going to school, work, leisure activities, medical appointments, a supermarket, and social contacts.

On a scale of 1 (no influence) to 10 (very important influence), all professionals (n = 115) placed the effect of Individuals with ID motivation on learning to travel independently at an average of 7.75 (SD = 1.39). In other words, this factor contributes significantly to learning and achieving independent travel.

Regarding signage, 85.22% of respondents consider that the signs, maps, etc., available in the environment are not adapted to people with ID because of their complexity. Furthermore, family fears and reticence have a significant influence on learning to travel independently. In fact, on a scale of 1 (no effect) to 10 (significant effect), all the respondents placed the average effect at 7.24 (SD = 1.62).

3.2 Skills Related to Independent Travel

Creating Spatial Representations About the Environment, Planning One's Movements. 94.78% of the professionals considered that people with ID use various strategies to move around in the environment in general. In response to the open-ended question asking them to specify the different strategies used by people with ID, the professionals formulated 213 proposals that were recorded. In particular, they mentioned strategies relating to finding landmarks (39.34% of the proposals related to the selection and memorization of landmarks), strategies for learning and memorizing the stages of the journey, in other words, the formation of routines (27.23% of the proposals formulated) and strategies for requesting human assistance, i.e., asking for help from professionals, parents, third parties or even traveling in a peer group (24.41%). Technological assistance (6.10% of the proposals made by professionals) and the use of reading (2.81% of the proposals cited) are two strategies that are little used by people with ID during their travel. According to the professionals, two tools are particularly used by these people during their travels: the support developed by the professionals (cited 98 times) and the cell phone to contact a person in case of need (cited 98 times).

Among the individual factors that cause the most problems (question, "In your opinion, which of the following factors are the most problematic for people with ID to travel alone in their neighborhood or city?"), all the professionals interviewed identified the factor "knowing how to plan a trip" (15.76% of the selected proposals) as the one that causes the most difficulties for people with ID when they have to travel alone. The second most difficult factor (12.57% of the proposals) concerns the factor "understanding the instructions given by outsiders" followed by the factor "memorizing the steps of the journey" (11.82%). These first three factors are completed by the factor "knowing how to read" (this factor represents 11.63% of the proposals), "paying attention to elements and information in the environment" (10.7% of the occurrences), and "knowing how to select fixed landmarks in the environment" (9.94% of the occurrences). Then come to the fact of not knowing one's left and right were mentioned 50 times (9.38%), and the factor "knowing how to resist distractions present in the environment", which is less identified as posing the most difficulties to ID people (8.25% of occurrences for each of these factors). These results are shown in Table 1.

Table 1. Distribution of the perceived barriers to independent travel

Factors that cause the most problems in achieving autonomous travel	Number of occurrences	Percentages
Knowing how to plan a trip	84	15.76
Understanding the information given by outsiders	67	12.57
Memorize the steps of the journey	63	11.82
Know how to read	62	11.63
Paying attention to environmental information	57	10.7
Knowing how to select fixed landmarks in the environment	53	9.94
Know your left and right	50	9.38
Knowing how to use landmarks to get around	44	8.25
Knowing how to resist distractions in the environment	44	8.25
Other factors	9	1.7
Total	533	100

Using Public Transportation. 83% of professionals support people with ID who use public transport. The most commonly used means of transportation is the bus (52%), followed by the train (25%), the tramway (14%), and the subway (9%). Table 2 presents the responses regarding the use of transportation. The results do not reveal any major difficulties in the use of transport by these people. Generally speaking, the purchase of a ticket is "sometimes" (54.74% of professionals selected this answer) or even "often" (28.42% of professionals selected this answer) carried out by the people themselves and they validate it "always" (49.48% of professionals selected this proposition) or "often" (34.74%) when they get on. The majority of people with IDs know "sometimes" (36.84%) or "often" (31.59%) that their ticket has a validity period. 42% of respondents believe that people with ID sometimes take transport schedules into account when organizing their trips. The identification of the direction of transport is "often" (44.21%) or "sometimes" (43.17%) correct according to the professionals. Regarding getting off the means of transportation, people with ID are "often" able to point out the desired stop (48.43%) and "often" getting off the means of transportation to the right stop (68.42%). 58.95%

of the professionals stated that the people accompanied "often" have a socially adapted attitude and behavior. 79% of the professionals mentioned difficulties (loss of reference points, planning difficulties, the complexity of the journey) during connections (changes of lines, directions, and combinations of different means of transport to continue the journey to the destination).

Table 2. Professional responses (n = 88) related to DI individuals' skills in using transportation (responses expressed as percentages).

People with intellectual disabilities	always	often	occasionally	never	Does not apply to
Purchase their own ticket	6.82	27.27	56.82	7.95	1.14
Validate or stamp their ticket	51.13	32.95	13.64	1.14	1.14
Know how to consider the schedules	20.45	31.82	40.91	6.82	0
Locate the direction of transport	6.82	44.32	44.32	2.27	2.27
Know how to signal the stop where they want to get off the means of transportation	23.86	46.59	26.14	1.14	2.27
Getting off the means of transportation at the right stop	18.18	69.32	12.5	0	0
Know the safety rules	22.72	39.80	32.95	2.27	2.27
Have a socially appropriate attitude and behavior	7.95	59.09	31.82	1.14	0

Resolve Unforeseen Events that Arise During Travel. 75.65% of the professionals consider that people with ID do not react appropriately if a problem occurs during a trip (detour, error on the part of the person, failure of public transport, etc.). According to the 114 professionals who answered the open question "in your opinion, what do these people do when they are confronted with a problem during their trips?", several strategies are observed in this type of situation: both "adapted" strategies (50.42% of the proposals formulated out of a total of 240 proposals put forward by the professionals) and non-adapted strategies (48.75% of the proposals formulated).

Adapted strategies correspond to behaviors that enable to persons with ID to resolve the unexpected by mobilizing their own cognitive and emotional resources and the social or material environment. In the event of an unexpected event, ID users predominantly turn to their social environment, whether it is their close circle (professional referral, family, friends), whom they ask for help by telephone, or third parties present at the scene (transport network professional, other users). This solicitation of the social environment is the first strategy used (39.17% of the proposals) compared to other strategies such as turning around, going back to a familiar place (home, establishment, etc.) which represents 5% of the proposals mentioned or reorienting oneself, taking the next means of transport (5% of the proposals). Finally, reading signs in the environment received 1.25% of the proposals mentioned by the professionals.

Non-adapted strategies refer to reactions and behaviors that do not make it possible to resolve the unexpected, endanger the person, or lead to the renunciation of the journey. According to professionals, people with ID suffer from anxiety and let themselves be overwhelmed by emotions when they encounter a problem during the trip (23.75% of the suggestions made). Professionals also report other behaviors such as not reacting, and waiting on the spot without taking any initiative (10.42% of suggestions). 6.25% of the proposals refer to the fact of being disoriented, or "lost" in these cases. Finally, some of the professionals' responses, which were in the minority, concerned giving up the trip (3.33%), showing behavioral problems (2.5%), walking the route (2.08%), or following a person they met on the route (0.42%). All of these results are shown in Table 3.

Table 3. Strategies used by people with ID during an unexpected event, according to professionals

Strategies in case of unforeseen events	Number of proposals	Percentages
Adaptive strategies	**121**	**50.42**
Solicit the people around you by phone (referring professionals, family, and friends…)	49	20.42
Ask for help from others in the vicinity (transportation professionals or other users)	45	18.75
Turn around, and go back to a familiar place (home, institution, etc.)	12	5
Redirect yourself, take the next bus	12	5
Read the signs	3	1.25
Non-adaptive strategies	**117**	**48.75**
Panic, anxiety, stress, being overwhelmed by anxiety (screaming, crying, getting upset)	57	23.75
Do not react, wait on the spot without asking for help	25	10.42
Being disoriented, lost	15	6.25
Staying at home, giving up travel	8	3.33
Behavioral disorders (including stereotypies, mutism, etc.)	6	2.5
Walk the route regardless of the distance	5	2.08
Following someone thinking they are going to the same place	1	0.42
Other strategies		
Getting off the means of transport	2	0.83
Total	240	100

3.3 The Use of a Technological Aid Dedicated to Navigation in People with ID and Professionals

We asked professionals about the tools used by people with ID when they travel. According to the answers obtained, two tools are particularly used by people with ID during their travels: the material developed by the professionals accompanying the people (cited 98 times) and the cell phone to contact a person in case of need (cited 98 times). Eight professionals gave an "other" answer to this question, mentioning the following tools: "Identifying journeys beforehand, accompanied concrete practice, locomotion lesson", "Observations/habits", and "Word of mouth".

Table 4 shows the tools used by professionals for learning to move independently with people with ID and the number of responses obtained for each tool. Among the proposed tools, the most used by professionals are tools specifically developed for the person (cited 97 times), the cell phone (cited 79 times), photographs of the environment (cited 61 times), and maps (cited 44 times). Navigation applications for smartphones such as Google Maps or Mappy (18.26%) as well as public transport network websites (17.39%) can also be used by professionals. On the other hand, navigation websites (8.69%) and smartphone applications for public transport networks (6.95%) are lesser used by carers. Three other answers were proposed by the professionals. It concerns the repetition of the journey, the chaining.

Table 4. Frequency of use of tools by professionals when learning to move with people with ID.

Tools preferred by professionals for travel	Number of occurrences	Percentages
Tools you have developed for the person (plans, maps, route cards, etc.)	97	84.35
Cell phone	79	68.70
Photographs of the environment	61	53.03
Maps (city, subway, bus, etc.)	44	38.26
Navigation applications for smartphones (Google Maps, Mappy, etc.)	21	18.26
Website of the public transport network	20	17.39
Navigation website (Mappy, Google Maps, etc.)	10	8.69
Smartphone application of the public transport network	8	6.95
Other (repeating the route, chaining)	3	2.61

To the question, "In your opinion, if a smartphone travel assistance application for people with intellectual disabilities were developed, what important principles should it respect (presentation, content, proposed aids, functionalities, way of transmitting information, etc.)?", the content analysis made it possible to distinguish several categories of answers. The professionals' responses advocate a simplified presentation of information, using auditory and visual aids. They also favored European standards for making information Easy to Read and understand[1]. Concerning the functionalities that such an application could offer, the answers suggested that the system should allow to establish itineraries and guide the person, ask for help in case of need, be geolocated, control the application by voice, manage unexpected events, give an indication of the duration of the journey (time of the journey, calculation of the departure time according to the desired arrival time, etc.) and integrate the data of public transport. Finally, the proposed application must be accessible and ergonomic, it must adapt to the user's preferences.

4 Discussion

This study carried out among professionals accompanying people with ID has enriched our knowledge of the characteristics of the mobility of these people. They very often make short, routine trips, limited to familiar environments. Although their motivation to learn to move around on their own contributes significantly to achieving this objective, the people accompanied nevertheless encounter difficulties in their daily movements. Such difficulties may be exacerbated by, among other things, inadequate signage, and reluctance on the part of their family and support staff. These reluctances are due to difficulties perceived by their relatives in planning and solving problems arising during travel. These results are in line with previous studies [6, 7, 28].

In addition, this descriptive study takes up some skills that are essential for learning to move about on one's own and sheds particularly rich light on the support practices of professionals and also their opinions and knowledge about the movements of people with ID:

- Creating spatial representations: the results show that people with ID select elements of the environment such as landmarks and memorize the stages of the journeys they must make. This result confirms those obtained in other previous works, adopting a comparative approach and showing learning of the environment from landmarks, and memorization of routes that requires many trials and includes several errors [29–31]. Throughout the literature, landmarks and routes form a first level of representation of the environment that continues to become more complex and subsequently promotes other more elaborate behaviors such as shortcuts or detours that require recourse to a more accomplished level of spatial representation namely knowledge of the configuration [32, 33]. In people with ID, the privileged use of landmarks and routes as tools for representing and navigating the environment is justified in more than one respect: very few of these people have a satisfactory command of reading, and those who can read make little use of reading to orient themselves or follow directions [34, 35]. This finding holds true during routine travel and during unexpected events as shown in this study.

- Using transport: Professionals consider that the various sub-skills that make up the ability to use public transport (planning departure times, buying tickets, understanding the concept of direction in transport, etc.) are mostly acquired despite the varying levels of mastery among ID people. This variability is partly explained by the fact that using transport requires making specific decisions (e.g. deciding on the appropriate means of transport, getting on the bus, looking for a seat, finding out where to get off, signaling

---

[1] https://www.inclusion-europe.eu/easy-to-read-standards-guidelines.

the stop, and getting off, etc.), but also mobilizing other resources not directly related to travel as such, such as reading, time management, use of money, lateralization, etc. However, mastery of these skills in general is very variable among people with ID. Finally, the results related to changes (connections) in transportation underline a point of increased vigilance for professionals and persons accompanied who may encounter significant difficulties with these changes.

- Resolving unforeseen events during travel: a large majority of the professionals interviewed recognize the negative influence of emotions (fear of getting lost, of making mistakes, of other travelers) on travel in the environment in general. In addition, a majority of professionals believe that people with ID have poor adaptation in unexpected situations. In the same proportions, they report strategies adapted to the situation, but also difficulties for the persons accompanied to regulate their emotions in these particular situations in order to solve the problem. These observations in relation to new, conflicting, complex situations during travel are pointed out in some works [26, 34, 35]. Addressing this topic during learning is recognized by the majority of professionals as having a high to very high priority. To do so, they mainly advocate mobilizing one's social environment (asking the referring professional for help by phone, calling the family, or asking a third party). In addition to this strategy, they mobilize their cognitive resources to solve the problem (find another way, etc.) and finally regulate their emotions by asking the people involved to calm down.

The results of this survey show a restrictive use of navigation aid systems by people with ID and the professionals who accompany them. In fact, during training, professionals make little use of this type of tool and prefer to carry out journeys directly in the field with the person with an ID or to draw up itineraries based on photographs of the environment, the stages of the journey, a simplified map of the town, the journey, etc. Similarly, people with ID make very little use of technological tools to plan their journeys but also during the execution of the journey. They reserve the use of their smartphone, on which an application dedicated to navigation can be implemented, to call a relative in case of need or unexpected event. Human assistance is preferred in these emergency situations. Nevertheless, the professionals interviewed recognized the importance of a technological assistance system dedicated solely to navigation and adapted to the cognitive profile of the users. Although this survey does not focus on the acceptability and acceptance of technological solutions for independent travel, interviewees outlined the characteristics and functions that such a tool should fulfill.

One of the limitations we can formulate here concerns the fact that these results are the perceptions of professionals, who despite their long experience and expertise developed on this subject, may have different perceptions of the issue from people with ID themselves. Overall, this descriptive study consolidates and enriches our knowledge about the mobility of people with ID. Indeed, the majority of work related to this topic relies on a research design that makes it difficult to generalize the findings obtained to other cases due to samples with very few participants [11, 15, 36]. In addition, we have knowledge pertaining to mobility learning methods, materials, and tools used by professionals and DI individuals. Thus, the information obtained allows us to distinguish axes of work that seem particularly important when learning mobility: planning trips and managing unexpected events. With regard to adapting to the environment, they highlight the need to simplify information (signs, maps, timetables, etc.) to make it easier for people with ID to find their way around and take information when using transport and making connections. In addition, the results provide information that is important to consider in the creation of technological aids. Indeed, as far as disruptions, deviations, and unforeseen events related to the transport system are concerned, we can wonder about the opportunity that people with ID and their relatives have to know about these events in advance and the need to have a reliable, available and adapted communication system so that they can consider other solutions for the continuation of the journey. Several collaborations between researchers stemming from various disciplines (psychology, education, computer science) as in Letalle and her colleagues [37], persons with ID, their families, and support staff seem to be necessary to define conceptual and practical aspects of technological solutions dedicated to navigation for the targeted population.

5   Conclusion

This paper presents a survey study among professionals accompanying people with intellectual disabilities (ID). The survey revolves around the mobility of people with ID, their mode of transportation, the challenges facing them as well as possible ways of dealing with these challenges. The survey reports how the mobility of these people is generally limited to a few, well-known trajectories in familiar environments. They are, in many cases able to use public transportation. However, among their biggest challenges are unforeseen events and incidents on their journey. Human assistance seems to be the privileged solution adopted by persons with ID as well as their family/accompanying professionals. The use of technology is very limited but might be of value if adopted, with the condition of it being adapted to the characteristics of the target population and the local environment.

Acknowledgement. The work presented in this paper is the result of a collaboration between researchers from the PSITEC Laboratory of the University of Lille and the LAMIH of the Univ. Polytechnique Hauts-de-France within the framework of two projects: TSADI (Technologie de Soutien à l'Apprentissage des Déplacements Indépendants) and SAMDI (Système d'aide à la mobilité pour les personnes présentant une déficience intellectuelle). We would like to thank the Maison Européenne des Sciences de l'Homme et de la Société (MESHS) Lille Nord de France for its financial support within the framework of the TSADI project. We would also to thank the Région Hauts de France for their financial support within the framework of the SAMDI.

References

1. Sandjojo, J., Gebhardt, W.A., Zedlitz, A.M.E.E., Hoekman, J., den Haan, J.A., Evers, A.W.M.: Promoting independence of people with intellectual disabilities: a focus group study perspectives from people with intellectual disabilities, legal representatives, and support staff. J. Policy Pract. Intellect. Disabil. **16**, 37–52 (2019)
2. Haveman, M., Tillmann, V., Stöppler, R., Kvas, Š, Monninger, D.: Mobility and traffic abilities. J. Policy Pract. Intellect. Disabil. **10**, 289–299 (2013)

3. Crocq, M.A.: Mini DSM-5® : Critères diagnostiques/American Psychiatric Association ; coordination générale de la traduction française Marc-Antoine Crocq et Julien Daniel Guelfi ; directeurs de l'équipe de la traduction française Patrice Boyer, Marc-Antoine Crocq, Julien Daniel Guelfi... [et al.]. Elsevier Masson (2016)
4. American association on intellectual and developmental disabilities (AAIDD): Homepage, https://www.aaidd.org/intellectual-disability/definition. Accessed 15 Nov 2022
5. INSERM, I. national de la santé et de la recherche. Déficiences intellectuelles. EDP Sciences (2016). https://www.ipubli.inserm.fr/handle/10608/6816
6. Alauzet, A., Conte, F., Sanchez, J., Velche, D.: Les personnes en situation de handicap mental, psychique ou cognitif et l'usage des transports, p. 145. INRETS, CTNERHI (2010). https://www.lescot.ifsttar.fr/fileadmin/redaction/1_institut/1.20_sites_integres/TS2/LESCOT/documents/Projets/Rapp-finalPOTASTome2.pdf
7. Mengue-Topio, H., Courbois, Y.: L'autonomie des déplacements chez les personnes ayant une déficience intellectuelle: Une enquête réalisée auprès de travailleurs en établissement et service d'aide par le travail. Revue francophone de la déficience intellectuelle 22, 5–13 (2011)
8. Alauzet, A.: Mobilité et handicap: Une question de point de vue. Transp. Environ. Circ. 235, 32–33 (2017)
9. Mengue-Topio, H., Letalle, L., Courbois, Y.: Autonomie des déplacements et déficience intellectuelle: Quels défis pour les professionnels? Revue Alter 14(2), 99–113 (2020)
10. Dever, R.B.: Habiletés à la vie communautaire : Une taxonomie. Presses Inter universitaires (1997)
11. Davies, D.K., Stock, S.E., Holloway, S., Wehmeyer, M.L.: Evaluating a GPS-based transportation device to support independent bus travel by people with intellectual disability. Intellect. Dev. Disabil. 48(6), 454–463 (2010)
12. Stock, S.E., Davies, D.K., Hoelzel, L.A., Mullen, R.J.: Evaluation of a GPS-based system for supporting independent use of public transportation by adults with intellectual disability. Inclusion 1(2), 133–144 (2013)
13. Mechling, L., O'Brien, E.: Computer-based video instruction to teach students with intellectual disabilities to use public bus transportation. Educ. Training Autism Dev. Disabil. 45(2), 230–241 (2010)
14. Price, R., Marsh, A.J., Fisher, M.H.: Teaching young adults with intellectual and developmental disabilities community-based navigation skills to take public transportation. Behav. Anal. Pract. 11(1), 46–50 (2017). https://doi.org/10.1007/s40617-017-0202-z
15. Gomez, J., Montoro, G., Torrado, J.C., Plaza, A.: An adapted wayfinding system for pedestrians with cognitive disabilities. Mob. Inf. Syst. 2015, e520572 (2015)
16. McCreadie, C., Tinker, A.: The acceptability of assistive technology to older people. Ageing Soc. 25(1), 91–110 (2005)
17. Williamson, B., Aplin, T., de Jonge, D., Goyne, M.: Tracking down a solution: exploring the acceptability and value of wearable GPS devices for older persons, individuals with a disability and their support persons. Disabil. Rehabil. Assist. Technol. 12(8), 822–831 (2017)
18. McShane, R., Skelt, L.: GPS tracking for people with dementia. Working Older People 13(3), 34–37 (2009)
19. Kearns, W.D., Rosenberg, D., West, L., Applegarth, S.P.: Attitudes and expectations of technologies to manage wandering behavior in persons with dementia. Gerontechnology 6, 89–101 (2007)
20. Robinson, L., et al.: Balancing rights and risks: conflicting perspectives in the management of wandering in dementia. Health Risk Soc. 9, 389–406 (2007)
21. Landau, R., Werner, S.: Ethical aspects of using GPS for tracking people with Dementia: recommendations for practice. Int. Psychogeriatr. 24(3), 358–366 (2012)

22. Landau, R., Werner, S., Auslander, G.K., Shoval, N., Heinik, J.: Attitudes of family and professional care-givers towards the use of GPS for tracking patients with Dementia: an exploratory study. Br. J. Soc. Work **39**(4), 670–692 (2009). http://www.jstor.org/stable/237 24323
23. Pot, A.M., Willemse, B.M., Horjus, S.: A pilot study on the use of tracking technology: feasibility, acceptability, and benefits for people in early stages of dementia and their informal caregivers. Aging Ment. Health **16**(1), 127–134 (2012)
24. McShane, R., Gedling, K., Kenward, B., Kenward, R., Hope, T., Jacoby, R.: The feasibility of electronic tracking devices in Dementia: a telephone survey and case series. Int. J. Geriatr. Psychiatry **13**(8), 556–563 (1998)
25. van Holstein, E., Wiesel, I., Bigby, C., Gleeson, B.: People with intellectual disability and the digitization of services. Geoforum **119**, 133–142 (2021). https://doi.org/10.1016/j.geoforum.2020.12.022
26. Delgrange, R., Burkhardt, J.M., Gyselinck, V.: Difficulties and problem-solving strategies in wayfinding among adults with cognitive disabilities: a look at the bigger picture. Front. Hum. Neurosci. **14**, 46 (2020). https://doi.org/10.3389/fnhum.2020.00046
27. LaGrow, S., Wiener, W., LaDuke, R.: Independent travel for developmentally disabled persons: a comprehensive model of instruction. Res. Dev. Disabil. **11**(3), 289–301 (1990)
28. Slevin, E., Lavery, I., Sines, D., Knox, J.: Independent travel and people with learning disabilities: the views of a sample of service providers on whether this need is being met. J. Learn. Disabil. Nurs. Health Soc. Care **2**(4), 195–202 (1998)
29. Courbois, Y., Mengue-Topio, H., Blades, M., Farran, E.K., Sockeel, P.: Description of routes in people with intellectual disability. Am. J. Intellect. Dev. Disabil. **124**(2), 116–130 (2019)
30. Purser, H.R., et al.: The development of route learning in Down syndrome, Williams syndrome, and typical development: investigations with virtual environments. Dev. Sci. **18**(4), 599–613 (2015)
31. Farran, E.K., et al.: Route knowledge and configural knowledge in typical and atypical development: a comparison of sparse and rich environments. J. Neurodev. Disord. **7**, 37 (2015)
32. Siegel, A.W., White, S.H.: The development of spatial representations of large-scale environments. In: Reese, H.W. (ed.) Advances in Child Development and Behavior, pp. 9–55. Academic Press (1975)
33. Poucet, B.: Spatial cognitive maps in animals: new hypotheses on their structure and neural mechanisms. Psychol. Rev. **100**(2), 163–182 (1993)
34. Courbois, Y., Blades, M., Farran, E., Sockeel, P.: Do individuals with intellectual disability select appropriate objects as landmarks when learning a new route? J. Intellect. Disabil. Res. **57**(1), 80–89 (2013)
35. Mengue Topio, H., Bachimont, F., Courbois, Y.: Influence des stimuli sociaux sur l'apprentissage de l'utilisation des transports en commun chez les personnes avec une déficience intellectuelle. Revue suisse de pédagogie spécialisée **3**, 7–13 (2017)
36. Kelley, K.R., Test, D.W., Cooke, N.L.: Effects of picture prompts delivered by a video iPod on pedestrian navigation. Except. Child. **79**(4), 459–474 (2013)
37. Letalle L., et al.: Ontology for mobility of people with intellectual disability: building a basis of definitions for the development of navigation aid systems. In: Krömker, H. (eds.) HCI in Mobility, Transport, and Automotive Systems. Automated Driving and In-Vehicle Experience Design, HCII 2020. LNCS, vol. 12212, pp. 322–334. Springer, Cham (2020). https://doi.org/10.1007/978-3-030-50523-3_23

Pedestrian Mobility Contexts of People with Intellectual Disabilities: The Role of Personalization

Léa Pacini[1,2,3], Sophie Lepreux[1]([☒]), and Christophe Kolski[1]

[1] Laboratoire d'Automatique, de Mécanique et d'Informatique industrielles et Humaines, UMR CNRS 8201, Université Polytechnique Hauts-de-France, Valenciennes, France
sophie.lepreux@uphf.fr

[2] Conservatoire Nationale des Arts et Métiers (CNAM), Centre d'Etudes et de Recherche en Informatique (CEDRIC, EA 4629), 292 rue St-Martin, 75003 Paris, France

[3] Université Paris Cité, Institut National de la Santé et de la Recherche Médicale (Inserm, U1284), System Engineering and Evolution Dynamics (SEED), 8 bis rue Charles V, 75004 Paris, France

Abstract. Everyday trip is essential for integration into society and for independence. The task of wayfinding is the process of determining and following a path between an origin and a destination. People with intellectual disabilities have difficulties in this task: fear of getting lost, difficulties in learning routes, in memorizing relevant landmarks, difficulty in questioning others to help them. Some of these people have expressed the desire to be more autonomous in their movements. This paper proposes to analyze two trip contexts as well as the design of navigation aids dedicated to people with intellectual disabilities, specific for each of these contexts. In order to address these important problems encountered by people with disabilities, two contexts are particularly highlighted: the monitoring of an already known route and the step-by-step guided learning of routes with selection of relevant landmarks. For each of these contexts, the importance of personalization is demonstrated. Perspectives in terms of validation of the proposals are proposed.

Keywords: Wayfinding · Personalization · Adaptation · People with intellectual disabilities · Navigation aid system

1 Introduction

Navigation Assistance Systems (NAS) are very common nowadays. The objective of their users is to have indications to reach a destination. The location of this destination is often not known to the user and in general, the environment along the route is not known either. Navigation, whether assisted or not, is a relatively complex cognitive task. Most NAS provide directions on top-down map representations, as well as verbal instructions. This information is not well suited to

H. Krömker (Ed.): HCII 2023, LNCS 14049, pp. 289–301, 2023.
https://doi.org/10.1007/978-3-031-35908-8_20

human cognitive functioning. Some people may therefore find themselves in great difficulty during a wayfinding task, especially people with intellectual disabilities (ID). We are interested in wayfinding, defined as the process of determining and following a path or route between an origin and a destination [19,27]. Based on the study of this process, this work focuses on the stages of navigation on a path known by the user and navigation on an unknown path. The objective is to design and evaluate NAS adapted to people with intellectual disabilities during their pedestrian trip.

Existing NAS (i.e. Google Maps) are considered useful by people without disabilities. But several studies have shown that this form of passive assistance does not help pedestrians memorize trips, nor do they help them become familiar with the environment [11,14]. As shown by Gardony and colleagues, such systems do not provide opportunities to develop spatial skills, and they tend to reduce spatial awareness [8]. Our proposition is to adapt the system to guide the users only when necessary in order to make them independent on the technology.

The following section presents a state of the art on wayfinding, as well as on the behaviors and difficulties encountered by people with intellectual disabilities. It also presents a short state of the art on NAS for ID people. Then, a global architecture is proposed in Sect. 3 in order to articulate the adaptation of the system to the user's wayfinding states. The two navigation contexts in known or unknown environment are detailed in Sects. 4 and 5. In each of these parts different mock-ups illustrate this proposal. The paper ends with a conclusion and research perspectives.

2   State of the Art

In this state of the art we consider representative works on wayfinding in a first step and then focus on the specificities of people with ID when they move. The second part positions the works on the proposed NAS for this population.

2.1   Wayfinding for ID

Spatial navigation is divided into two components (locomotion and wayfinding) [19,27]. Each component interacts with the environment in a different way. Concerning the first component, the locomotion, more considerations are given to accessibility level of a sidewalk or a road for example. Concerning the cognitive component, named wayfinding, the focus is given to path details as landmarks and salient elements of the surroundings. The paper focuses on the wayfinding component.

People with ID have specific functioning, needs and difficulties. In particular, they report significant differences in the urban mobility situations they encounter and in their problem-solving strategies, compared to a control group. They seem to recognize relevant landmarks less effectively than typical subjects, which leads to an impact on their wayfinding performance [3]. They more often choose to ask another person for help rather than change their route [5]. In consequence, people with ID are so not encouraged to go out, even more for trips they do not really know much about or not at all. Whether it comes from themselves or from their relatives, the fear of not being able to find their way back is an obstacle to their independence.

However, it is quite possible for them to learn specific routes through individualized and supported learning [4,20]. Unfortunately, this learning based on regular training is particularly time consuming. Literature provides guidelines for improving ID people navigation through spatial cognition. Sohlberg and colleagues [25] recommend to: (1) use landmarks, (2) give short and clear instructions, (3) use written and auditory modalities, (4) link with a caregiver if needed, (5) allow the repetition of the instructions to overcome memory issues and parasite noises in urban environment, and (6) to include user's notes and landmarks personally chosen. These recommendations address two main points: taking into account information coming from the user (behavioral information) as well as from the outside (contextual and environmental information).

Some researchers have further demonstrated the importance of emotion in spatial representations. Indeed, positive emotions will improve spatial information retention [24,26]. According to Delgrange [5], proposing emotional routes could therefore improve the performance of people with cognitive disabilities, who are prone to experiencing negative emotions in complex navigation situations.

2.2   NAS for ID People

There is a set of studies, in particular systematic literature reviews, on accessibility or NAS [6,13,18,22]. In these works, we remark that few papers in the literature propose NAS for ID people. For example, one work [7] concerns the dementia as cognitive disability. It addresses the problem of monitoring that a person does not get lost. This goal is interesting and can also be used for IDs but the system is not dedicated to guidance and learning and is not adapted to this population. In [10], Gong and colleagues focus on equipping large transportation hubs to offer indoor navigation services to people with special needs but without focusing on ID people. The work proposed by [23] aims at the design of a complete platform for universal access to the guidance service. The architecture is presented as well as its constitutive components. However, it is difficult to conclude on the usability of the system with people with ID because they were not involved in the design nor the evaluation.

Davies and colleagues [4] propose a step-by-step system allowing ID people to take the bus for the first time on a new route but does not emphasize the learning of these routes. As for the system proposed by Gomez and colleagues [9], we find it somewhat confusing in terms of user interface, and not taking into account users who would not be lateralized (i.e. do not know their left and right).

Despite their respective qualities, these systems do not offer any adaptation to ID people's behavior: They do not plan to react to potential changes in the route or in case of unexpected events; They do not reassure users in such cases. In order to take into account the context and specificities of such users during their mobility, we will use the definitions proposed in the global Ontology on the wayfinding by Letalle et al. [17].

Our proposition relies on the work proposed by Lakehal et al. [15,16]. These authors proposed Augmented Reality glasses-based NAS considering different wayfinding states. Indeed, we want to base the real-time adaptation on user's behavior and on environment information by using Wayfinding Model. Concerning the personalization, we aim at following the guidelines expressed in [25] going further in personalization and pushing towards learning new routes as it is a wish expressed by some people with ID. This work comes in the continuity of an ongoing work presented by Pacini et al. [21] in which the focus was on route learning and centered on the emotionality of ID people as it is a wish expressed by some people with ID.

3 Proposal: Architecture Focused on Two Navigational Contexts

The proposal consists in the principles of adaptation of a NAS dedicated to people with intellectual disability.

Considering that the system must be adapted and personalized both on the form (human-computer interaction) and on the content (personalized proposal of the proposed path), the architecture visible in Fig. 1 allows to show the importance of the user profile and its role in the generation of the two personalization aspects.

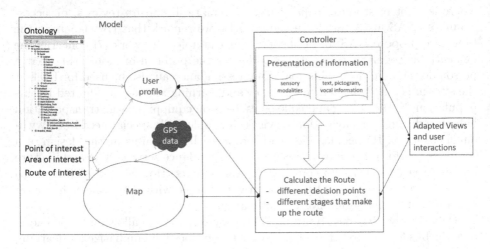

Fig. 1. Global architecture of the navigational assistive system

The proposed architecture follows the MVC design pattern which consists in separating the concerns [2]. The Model concerns the data that are stored or captured. In this part, we position the information that is necessary for the personalization. We can note the importance of the choice of the points, routes and decision areas that will be useful to personalize the route. Indeed, the system will propose to use the points of interest (POI), Route of interest (ROI) and Area of Interest (AOI), as proposed by [12]. The ontology proposed by Letalle et al. [17] allows to help to structure and collect data which are important to this domain. The ontology allowed us to establish a set of concepts and relationships between these concepts. It was proposed with the aim of adapting the system to the needs of users in the context of their mobility, which corresponds exactly to our need. The objective here is to collect the data corresponding to the elements when they exist. In this Model part, we have also positioned the GPS data. Indeed, the GPS data are captured to ensure the tracking of the route.

The controller is the part that concentrates the processing using the data provided by the model. It is in this part that the algorithms will operate to provide useful specifications for the interactions with the user. The result of these processes is visible in the View part. In the controller, we propose to distinguish two components: (1) to compute the personalization with the point of view of the human-computer interaction and (2) from the point of view of the route to be chosen according to the user's preferences in terms of choice of route, duration, modes of transport, etc.

The View part is not detailled here, we just precise that views and interactions must be adapted. The following sections will detail the use of these data according to the contexts in which they are used and the proposed mock-up as adapted views.

Then, the objective is to encourage the autonomy of people by reassuring these users (but also their entourage) when they are on the move. On the one hand, they are monitored during known trips in order to reassure both the user and relatives. On the other hand, the system allows people who master the safety rules to go outside in unknown areas in order to encourage them to be a little more independent. The assistance will then be done in two different ways: undirected assistance, and directed assistance.

To begin with, the system will propose a main route (route planning in Fig. 2) from a starting point to a destination, divided into different sequences. Each sequence is determined according to different landmarks (POI, ROI, AOI) along the route, which will constitute the stages of the route. Each route is also made up of decision points, where the user will have to make a choice along the way. A decision point can be a stage of the trip. Whether or not the user knows a sequence will determine the type of guidance.

In order to orchestrate the passages between the two contexts, a transition state diagram, inspired by the Wayfinding model [27] and the State model [16] is proposed on Fig. 2. We focus on the Direct navigation assistance with the specific known destination. The change between states is centered on the knowledge of route and environment, by using POI, ROI, AOI, in other words the landmarks.

For each state it is indicated which are the important activities to be carried out by the system to reassure and guide the user. For example, in the most

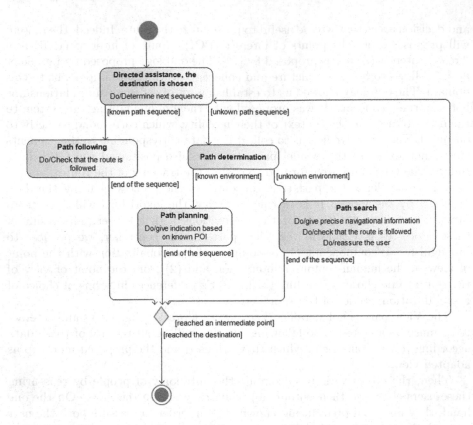

Fig. 2. State transition diagram of the proposed Navigational Assistive System

sensible case where the user knows neither the path nor the environment (state named *Path search*), there are three activities to be carried out:

- guide the user with the landmarks the user does not know by giving precise indications,
- check that the user follows the directions,
- reassure the user.

The system also plans to allow the user to choose one mean of transport rather than another. Indeed, for instance, some user are more at ease with buses than with subway. The ontology also covers this aspect by data relative to the transport modality. Furthermore, users will be able to integrate pictures and photography of well-know and preferred frequently visited places, such as a bakery or a friend's house, as destination choices. These features address the personalization aspect issues, while including in the NAS the emotional aspect. Concerning the management of the Points of Interest, we propose to question the user during the trip to know impressions on the landmarks, roads or areas. The Fig. 3 allows to collect the user's impression about a building.

The following section will focus on the context where the path is known.

Fig. 3. Proposed mock-up to obtain personalized data about environment (in this case, on a building that may or may not be used as a landmark)

4 Context of Known Route

4.1 Context Analysis

Autonomy mode (Fig. 4) allows the users to be monitored remotely on a route they already know. This Figure represents a state diagram in UML. The states correspond to the user states and for each state the system's actions. The global purpose would be to increase the number of routes/sequences known by the users and thus, to reinforce their autonomy.

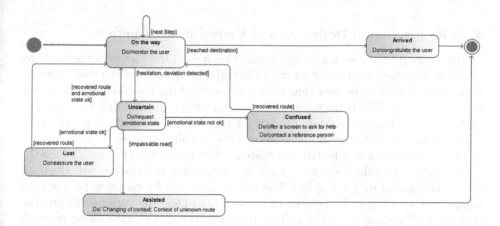

Fig. 4. State transition diagram in the context of known route

In this context, the users know the path but can be encouraged by displaying congratulatory messages when they reach a step or the destination. If the destination is reached, the state of the system will be "Display of congratulations". During the trip, the system anticipates when the users may no longer follow a planned path, for example by hesitating, turning back or deviating from the route. These behaviors may indicate a mental state that would make them less likely to continue the trip. In this case, the system must first inform them that they are deviating from the planned path to know if it is voluntary or not. If it is not voluntary, they are then guided back to the planned route, while reassuring them with written or auditory messages, in order to encourage the users to continue the trip (Reassuring the user and displaying instructions).

As proposed in [21], in case these behaviors seem to be repeated, the user has the opportunity to decide whether to continue the trip or not, without endangering them. After analyzing the path-following, the system will first self-assess their emotional state. If the self-assessment is positive, the user will then be allowed to continue the trip, and a caregiver will be informed at the same time, as a precaution. If the self-assessment is negative, the system will display some help. The user will be able to choose between an automated form, allowing assistance to explain the situation to a person present in the immediate environment, and a direct call to a caregiver, to whom the GPS location will be shared in order to remotely help the user in need. This help feature will be accessible at any time during the trip.

It would be possible to decide on certain sensory modalities for displaying route instructions (for example, only auditory instructions, or visual with vibrations), event if only mobile mock-ups are shown in this paper.

To overcome memory and attentional difficulties met by people with ID, the system will regularly ask a first feedback from the user to indicate good reception of a notification, then a second feedback to indicate they understood the instruction or the message displayed.

4.2   Proposition of Design for the Known Route Context

In this context, the user knows the path. The system watches over the users by checking that they are consistent with the usual path. Most of the time, the system will just inform the user that they have reached the intermediate objectives of the sequence. This helps to reassure and encourage the user (cf. Fig. 5a). The system keeps giving information about landmarks encountered during the trip. This will help memorization and reinforce the learning process.

When hesitation is detected, the system informs the user (Fig. 5b). When the users hesitate or take the wrong path at least twice, the system will then ask their emotional state (cf. Fig. 5c). The users can answer by choosing between the happy or the sad face. If the user chooses the sad face, the system will propose two ways of getting help: the call to the caregiver or the automated form. A template of this form that the user can display and show to a third-person to ask for help is visible on Fig. 5d). On top of all frames, we can see on the left a speaker icon, allowing to hear or repeat an instruction or a message, and on the right a shelter icon that gives access to the help screen (with the form or the direct call, as mentioned previously).

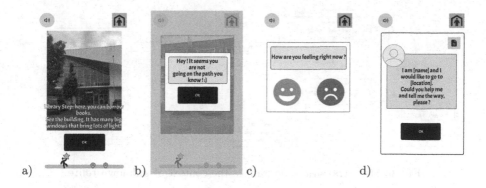

Fig. 5. Mock-ups in the context of known route

The following section focuses on the description in the context of the unknown route in order to provide the user with the adapted information in the adapted format.

5   Context of Unknown Route

5.1   Context Analysis

The objective here is to guide the users through a sequence that it is not known. This is unusual for unaccompanied ID users. It can be useful in case the usual path is not accessible (for example if there are road works that prevent the movement [1]). The system will adapt the proposed route according to the obstacles present and the user's preferences. The system should display each step, step by step, giving indications based on the environment. The users can then memorize the route. In this context, the system must also be careful with the emotional state. In case the users do not find the proposed way correctly, two suggestions are given to ask for help to someone close to their or to contact by phone or SMS someone close to their.

The succession of states can be seen on Fig. 6. As for Fig. 4, this Figure represents the user's states and the system's actions. The nominal scenario is represented by the horizontal path of the states and the alternative scenarios are added below. Depending on the problem and the user's state, different options are proposed. Among these options we find the help to ask for the way but also the right to turn back.

5.2   Proposition of Design for the Unknown Route Context

From the proposition of adaptive system, modeled by UML diagrams, mock-ups are proposed to allow a better visualization of the system and the possible interactions with the user. From works [4,24], the mock-ups of user interaction contains (1) the progress bar, (2) the step-by-step process with interaction

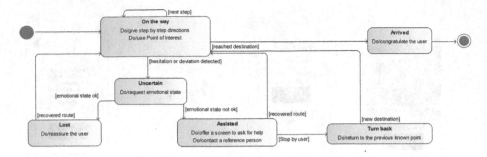

Fig. 6. State transition diagram in the context of unknown route

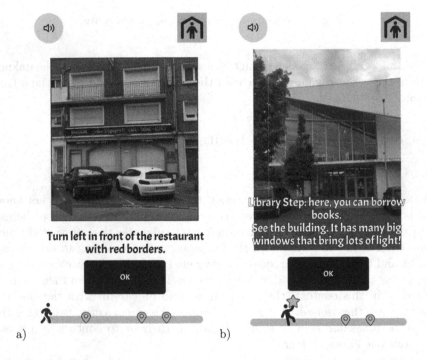

Fig. 7. Mock-ups in the context of unknown route (Color figure online)

needed with the user, (3) the always available help function, (4) the brief user interface, (5) the use of precise and street-level pictures and photos, (6) the personalization in the management of landmarks and photos.

Figure 7 shows different screens of the prototype. Figure 7a represents a street-level photography with a textual indication of the direction to take, also indicated by the blue arrow showing the left direction. This will be the kind of instruction provided to the users. If they cannot read, the message will be displayed as an audio message. Figure 7b shows what happens when the user reaches a step on the route. We can see a photo of the landmark, the library, and a physical description to help memorization.

A mock-up also shows textual indication explaining what the purpose of the building is. At the bottom of both of Figs. 7a) and 7b), we can see a progress bar, where the gray pin icons represent the steps not yet reached. On Fig. 7b), the first pin has a star shape because the user has just reached it. Once the user has moved forward, the pin will return to its original shape and be colored yellow, to indicate that the step has already been reached.

We showed the mock-ups and talked about the proposal to two psychologists working with people with intellectual deficiency. To their knowledge, the proposal seemed well fitted to this kind of end-users and were enthusiastic about the personalization features. They validated the mock-ups and argued that we should maybe reduce the text displayed with landmarks, because it may be too much information. According to them, we should stick to the name of the landmark and what it used for, since the rest seems unnecessary.

6   Conclusion

The principles of user context personalization that we have proposed aim at improving urban mobility for people with intellectual disabilities. Our proposal is based on two features. On one hand, it is based on the wayfinding model to consider the travel context, and on the other hand, on the user profile to propose adapted human-computer interactions. Our goal is to respond to the needs and desire of people with intellectual disabilities to be more independent in mobility contexts. For that, the users can be reassured during current trips because the system proposes known routes and they are helped in case of unforeseen event. When they have to go to unknown areas/paths, they are guided by relevant landmarks. Finally, the personalization of the tool by including recurrent routes but also places appreciated by the user as well as adapted interactions should facilitate acceptance. The paper described our work in progress. The mock-ups have already been shown to experts (psychologists, educators). Yet we still need to validate the interactions and the principles with caregivers as much as people themselves in a user-centered approach. It remains to validate the interactions and the principles with the people themselves in a user-centred design approach. Once validated a priori, it will be necessary to evaluate the ease of use and acceptability of our system through user studies covering a variability of profile allowing to test all the stated contexts.

Acknowledgments. This works has been partially supported by PIA Accroche Active in particular Valmobile action and the SAMDI Project supported by the Region Hauts-de-France. They also thank PRIMOH, PSITEC (Univ. Lille), UDAPEI, APEI de Valenciennes, APEI de Denain and the "Nous Aussi" association.

References

1. Blanchard, E., Duvivier, D., Kolski, C., Lepreux, S.: Towards a framework for detecting temporary obstacles and their impact on mobility for diversely disabled users. In: Krömker, H. (ed.) HCI in Mobility, Transport, and Automotive Systems - 4th International Conference, MobiTAS 2022, Held as Part of the 24th HCI International Conference, HCII 2022, Virtual Event, 26 June–1 July 2022, Proceedings. LNCS, vol. 13335, pp. 475–488. Springer, Cham (2022). https://doi.org/10.1007/978-3-031-04987-3_32
2. Bucanek, J. (ed.): Model-View-Controller Pattern, pp. 353–402. Apress, Berkeley (2009). https://doi.org/10.1007/978-1-4302-2370-2_20
3. Courbois, Y., Blades, M., Farran, E.K., Sockeel, P.: Do individuals with intellectual disability select appropriate objects as landmarks when learning a new route? J. Intellect. Disabil. Res. **57**(1), 80–89 (2013). https://doi.org/10.1111/j.1365-2788.2011.01518.x
4. Davies, D., Stock, S., Holloway, S., Wehmeyer, M.: Evaluating a GPS-based transportation device to support independent bus travel by people with intellectual disability. Intellect. Dev. Disabil. **48**, 454–63 (2010). https://doi.org/10.1352/1934-9556-48.6.454
5. Delgrange, R.: Améliorer la mobilité urbaine des personnes ayant un handicap cognitif en assistant la cognition spatiale: analyse exploratoire de la chaîne du déplacement et évaluation expérimentale d'un prototype d'aide à la navigation. PhD Thesis, Université Paris Cité (2020)
6. El-taher, F.E.Z., Taha, A., Courtney, J., Mckeever, S.: A systematic review of urban navigation systems for visually impaired people. Sensors **21**(9) (2021). https://doi.org/10.3390/s21093103
7. Freina, L., Caponetto, I.: A mobile guardian angel supporting urban mobility for people with dementia - an errorless learning based approach. In: Proceedings of the 1st International Conference on Information and Communication Technologies for Ageing Well and e-Health - SocialICT, (ICT4AgeingWell 2015), pp. 307–312. INSTICC, SciTePress (2015). https://doi.org/10.5220/0005502503070312
8. Gardony, A.L., Brunyé, T.T., Mahoney, C.R., Taylor, H.A.: How navigational aids impair spatial memory: evidence for divided attention. Spatial Cogn. Comput. **13**(4), 319–350 (2013). https://doi.org/10.1080/13875868.2013.792821
9. Gomez, J., Montoro, G., Torrado, J.C., Plaza, A.: An adapted wayfinding system for pedestrians with cognitive disabilities. Mobile Inf. Syst. (520572) (2015). https://doi.org/10.1155/2015/520572
10. Gong, J., et al.: Building smart and accessible transportation hubs with Internet of Things, Big Data analytics, and Affective computing, pp. 126–138 (2017). https://doi.org/10.1061/9780784481219.012
11. Ishikawa, T., Fujiwara, H., Imai, O., Okabe, A.: Wayfinding with a GPS-based mobile navigation system: a comparison with maps and direct experience. J. Environ. Psychol. **28**(1), 74–82 (2008). https://doi.org/10.1016/j.jenvp.2007.09.002
12. Karimi, H.A., Dias, M.B., Pearlman, J., Zimmerman, G.J.: Wayfinding and navigation for people with disabilities using social navigation networks. EAI Endorsed Trans. Collaborative Comput. **1**(2) (2014). https://doi.org/10.4108/cc.1.2.e5
13. Khan, S., Nazir, S., Khan, H.U.: Analysis of navigation assistants for blind and visually impaired people: a systematic review. IEEE Access **9**, 26712–26734 (2021). https://doi.org/10.1109/ACCESS.2021.3052415

14. Konishi, K., Bohbot, V.D.: Spatial navigational strategies correlate with gray matter in the hippocampus of healthy older adults tested in a virtual maze. Front. Aging Neurosci. **5**(FEB), 1–8 (2013). https://doi.org/10.3389/fnagi.2013.00001
15. Lakehal, A., Lepreux, S., Letalle, L., Kolski, C.: Modélisation des états de la tâche de wayfinding dans un but de conception de système d'aide à la mobilité des personnes présentant une déficience intellectuelle. In: Proceedings of the 30th Conference on l'Interaction Homme-Machine, IHM 2018, pp. 202–208. Association for Computing Machinery, New York, NY, USA (2018). https://doi.org/10.1145/3286689.3286710
16. Lakehal, A., Lepreux, S., Letalle, L., Kolski, C.: From wayfinding model to future context-based adaptation of HCI in urban mobility for pedestrians with active navigation needs. Int. J. Human-Comput. Interaction **37**(4), 378–389 (2021). https://doi.org/10.1080/10447318.2020.1860546
17. Letalle, L., et al.: Ontology for mobility of people with intellectual disability: building a basis of definitions for the development of navigation aid systems. In: Krömker, H. (ed.) HCI in Mobility, Transport, and Automotive Systems. Automated Driving and In-Vehicle Experience Design - Second International Conference, MobiTAS 2020, Held as Part of the 22nd HCI International Conference, HCII 2020, Copenhagen, Denmark, 19–24 July 2020, Proceedings, Part I. LNCS, vol. 12212, pp. 322–334. Springer, Cham (2020). https://doi.org/10.1007/978-3-030-50523-3_23
18. Marques, V.L., Graeml, A.R.: Accessible maps and the current role of collective intelligence. GeoJournal **84**(3), 611–622 (2019)
19. Montello, D.R.: Navigation. Cambridge Handbooks in Psychology, pp. 257–294. Cambridge University Press, Cambridge (2005). https://doi.org/10.1017/CBO9780511610448.008
20. Newbigging, E.D., Laskey, J.W.: Riding the bus: teaching an adult with a brain injury to use a transit system to travel independently to and from work. Brain Injury **10**(7), 543–550 (1996). https://doi.org/10.1080/026990596124250
21. Pacini, L., Lepreux, S., Kolski, C.: Towards behavioral adaptation for people with intellectual disabilities in a mobility context. In: Proceedings of the 19th International Conference on Human-Computer Interaction - RoCHI 2022 (6–7 October), Craiova, Romania, pp. 126–129 (2022)
22. Prandi, C., Barricelli, B.R., Mirri, S., Fogli, D.: Accessible wayfinding and navigation: a systematic mapping study. Univ. Access Inf. Soc. (2021). https://doi.org/10.1007/s10209-021-00843-x
23. Rodriguez-Sanchez, M., Martinez-Romo, J.: GAWA - manager for accessibility wayfinding apps. Int. J. Inf. Manage. **37**(6), 505–519 (2017). https://doi.org/10.1016/j.ijinfomgt.2017.05.011
24. Ruotolo, F., Claessen, M.H.G., van der Ham, I.J.M.: Putting emotions in routes: the influence of emotionally laden landmarks on spatial memory. Psychol. Res. **83**(5), 1083–1095 (2018). https://doi.org/10.1007/s00426-018-1015-6
25. Sohlberg, M.M., Todis, B.J., Fickas, S., Hung, P.F., Lemoncello, R.: A profile of community navigation in adults with chronic cognitive impairments. Brain Injury **19**, 1249–1259 (2005)
26. Storbeck, J., Maswood, R.: Happiness increases verbal and spatial working memory capacity where sadness does not: emotion, working memory and executive control. Cogn. Emotion **30** (2015). https://doi.org/10.1080/02699931.2015.1034091
27. Wiener, J.M., Büchner, S.J., Hölscher, C.: Taxonomy of human wayfinding tasks: a knowledge-based approach. Spatial Cogn. Comput. **9**(2), 152–165 (2009)

The SOLID Model of Accessibility and Its Use by the Public Transport Operators

Gérard Uzan[1], Caroline Pigeon[2(✉)], and Peter Wagstaff[2]

[1] Laboratoire CHart-THIM, Université Paris 8, Saint-Denis, France
[2] Lescot, Université Gustave Eiffel, Campus de Bron, Bron, France
Caroline.Pigeon@univ-eiffel.fr

Abstract. Public transport could be the only transport means for some people, in particular those with disabilities. Accessibility of public transports is thus essential to enable everyone to participate in the society, for example by accessing to employment, healthcare and leisure activities. Legislative framework and regulations ensure minimum accessibility standards, but are not always enough to ensure effective accessibility. In this paper, the SOLID model of accessibility and its main components are presented. This model has been developed through public consultations and experiments with people with visual impairment, and then refined through experiments with other populations and discussions with public transport companies. Then the paper presents examples of usage of the model for improvements in public transports in and out of the disability context. This illustrates how the model is a useful tool to identify accessibility issues and propose solutions in different domains such as architecture, vehicle design, traveller information or smartphone application. The paper also demonstrates how research on disability is relevant to work on the general human condition.

Keywords: Accessibility · Passenger information · Public transport · Disabled passengers

1 Introduction

Mobility, defined as the ability to move within community environments that expand from home [1] is an essential requirement to participate in society, i.e. access to employment, education, healthcare and social and leisure activities. Public transports can be the only transport means at a distance from home for some people, especially for those who cannot or no longer drive, for example people with impairments [2, 3]. A lack of access to a reliable transport means lead to isolation and social participation restrictions [3].

According to social models of disability, for example the International Classification of Functioning, Disability and Health (ICF) by the WHO [4], or the Disability Creation Process (DCP) of Fougeyrollas and colleagues [5], disability results of the restriction of participation in activities due to the interaction of environmental barriers and personal

---

P. Wagstaff—Deceased

difficulties. Thus, if public transports are not adapted, the mobility of people with physical or sensorial impairments may be hindered, but also that of those with more temporary difficulties, such as people with bulky luggage, pregnant women, or people with young children, including strollers [6]. However, mobility and therefore social participation can be facilitated for all, when environment is optimized, for example when access to public transports is barrier-free.

In France, the law on accessibility introduced on the 11th of February 2005 aiming to achieve accessibility for all in 2015 lead to improvements in several areas, such as education, employment, built environment, transportation, and internet. However, its implementation has been insufficient, so much so that a revised timeframe has been put into place. Indeed, despite the immediate needs of users with impairments, accessibility can take a long time to implement, especially in the context of transport infrastructures (tracks, stations, stations), whose renewal rate is around 50 years. A distinction must therefore be made between improvements that can be made quickly and at low cost (particularly on vehicles), modifications made as part of renovation, and accessibility application as part of the implementation of new transports. Furthermore, although legislative framework and regulations ensure minimum accessibility standards, it is necessary to go further, and think about an effective accessibility, which meets the real needs of individuals.

Nonetheless, it is not only the physical accessibility of the vehicle that guarantees access to public transports, but the accessibility of the entire chain of mobility (Fig. 1), from the planning of the journey to the exit of the vehicle, including the management of disturbed and emergency situations [7]. A journey using public transport is a succession of phases in different environments, where the users are either walking, in a stationary infrastructure or immobile (or almost) in a moving vehicle. Each of these phases is

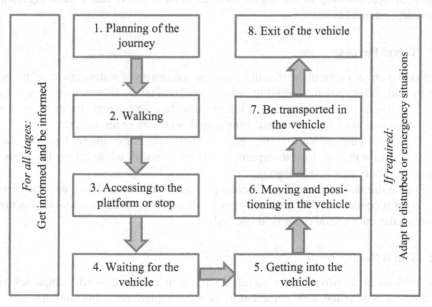

Fig. 1. Stages of the mobility chain of the public transport users

accompanied by different levels of physical, perceptual and mental effort [7], which also varied according the abilities and impairments of the users. For example, by using the think aloud method with participants with visual impairments in journeys including the use of public transport, we found that it was in multimodal hubs poles that the participants expressed the most difficulties [8, 9]. In addition, stress was more expressed in the three following situations: large spaces, crossings (road or tram track) and stairs [8, 9].

Ensuring effective accessibility for all users of public transport and enabling them to reach their destination safely requires considering the physical aspects, but also providing them with the information necessary for their orientation, at all stages of the mobility chain. These different elements will be developed in the following section, which describes a model of accessibility called SOLID, with a focus on public transport and data transport.

2 The SOLID Model

SOLID is an accessibility model, which describes the representations of transport users in the entire chain of mobility. SOLID is an acronym for five elements ensuring an optimal accessibility of transport systems: Safety, Orientation, Localization, Information, Displacements. This model describes the travel purposes, the travel zones, the user needs and information nature. This model has been developed through public consultations and experiments with people with visual impairments, aiming at developing solutions to improve public transport accessibility [7, 10]. Then the model has been refined through experiments with other populations and discussions with public transport companies. This model lets the analysis of problems and solutions or validates the accessibility of system by decomposing all the phases involved in the mobility chain including a public transport system [11].

2.1 Travel Purpose

Depending on the cause of the trip, there are four categories of pedestrian travel purpose: 1) physical activity, such as walking or running, 2) social relations with others, such as going for a walk while talking, 3) getting to know the environment (for example, a new neighbourhood, touring), and 4) reaching a goal, a person, or an object.

The travail purpose has an effect on the experience. For example, taking the stairs to reach a train that is close to departure may be stressful, while taking the stairs for physical activity may be motivating.

During the same travel, these different purposes can be associated, they can also be priority or secondary. In the context of public transport use, the last purpose is prioritized, although the others should not be neglected.

2.2 Travel Zones

The traditional division of the travel environment from transport companies view includes 3 zones: the ticket validation zone, the platform zone and the zone of transportation.

The SOLID model proposes another division of the travel environment, in 4 zones. The first one is the **open zone**, which is multifunctional: not all pedestrians who come here will take public transport, and there are commercial, cultural and residential activities. Even if it is not dedicated to transportation, vehicles (bus, streetcar, scooter, etc.) can be found there. This open zone is traditionally non-considered by transport companies. The second zone is the **access zone**, where the main activity of the people in this zone is to go to the transfer zone (which will be described next) or to return from it. The access zone is equivalent of the ticket validation zone of companies. The third zone is the **transfer zone**, which consists of the platform (or curb, for buses) and the vehicle entry/exit platform. This is where the vehicle docks and departs and where passengers are transferred to and from the vehicle. It is a time pressure zone where safety can be engaged. Finally, the **zone of transportation** is the interior of the vehicle, with the exception of the vehicle entry/exit platform, where the user is moved while remaining immobile. In the traditional division of transport companies, the vehicle entry/exit platform is included in the zone of transportation.

2.3   User Needs: Safety, Orientation, Localization, Information and Displacements

The acronym of the model refers to the different need of transport users. Firstly, the main responsibility of the transport systems is to ensure the **safety** and the security of the users at all phases of the journey, and in all circumstances.

The safety of transport users may be hindered in multiple ways, especially if they have disabilities. The main several kinds of risks are strokes, collisions between users, falls, and theft or assault. In the case of visual impairments, falls can be caused by perceptual alteration, attentional saturation due to an excessive amount of information to be processed, or the absence of information concerning an empty space (for example stairs or an unclosed trap doors). The presence of posts, poles, and protruding elements at the head level increases the risk of strokes. Poor flow guidance can lead to collisions between different users, and some ambiences related to architecture favour the risk or the feeling of risk of theft or assault. For someone with balance disorders, sudden starts and stops of vehicles can be a source of falls, especially if the person has not had time or space to sit down.

Transport systems, infrastructures and human-machine interfaces should be designed to alleviated all these kinds of risks. In addition, in case of dysfunction, incident, or accident, special safety procedures should be adapted to all types of impairments.

The transport users have to maintain a correct **orientation** during the journey. It requires having indications of the direction to be taken at each journey portion. These indications can be stored in memory (mental image of a map or an itinerary) or timely provided by infrastructures and human-machine interfaces. Standard signs or display panels are usual for indicating directions, but standard signs may not be used by all the users (for example people who are visually impaired, foreigners or illiterates). In addition, maintaining a straight path can be difficult for blind people, and understanding abstract written information (e.g., acronyms) or pictograms and navigating space and time can be difficult for some people with intellectual impairments [12, 13].

Furthermore, in case of modification of the initial itinerary (disruption, works, detours), the users must be able to find an alternative route, which can be challenging if the information needed is not provided in an appropriate format.

The transport users need to know at any time their **localization** in relation to required points of interest, such as entrances and exits, ticket barriers, bus stop, station or platform. They also need to know the relative position between secondary points of interest, for example ticket offices, information displays, maps, stairs, escalators, lifts, toilets, and commercial installations. Systems that confirm to the users that they reached strategic points or intermediate milestones help visually impaired users locate themselves, and reassure users with anxiety.

Throughout the travel chain, users must be able to access the **information** necessary for their orientation and location, and for directing them to points of interest. This information can be obtained through active research by the users, for example to verify their position or other travel information, particularly crucial in the case of visual impairments or anxiety. Information can also be broadcast to users (they are informed). In order to be perceived and understood by all types of users, the information could be disseminated with different sensory channels. In addition, visual information (written language and pictograms) must be disseminated in adequate formats, in terms of font, contrast, size, luminosity, and auditory information must to be audible through the background noise and adequate in terms of structures and delivery speed.

Information about the presence of alternative routes, allowing the use of an escalator or elevator, will also facilitate the journey of users with walking difficulties. The information provided can also be enriched with train destinations, arrival times and disruptions or delays, as well as commercial or tourist information, useful for all types of users.

A journey involving public transportation use involves different types of **displacements**. There are pedestrian phases, phases in which users are immobile (or almost) in a moving vehicle, lift or escalator, and transitions between theses phases, such as entering and go through the station, finding the correct platform and entering the vehicle and leaving it, without forgetting the procedures to follow in case of an emergency. Generally, pedestrian phases require the greatest physical effort, while transition phases require the greatest perceptual and cognitive effort. However, this depends of the characteristics of the users and their difficulties; for example, in a wheelchair, it can be physically difficult to get in and out of a vehicle (a transition phase) when the horizontal and vertical gaps are high.

2.4   Information Nature

In the context of public transport, three types of information can be characterized, when classified according to their periodicity. **Event-based information** is short-lived, and informs about a specific event (arrival of a vehicle, traffic interruption) whose relevance for the transport user disappears after the event is over. The duration of this type of information is less than the daily cycle. **Contextual information** is temporary information that lasts for a certain period of time (e.g., maintenance), whereas **structural information** is asynchronous or planned (e.g., programmed schedules).

3   The SOLID Model, a Tool for Implementation of Accessibility in Public Transports

In France, public transport is organized by a multitude of stakeholders: authorities in charge are different according the areas concerned (national transport, regional transport, suburban transport and urban transport), and each public transport authority uses (private or public) operators [8]. In addition, urban infrastructures such as roadways, sidewalks, sidewalk sign are on the responsibility of regions or conurbations.

Hence accessibility of the mobility chain relies on a variety of stakeholders, and they do not share a common guideline, that leads to a lack of continuity and coherence between the different places travelled during a journey, in terms of physical architectures, information provided, ways to buy a ticket and their interoperability in the case of a multimodal journey, line characterizations and so on. This phenomenon is illustrated by the Fig. 2, which presents maps of three different transport networks (urban, suburban bus and suburban train), all focusing on the central railway Perrache station, in Lyon, France.

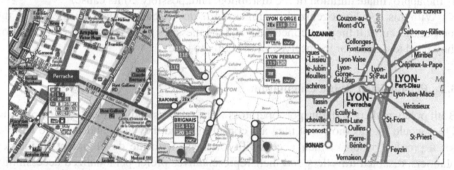

Fig. 2. Maps caption from urban, suburban bus and suburban train networks focusing in the Perrache station (Lyon, France).

The SOLID model was used in consultations/confrontations involving the first author of the present paper in the various departments of the transport operators of the Ile de France region, or in Lyon. These consultations took for example the form of Q&A sessions, organized by the department of disabled passengers with each of the services (two sessions per service). Some concrete application issues were then studied in depth in the framework of working groups, for example for the layout of a station with the architects' department or for the improvement of the lexicon and grammar of traveller messages with the information systems department. About 20 projects were conducted, for example AUTOMOVILLE, RAMPE, DANAM, Guide Urbain, INFOMOVILLE, SIV1, SIV2, Atlas Sonore, Besoin d'Humain, Ma Lanterne, and IMAGE [10, 14–18]. These different projects were the subject of partnership agreements with operators with confidentiality clauses that do not allow us to reveal all the details or to publish. Two examples of application of the SOLID model will be described in the following subsections.

3.1 Grammar and Lexicon of Traveller Information

Within the framework of the working group on the grammar and lexicon of traveller information, the SOLID model has allowed an optimization of the messages broadcasted in case of traffic disruptions on the metro lines [19].

Initially, traveller information was built on the basis of maintenance and traffic management messages. The messages were not homogeneous, since they partly depended on the line managers, but the cause of the disruption always came first, the line number always came late, and action to be taken for users (when there was one) always came last.

An underlying element of the SOLID model is that transport users can become contributors to traffic regulation. Indeed, if transport users have the right information, they can implement the most optimal strategy for themselves, e.g. decide to wait or to look for an alternative route, depending on their own constraints (safety, travel time). The working groups thus led to a homogenization of the lexicon used, and of the proposed order, by presenting, and in this order: 1) the line concerned, 2) the type of disruption (e.g.: interruption or slowdown), 3) the section concerned (e.g.: between two stations), 4) the approximate duration of the disruption, and, possibly, 5) its cause.

With this new homogenized grammar, users do not have to wait for the end of the message to know if they are concerned, and the message gives them all the information they need to decide for themselves what action to take.

3.2 Personalized Route Calculation

Both policy makers and transport operators want an accessibility label for legal reasons. They would like to show that they are consulting (and they are), and they are looking for a proof of result of their accessibility actions. With the development of journey planners, disabled travellers and their representatives prefer personalized accessibility in which the calculation takes account of their disability. The SOLID model was used in a project of an application for guidance of people with all types of impairments in a complex train station (Paris Gare de Lyon, France; *Ma Lanterne* project [16]). In this project, a weighting matrix that considers the nature of the layout and the nature of the disability was developed for personalized route calculation inside the station. This project has shown the relevance of this personalized approach, and at the same time has been one of the supports for the development of the NeTEx (Network Timetable Exchange) database of the Transmodel standard [20]. The NeTEx database support information exchange of public transports data such as network, timetable and fare information (including multimodal fares) for passenger information and Automated Vehicle Monitoring Systems. The database collects and integrates data from different transport operators, and reintegrates it as it evolves through successive updates, and data considers the accessibility.

The example of the staircase will allow us to illustrate because it links different SOLID elements: the construction of the NeTEx standards database, the concerns of route calculation and the issues between the political-administrative transport authorities and the public transport management companies on the one hand, and the disabled travellers and their representatives on the other. A staircase is a blocking element for wheelchair users and an element of hardship for people with walking difficulties, with a cane, or people with cardiac or respiratory difficulties.

In the case of a person with two crutches, the difficulty is not the same when going up or down the stairs. Stairs are therefore vectored. The difficulty in going up (a hardship factor in the Displacement section of the SOLID model) becomes the risk of falling and colliding with other travellers in the descent (i.e. the Safety section of the model).

In the case of a person with two crutches on the stairs, the existence of a handrail to assist the ascent and to secure oneself on the descent is a design component that improves the accessibility of the stairs. This handrail must therefore appear in the NeTEx descriptor base. The SOLID model is therefore a contributor to the database.

The presence of this handrail is important not only because it reduces both the effort to climb the stairs (hardship) and the risk of falling when descending, but also because it makes it possible to define a weighting gradient in an itinerary calculation that takes the disability into account. Thus, the precision of the calculation can integrate a time requirement (train at departure for example) which is found in the fourth travel purpose (reaching an object).

We should specify that in our example, the hardship is physical, but in other situations, this hardship can be cognitive (in particular for people with cognitive or sensory difficulties) or of communication (for example for the deaf people).

4 Conclusion and Future Directions

The SOLID Model was initially created based on the issues faced by blind people in public transports. It was then generalized to other sensory impairments, then to cognitive impairments, then to all impairments, then tourists and finally to all transport users. While ensuring the accessibility of the mobility chain provides comfort for all, it is crucial for people with impairments. Indeed, travellers with impairments can avoid the non-adapted or uncomfortable environments by lengthening their journey by taking a longer alternative route, or even by giving up going somewhere. The present paper illustrates how a conceptual model of accessibility, the SOLID model, can be used by operators to identify accessibility issues in the mobility chain. However, the model does not define how to ensure this accessibility. The elaboration of specifications based on the SOLID model, developed in consultation with transport operators, users and researchers, would allow the implementation of concrete, coherent and common solutions for the accessibility of public transport.

The SOLID model can help to achieve a minimum level of comprehension of the real needs of different types of users for other transport means and in other contexts. For example, although autonomous vehicles are often presented as a solution for those who cannot drive, our recent literature review on non-rail autonomous public transport vehicles shown that few studies investigated needs, expectations and concerns of people with impairments toward autonomous vehicles [10]. Moreover, lack of awareness of disabled situations of from municipalities stakeholders' authorities and employees is one of the barriers to the implementation accessibility measures in municipalities identified in a recent scoping review [21].

Finally, this paper illustrates how research can be used by stakeholders to improve transport accessibility and comfort, and how interaction with transport stakeholders can refine research models. It also shows how research on disability is relevant to work on the general human condition.

Acknowledgments. The authors gratefully acknowledge the support of the RATP, the SNCF, TCL and numerous partners from other institutions that have collaborated in the projects, which have contributed to the development and validation of the ideas presented in this paper.

References

1. Webber, S.C., Porter, M.M., Menec, V.H.: Mobility in older adults: a comprehensive framework. Gerontologist **50**, 443–450 (2010)
2. Penfold, C., Cleghorn, N., Creegan, C., Neil, H., Webster, S.: Travel behaviour, experiences and aspirations of disabled people. London, England: National Centre for Social Research (2008)
3. Jansuwan, S., Christensen, K.M., Chen, A.: Assessing the transportation needs of low-mobility individuals: Case study of a small urban community in Utah. J. Urban Plann. Developm. **139**, 104–114 (2013)
4. World Health Organization: International Classification of Functioning, Disability and Health. WHO, Geneva, Switzerland (2001)
5. Fougeyrollas, P., Noreau, L., Bergeron, H., Cloutier, R., Dion, S.A., St-Michel, G.: Social consequences of long term impairments and disabilities: Conceptual approach and assessment of handicap. Inter. J. Rehabil. Res. **21**, 127–141 (1998)
6. Hallet, P.: Les personnes à mobilité réduite (PMR). FICHE n 04. CERTU, Lyon, France (2010)
7. Uzan, G., Wagstaff, P.: Solid: a model to analyse the accessibility of transport systems for visually impaired people. In: Pissaloux, E., Velazquez, R., (eds.), Mobility of Visually Impaired People. pp. 353–373. Springer (2018). https://doi.org/10.1007/978-3-319-54446-5_12
8. Grange-Faivre, C., Pigeon, C., Pagot, C., Cosma, I., Chateauroux, E., Marin-Lamellet, C.: Design for all: Multimodal transport hubs and travelers with visual impairments (TIMODEV project). In: Proceedings of the 14th Conference on Mobility and Transport for Elderly and Disabled Persons, pp. 838–853, Lisbon, Portugal (2015)
9. Pigeon, C., Grange-Faivre, C., Marin-Lamellet, C.: Difficulties and strategies of visually impaired people in multimodal transport hubs. In: International Mobility Conference, Montréal, Québec (2015)
10. Pretorius, S., Baudoin, G., Venard, O.: Real time information for visual and auditory impaired passengers utilising public transport-technical aspects of the infomoville project. In: AMSE Modelling, Measurement and Control, Series C: Chemistry, Geology, Environment and Bioengineering. (2010)
11. Uzan, G., Hanse, P.C., Seck, M., Wagstaff, P.: Solid: a model of the principles, processes and information required to ensure mobility for all in public transport systems. In: Proceedings 19th Triennial Congress of the IEA, Melbourne, pp. 9–14 (2015)
12. Guth, D., LaDuke, R.: The veering tendency of blind pedestrians: An analysis of the problem and literature review. J. Vis. Impairment Blindness **88**, 391 (1994)
13. Leyrat, P.-A., Mathon, S.: COGI to Access (Cognition et information pour l'accessibilité). Rapport de recherche Cognition et réseau d'information : Une condition du niveau de service du système de mobilité multimodale pour tous. CEREMH, Vélizy-Villacoublay, France (2016)

14. Badillo, P.-Y., Tarrier, F.: Mobilité et ubiquité : vers le nomadisme numérique. Les Cahiers de l'ANR (2009)
15. Baudoin, G., Venard, O., Uzan, G., Paumier, A., Cesbron, J.: Le projet RAMPE: système interactif d'information auditive pour la mobilité des personnes aveugles dans les transports publics. In: Proceedings of the 2nd French-Speaking Conference on Mobility and Ubiquity Computing, pp. 169–176 (2005)
16. Kahale, E., Hanse, P.C., Destin, V., Uzan, G., Lopez-Krahe, J.: Optimisation de Parcours Pour le Deplacement Indoor Selon les Caracteristiques des Usagers. In: La recherche au service de la qualité de vie et de l'autonomie, pp. 219–224, Paris, France (2016)
17. Baudoin, G., Venard, O., Dessaigne, M.-F., Uzan, G., Le Maître, Y.: INFOMOVILLE: Environnement temps réel pour l'information et l'orientation des voyageurs à handicap sensoriel dans les transports collectifs. Génie logiciel (1995)
18. Destin, V., Guérin, C., Soulivong, P., Uzan, G.: Aides humaine et technologique dans l'assistance aux voyageurs handicapés : quelles interactions ? quels besoins ? In: La recherche au service de la qualité de vie et de l'autonomie, pp. 121–126, Paris, France (2016)
19. Uzan, G., Wagstaff, P.: Audio-based interface of guidance systems for the visually impaired in the paris metro. In: Krömker, H. (ed.) HCII 2021. LNCS, vol. 12791, pp. 563–576. Springer, Cham (2021). https://doi.org/10.1007/978-3-030-78358-7_39
20. NF EN 12896 - Systèmes de transport intelligents - Transports en commun - Identification des objets fixes dans les transports publics (IFOPT)
21. Corcuff, M., Ruiz Rodrigo, A., Mwaka-Rutare, C., Routhier, F., Battalova, A., Lamontagne, M.-E.: Municipalities' Strategies to Implement Universal Accessibility Measures: A Scoping Review (2023). https://papers.ssrn.com/abstract=4327401, https://doi.org/10.2139/ssrn.432 7401

Correction to: Research on Interactive Interface Design of Vehicle Warning Information Based on Context Awareness

Fusheng Jia, Yongkang Chen, and Renke He

Correction to:
Chapter 12 in: H. Krömker (Ed.): *HCI in Mobility, Transport, and Automotive Systems*, **LNCS 14049,**
https://doi.org/10.1007/978-3-031-35908-8_12

In the original version of this chapter the affiliation of the second author, Yongkang Chen, was published incorrectly. This has now been corrected to read as: "College of Design and Innovation, Tongji University, Shanghai 200092, China".

The updated version of this chapter can be found at
https://doi.org/10.1007/978-3-031-35908-8_12

Author Index

H. Krömker (Ed.): HCII 2023, LNCS 14049, pp. 313–315, 2023.
https://doi.org/10.1007/978-3-031-35908-8

Printed in the United States
by Baker & Taylor Publisher Services